Biotransformations in Preparative Organic Chemistry

The Use of Isolated Enzymes and Whole Cell Systems in Synthesis

BEST SYNTHETIC METHODS

Series Editors

A. R. Katritzky
University of Florida
Gainesville, Florida
USA

O. Meth-Cohn
Sterling Organics Ltd
Newcastle upon Tyne
UK

C. W. Rees
Imperial College of Science
and Technology
London, UK

R. F. Heck, *Palladium Reagents in Organic Syntheses*, 1985
A. H. Haines, *Methods for the Oxidation of Organic Compounds: Alkanes, Alkenes, Alkynes, and Arenes*, 1985
P. N. Rylander, *Hydrogenation Methods*, 1985
E. W. Colvin, *Silicon Reagents in Organic Synthesis*, 1988
A. Pelter, K. Smith and H. C. Brown, *Borane Reagents*, 1988
B. Wakefield, *Organolithium Methods*, 1988
A. H. Haines, *Methods for the Oxidation of Organic Compounds: Alcohols, Alcohol Derivatives, Alkyl Halides, Nitroalkanes, Alkyl Azides, Carbonyl Compounds, Hydroxyarenes and Amino-arenes*, 1988
H. G. Davies, R. H. Green, D. R. Kelly and S. M. Roberts, *Biotransformations in Preparative Organic Chemistry: The Use of Isolated Enzymes and Whole Cell Systems*, 1989

In preparation

I. Ninomiya and T. Naito, *Photochemical Synthesis*, 1989

Biotransformations in Preparative Organic Chemistry

The Use of Isolated Enzymes and Whole Cell Systems in Synthesis

H. G. Davies and R. H. Green

Medicinal Chemistry Department
Glaxo Group Research Ltd
Greenford, Middlesex, UK

D. R. Kelly

Department of Chemistry
University College, Cardiff
PO Box 78
Cardiff, UK

Stanley M. Roberts

Department of Chemistry
University of Exeter
Stocker Road
Exeter, UK

Academic Press

Harcourt Brace Jovanovich, Publishers
London   San Diego   New York   Berkeley
Boston   Sydney   Tokyo   Toronto

ACADEMIC PRESS LIMITED
24–28 Oval Road
London NW1 7DX

US Edition published by
ACADEMIC PRESS INC.
San Diego, CA 92101

This book is a guide providing general information concerning its subject matter; it is not a procedural manual. Synthesis of chemicals is a rapidly changing field. The reader should consult current procedural manuals for state-of-the-art instructions and applicable government safety regulations. The publisher and the authors do not accept responsibility for any misuse of this book, including its use as a procedural manual or as a source of specific instructions

British Library Cataloguing in Publication Data
Biotransformations in preparative organic
  chemistry: the use of isolated enzymes and
  whole cell systems in synthesis
  1. Organic compounds. Synthesis.
  Biochemical techniques
  I. Davies, H. G. II. Green, R. H. III. Kelly, D. R.
  IV. Roberts, S. M. V. Series
  547.'2

ISBN 0-12-206230-2

Typeset by Bath Typesetting Ltd., London Road, Bath
and printed in Great Britain by T. J. Press (Padstow) Ltd., Padstow, Cornwall

Contents

Foreword

There is a vast and often bewildering array of synthetic methods and reagents available to organic chemists today. Many chemists have their own favoured methods, old and new, for standard transformations, and these can vary considerably from one laboratory to another. New and unfamiliar methods may well allow a particular synthetic step to be done more readily and in higher yield, but there is always some energy barrier associated with their use for the first time. Furthermore, the very wealth of possibilities creates an information-retrieval problem. How can we choose between all the alternatives, and what are their real advantages and limitations? Where can we find the precise experimental details, so often taken for granted by the experts? There is therefore a constant demand for books on synthetic methods, especially the more practical ones like *Organic Syntheses*, *Organic Reactions*, and *Reagents for Organic Synthesis*, which are found in most chemistry laboratories. We are convinced that there is a further need, still largely unfulfilled, for a uniform series of books, each dealing concisely with a particular topic from a *practical* point of view—a need, that is, for books full of preparations, practical hints and detailed examples, all critically assessed, and giving just the information needed to smooth our way painlessly into the unfamiliar territory. Such books would obviously be a great help to research students as well as to established organic chemists.

We have been very fortunate with the highly experienced and expert organic chemists, who, agreeing with our objective, have written the first group of volumes in this series, *Best Synthetic Methods*. We shall always be pleased to receive comments from readers and suggestions for future volumes.

A.R.K., O.M.-C., C.W.R

Preface

This book was written by chemists for use by other chemists working in an organic chemistry laboratory and for all scientists with an interest in biotransformations. It is an attempt to summarize the important aspects of work in the burgeoning field of biotransformations, i.e. the use of micro-organisms and/or isolated, partially purified enzymes for the conversion of a given substrate into a useful product. The emphasis within the book is on the description of processes and transformations which should be easy to reproduce within a chemistry laboratory without the requirement of special equipment. Many of the transformations that are described can be conducted on a multigram scale, and in a good number of cases there is no reason to believe that the corresponding kilogram scale reaction would be impractical. This review of the field of biotransformations is not intended to be comprehensive. Many enzyme-catalysed transformations that have been conducted only on a very small scale have been omitted. In addition the vast majority of the work involving the synthesis of labelled materials for biosynthetic studies has been omitted. Similarly, specific biotransformations that are of crucial importance to the work of one laboratory but are not, at least at the present time, available to other chemists are not discussed in detail. Thus the use of penicillin synthetase to produce novel penicillins and cephalosporins from tripeptides is hardly mentioned in this text although the scientific merit of the work is enormous. The use of restriction endo-nucleases and ligases is routine in laboratories concerned with DNA synthesis. However, these enzymes are not used as catalysts for a wider range of other processes and hence a full discussion of the impact of enzyme controlled reactions on recombinant DNA technology is considered to be outside the scope of this book. At the appropriate point interested readers are referred to relevant texts covering this area.

This book does not attempt to emphasize the use and importance of enzymes in a number of industrial applications. For example, proteases are the most widely utilized enzymes, being important additives in "biological" washing powders. The uses of the amylases in the sugar industry, rennet in cheese production, isomerase enzymes for the conversion of glucose to fructose, and lipases for *trans*-esterification reactions in fats are all vitally important processes in the food industry. Equally important processes using whole cells include sewage treatment on the one hand, and the production of solvents such as ethanol and butanol on the other. We believe that the study of such industrial processes has given a basis to a number of the areas of research of current importance as described hereafter, but a full description of the historical perspective concerning biological catalysis and biotechnology in the chemistry industry is best sought elsewhere (*1*).

In summary, the authors have concentrated their efforts on distilling from the literature those biotransformations which should be of interest to a wide range of scientists.

We thank the following persons for contributing experimental procedures for incorporation in this book: Professor D. W. Brooks, Professor C. Fuganti, Dr O. Ghisalba, Professor J. B. Jones, Professor A. M. Klibanov, Professor K. Mori, Professor A. Ohno, Professor M. Ohno, Professor D. Seebach, Professor C. J. Sih, Professor C. Wandrey, Professor G. M. Whitesides, Professor C.-H. Wong, Professor H. Wynberg and Dr D. W. Young. Thanks are also due to S. Green, S. D. Bull, R. Collier and J. W. Kelly for help in the preparation of the manuscript.

REFERENCE

1. See, for example, M. K. Turner, *The Chemical Industry*, ed. C. A. Heaton, Blackie, Glasgow, 1986.

Detailed Contents

Biotransformations—Introduction and Background Information

1.1. THE PAST: A SELECTION OF THE IMPORTANT, WELL-ESTABLISHED USES OF WHOLE-CELL AND ENZYME CATALYSED BIOTRANSFORMATIONS

The fermentation of sugar to ethanol by yeast is a process that has been known for a very long time and, indeed, the conversion featured in some of the earliest scriptures. The process has been modified extensively over the years: even today there is considerable interest in optimizing the conversion of sugar cane into alcohol so that the latter compound can be added to the gasoline needed for automotive power. This strategy is obviously of most importance in countries (e.g. Brazil) where sugar cane is relatively cheap.

Not only is the sugar → alcohol conversion of established commercial importance but, through Pasteur's studies, it helped to lay the foundation of the science of microbiology and contributed to our understanding of catalysis. During his work Pasteur noted that yeasts produced glycerol as a minor component of sugar fermentations. Later research demonstrated that when *Saccharomyces cerevisiae* is grown in a medium buffered to alkaline pH, the yield of glycerol increased. Thus manipulation of the yeast's metabolism allowed large-scale production of the triol. *Clostridium acetobutylicum* was shown to metabolize glucose in yet another way: under anaerobic conditions the solvents butanol and acetone are produced. This field was researched extensively during World War 1.

Just as the production of ethanol underlies the traditional process of brewing, so the fermentative production of acetic acid and lactic acid relate to vinegar manufacture and milk fermentation, respectively. The production of lactic acid was first demonstrated in 1880. Today lactic acid is produced from the lactose in whey using the organism *Lactobacillus bulgaricus*. Related organisms (*L. delbrueckii* and *L. pentosus*) convert glucose into the same product.

Citric acid is another important compound that is, for the most part, obtained by a fermentation process dating back over 70 years. The original process involved the degradation of sucrose by the micro-organism *Asper-*

gillus niger and a variant of this process is still employed today. The yeasts *Candida guilliermondi* and *C. lipolytica* allow the use of alternative feedstocks such as low molecular weight alcohols and n-alkanes. Today citric acid is used as a flavouring agent (for food and drink) and as an antioxidant.

The extraction of amino acids from protein is the oldest method for their production and L-cysteine is still obtained in this way. However, at the present time the manufacturing processes for L-glutamate, L-lysine, L-glutamine, and L-arginine are all based on microbiological processes. For example, L-glutamate is produced from starch or molasses using the organism *Corynebacterium glutamicum*. The titre of L-glutamate in the fermentation broth can reach 60% of the theoretical yield based on the amount of input sugar. A strain of *C. glutamicum* can also be used to manufacture L-lysine.

In addition to products resulting, by and large directly, from the primary metabolic cascade, many organisms have been shown to produce important secondary metabolites. Perhaps the best known secondary metabolites produced by fungi are the penicillins and cephalosporins produced by *Penicillium* spp. and *Cephalosporium* spp. The annual world production of penicillin-G (**1**) and -V (**2**) is ca. 12 000 tonne. Of no less commercial importance was the discovery of the hydrolysis of these penicillins to 6-aminopenicillanic acid (**3**) using the bacterium *Escherichia coli* or the appropriate amidase isolated from the micro-organism. This microbiological process rivalled the alternative chemical processes in terms of simplicity and cost and led to the preparation of "semisynthetic" penicillins such as methicillin (**4**) for use in the clinic as powerful anti-bacterial agents.

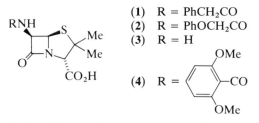

(**1**) R = $PhCH_2CO$
(**2**) R = $PhOCH_2CO$
(**3**) R = H

(**4**) R =

Steroids are derived from plants and mammals. Once again the biologically important materials (e.g. anti-inflammatory agents) must be synthesized from the natural products. Almost 40 years ago it was observed that progesterone was converted into 11α-hydroxyprogesterone (**5**) using *Rhizopus arrhizus*. This regioselective oxidation of the complex tetracyclic molecule was not only to become of historical interest, but also became commercially important for the manufacture of cortisone. Another spectacular, well-known transformation performed by mycobacteria, is the oxidation of β-sitosterol (**6**) into androstenedione (**7**).

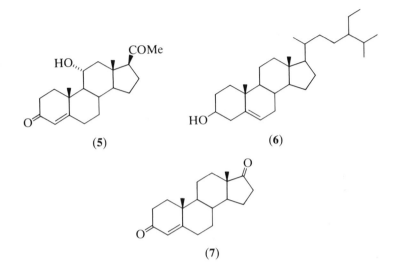

(5)   (6)

(7)

Since about 1960, the use of partially purified enzymes as specific catalysts for selected processes became of great importance to a number of industries. The enzymes which hydrolyse proteins (bacterial proteases) are important additives in "biological" washing powders and have also found applications in the food, textile and photographic industries. Enzymes such as bacterial α-amylase and amyloglucosidase are widely used in the manufacture of sugar. Glucose isomerase (e.g. from *Bacillus coagulans*) is used to convert glucose into the higher value, sweeter fructose. Rennet is used in cheese manufacture.

The manufacture of fine chemicals using enzyme catalysed reactions is also well established. Some optically active amino acids are produced on a tonne scale using this methodology and the methods for the preparation of compounds such as L-leucine from convenient starting materials such as L- or D-hydroxyisocaproate illustrate the complexity of the reaction cascades in which purified enzymes participate (Scheme 1.1).

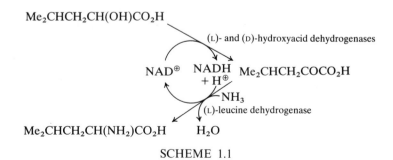

SCHEME 1.1

Most of the above well-established fermentation procedures are still in operation today: in addition the development of an increasing number of processes that depend on enzyme catalysed reactions point to continued steady growth in this direction. The large-scale conversions using whole cells and the enzymic transformations were established on the basis of continued research over 100 years. The vastly increased amount of effort put into this research area over the past 15 years should prepare the ground for more and more industrially important biotransformations.

1.2. THE PRESENT

In the following Chapters we have collated most of the important research work of the past 15 years (up to late 1987) that involves bioconversions using whole-cell systems or isolated enzymes.

For convenience, we have divided the text into the following sections: Chapter 2, hydrolysis and condensation reactions; Chapter 3, enzyme catalysed reduction reactions; Chapter 4, oxidation reactions; and Chapter 5, other biotransformations. In each Chapter and its sub-sections, the enzymic and microbiological methods for the transformation(s) under scrutiny are surveyed. Emphasis has been placed on reporting reactions that may be of general interest and which may be taken up by scientists initially inexperienced in the use of bio-catalysts.

A newcomer to the field is often faced with an initial dilemma: if wishing to repeat a "literature" process, or if wishing to try a known transformation on a new substrate, a scientist must sometimes decide whether to use a partially purified enzyme or a whole-cell system for the desired conversion. There are various "pros and cons" to be considered and some of these are summarized in Table 1.1. Except for the instances where common yeasts (bakers' yeast, brewers' yeast) are required for the transformation, scientists with little or no experience in handling or growing micro-organisms would have to enlist the help of a microbiologist, at least initially. The vast majority of the organisms mentioned in this text are available from culture collections and the ease of growth, handling and use of the cells is detailed in Appendix 1.1 [1].

Once the micro-organism has been obtained as a suspension in water or as a wet paste, the use of a shake-table to mix the substrate and the catalyst is preferred; an overhead stirrer can be employed if shear stress on the cells is avoided. After the reaction is complete, the cells can be removed by filtration or (better) by centrifugation. The whole-cell reactions often demand the use of large quantities of water. Obviously water soluble end products can be difficult to extract from the solution; equally, care should be taken to ensure that lipophilic products are washed from the cell debris.

Table 1.1   Some of the advantages and disadvantages of using a whole-cell system or an enzyme system for a biotransformation

Biotransformation system	Advantages	Disadvantages
Whole cell	Cheap, enzyme cofactors present	Fermentation equipment and/or microbiological expertise often required. Work up (involving, for example, removal of spent cells and extraction of products from a large volume of water) can be tedious. Side reactions can interfere or dominate. Substrate and/or product and/or organic co-solvent can affect transport of material into or out of the cell.
Isolated enzymes	Simple apparatus. Process easier to control and monitor. Simple work up. Specific for selected reaction. Cosolvents tolerated better.	Expensive. Addition of enzyme cofactors may be required; for processes other than very small scale reactions, cofactor recycling must be set up where necessary.

Micro-organisms catalyse a great number of biotransformations. The transformation (say A → B) required by the chemist may be overshadowed by a second metabolic pathway (A → C), or the first formed product may be rapidly turned into other compound(s)

$$[A \xrightarrow{k_1} B \xrightarrow{k_2} C (+ D + E \ldots); k_2 > k_1].$$

The above-mentioned disadvantages of using a whole-cell system are counter-balanced by certain advantages, namely that the use of micro-organisms is inexpensive and that all the entities necessary for the transformation [enzyme(s), cofactor(s), metal ion(s), etc.] are present within the cell.

The use of an isolated enzyme as the catalyst for the required biotransformation is often the alternative course of action that should be considered. Enzymes are categorized into six groups following the recommendations of the International Union of Biochemistry: the groups comprise the following enzymes:

(1) Oxidoreductases: enzymes catalysing reactions such as

$$-\overset{|}{C}HOH \; \rightleftharpoons \; C=O; \quad -\overset{|}{C}H-\overset{|}{C}H- \; \rightleftharpoons$$

$$\overset{\backslash}{\underset{/}{C}}=\overset{/}{\underset{\backslash}{C}} \quad \text{and} \quad -\overset{|}{\underset{|}{C}}-H \; \longrightarrow \; -\overset{|}{\underset{|}{C}}-OH$$

(2) Transferases: enzymes catalysing the transfer of groups (such as acyl or phosphoryl moieties) from one molecule to another.
(3) Hydrolases: enzymes catalysing the hydrolysis of esters, amides, glycosides, etc.
(4) Lyases: enzymes catalysing the addition of HX (X ≠ OH) to alkene, imine and carbonyl groups.
(5) Isomerases: enzymes catalysing isomerization processes such as racemization and epimerization.
(6) Ligases: enzymes which catalyse the connection of two molecules, mediating the formation of C–O, C–N and C–C bonds, coupled with the cleavage of a pyrophosphate unit from adenosine triphosphate (or a similar triphosphate).

The appropriate number from the above list is the first of a series of four numbers (the E.C. number) that is given to characterize a particular enzyme. The second number in the series indicates the subclass. For oxidoreductases it indicates the type of group in the donors which undergoes oxidation (1 denoting a secondary alcohol group, 2 denoting an aldehyde or ketone unit and so on); for transferases it gives an indication of the functional group which is transferred (1 indicates the transfer of a one-carbon unit); for the hydrolases it earmarks the functional group hydrolysed (1 is used when an ester group is hydrolysed); for the lyases the subclass number indicates the group HX (3 indicates the addition of ammonia); for the isomerases the second number shows the type of isomerization (2 indicates alkene *cis–trans* isomerization), while for ligases the type of bond formed is indicated (4 indicates carbon–carbon bond formation).

The third number in the series serves to allocate the enzyme a sub-subclass. For oxidoreductase enzymes this third figure shows the type of acceptor involved (for example, 1 denotes a co-enzyme {such as nicotinamide adenine dinucleotide (phosphate) [NAD(P)]}, 2 a cytochrome, 3 molecular oxygen, etc.); for the transferases the third figure allows a subdivision of the type of group transferred (thus the C_1 unit can be defined as methyl or carboxyl, etc.). For the hydrolases, the lyases, the isomerases and the ligases the third figure shows more precisely the type of bond hydrolysed, the nature of the added group, further detail of the isomerization, and the type of substance formed, respectively.

The fourth number in the series completes the serial number of the enzyme and allows each enzyme to have a unique four-figure number for reference purposes.

A key to the numbering and classification of the enzymes is available in text-books on biochemistry [2].

Well over 2000 enzymes have been identified (Table 1.2) and roughly 15%

are commercially available as partially purified protein (Appendix 1.2). (Note that the commercially available enzymes can contain small amounts of protein with catalytic properties totally different to those required. However, this is rarely a problem in the field of biotransformations unless a series of related enzymes (isozymes) are present in the mixture.)

Table 1.2. Numbers of enzymes that have been identified and the numbers that are commercially available

Enzyme type	Number of enzymes identified	Number of enzymes commercially available
Oxidoreductase	650	90
Transferase	720	90
Hydrolase	636	125
Lyase	255	35
Isomerase	120	6
Ligase	80	5

The cost of commercially available enzymes varies considerably and depends to a large extent on whether the enzyme in question is used for a commercial purpose (e.g. useful lipases are relatively cheap because of their use in effecting transesterification processes in the food industry) and on how difficult it is to isolate the enzyme from the natural source.

The use of isolated enzymes has a distinct advantage over the use of whole-cell systems in that the chosen enzyme will generally catalyse just one reaction and unwanted, totally different side reactions are rarely observed. The use of partially purified enzymes allows the reaction to be conducted in a small amount of water/solvent; the temperature and pH of the solution are easy to control and organic cosolvents are often well tolerated. Indeed some enzymes can tolerate almost complete exclusion of water from the reaction medium, a fact which would have been unthinkable for most scientists just a few years ago. Work up of an enzyme catalysed reaction is rarely problematic, but can be made even easier by attaching the enzyme to a solid support. The various processes for enzyme immobilization are extremely straightforward [3]; for example, the enzyme is stirred gently for some hours with the chosen support in a 1 M phosphate buffer. The solid is filtered and washed with buffer until no protein is present in the filtrate (monitored by UV absorption bands at 260 nm and 280 nm). As well as providing a useful reusable catalyst system, immobilization of an enzyme can act to stabilize the macromolecule. Thus it is often preferable to immobilize an enzyme before conducting transformations in organic solvents or at higher temperatures.

Typical solid supports for enzymes are Sepharose, silica, Fluorisil, neutral alumina, porous ceramics, and epoxy-acrylic polymers such as Eupergit-C. Other strategies for the immobilization and stabilization of enzymes are published; for example, Whitesides has recommended the enclosure of enzymes in a dialysis bag to protect the protein from damage by organic solvent in a two-phase system [4]. It is noteworthy that whole cells can also be trapped and immobilized in polymers such as alginate, but the techniques involved are not as simple and would require some practice by a newcomer to the field.

Enzymes such as esterases and lipases which need only the presence of water to promote hydrolysis reactions are very easy to use. The only control necessary is that of the pH of the solution; as an ester is hydrolysed the release of the carboxylic acid causes a decrease in the pH of the medium. The pH can be maintained at the desired value using an autotitrator; however, many esterases and lipases operate over a range of pH values so that, unless the hydrolysis is very rapid, the employment of a pH meter and the periodic addition of base may be preferred.

Taking other enzymes out of their natural habitat can give rise to problems which require more attention. For example, oxidoreductases require cofactors such as nicotinamide adenine dinucleotide (phosphate) $[NAD(P)^+]$ in order to accomplish a reduction: (Scheme 1.2).* Such cofactors are difficult to obtain and are also expensive to buy so that use of a

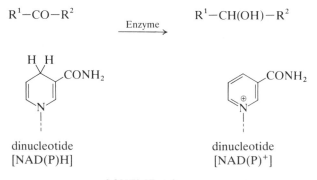

SCHEME 1.2

molar equivalent of cofactor in the reaction is unreasonable except when the reaction is a small-scale, "one-off" transformation. In all other cases cofactor recycling must be set up as shown in Scheme 1.3. Enzyme 1 and enzyme 2 may be the same or different. The details concerning cofactor recycling will be considered later in the relevant sections; suffice it to say here

* Prior to the 1960s NAD^+ and $NADP^+$ were referred to as diphosphopyridine dinucleotide (DPN) and triphosphopyridine dinucleotide (TPN), respectively.

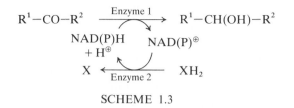

SCHEME 1.3

that the recycling of NAD(P)H and ATP has received a lot of attention and all three cofactors can be recycled very efficiently and effectively. Recycling NAD(P)$^+$ is not as straightforward and a simple, generally useful method for the regeneration of this cofactor has yet to be found.

The remaining, perhaps the most difficult, question to address in this introductory Chapter is when is it timely for a chemist to think of employing a biotransformation as a key step in a multistage synthesis? Perhaps the simplest answer is to consider the commercially available enzymes and the more readily obtained and easily handled micro-organisms as alternative synthetic methods to be considered alongside other more conventional reagents and catalysts. Enzyme and whole-cell transformations may be particularly favoured when the substrate and/or product is relatively un-stable, such that the desired reaction must be performed under mild conditions of temperature and pH. However, the greatest advantage gained by the use of biocatalysts is that, being chiral, enzymes often promote reactions in a stereoselective or enantioselective fashion. The obtention of optically active compound(s) from a biotransformation process is common. The other peculiarity of enzymes is their ability to promote reactions at ostensibly non-activated sites in the substrate molecule.

"Biotransformations" should, therefore, be thought of in terms of provid-ing a complementary set of reagents to be considered alongside other well-documented catalysts. It may be useful to consider some specific examples: if an ester such as (8) is needed as an intermediate and the compound is required in optically active form, one method of preparation to be con-sidered involves reacting the corresponding diol with benzoyl chloride in the presence of a chiral amine [5]. Another method would be to make the appropriate diester (9) and effect selective hydrolysis with an esterase or

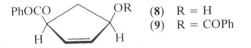

(8) R = H
(9) R = COPh

lipase (Chapter 2). The latter method would probably be the preferred strategy at this time. The preparation of optically active β-hydroxyesters can be achieved from the corresponding β-ketoesters using dehydrogenase enzymes (with cofactor recycling), or by using whole-cell systems (Chapter 3). The biotransformation processes are easy to operate and cheap to run. However, the possible utilization of such a process must be assessed alongside the alternative methodology employing excellent, albeit expensive, man-made chiral catalysts such as Noyori's reagents [RuX$_2$(BINAP) or Ru$_2$Cl$_4$(BINAP)$_2$pyr] which can be used for the same purpose [6].

Hydroxylation at an unactivated carbon atom remote from other functionality can be accomplished using whole-cell systems (a technique brilliantly exploited in the preparation of biologically active steroids), although the predictability of the point of oxidation in other alicyclic systems as well as acyclic systems leaves a lot to be desired (Chapter 4). Non-enzymic mimics of such a process are still very much in their infancy [7]. Other processes, such as the conversion of toluene into optically active 3-methylcyclohexane-1,2-diol (Chapter 4), simply cannot be matched by a chemical process. On the other hand, stereocontrolled aldol reactions [8] and the formation of optically active epoxides [9] are conversions best accomplished at the present time by chemical techniques, although the use of isolated enzymes and whole-cell systems to serve the same purposes and to extend the ranges of products available is being actively researched by a number of groups.

In a nutshell, the following Chapters in this book should serve to show:

(a) esterases, lipases and amidases are simple-to-use catalysts for the preparation of optically active alcohols, amines, and acids. The area is sufficiently well-researched so as to be of potential use to a wide range of chemists (Chapter 2);

(b) dehydrogenase enzymes and micro-organisms can be used to effect stereoselective or enantioselective reduction of a ketone to furnish the corresponding secondary alcohol. Further research work is needed to optimize these reduction processes before their use for large-scale processes becomes widespread (Chapter 3);

(c) the synthesis of optically active phosphate esters is now possible using enzyme catalysis and this tactic should be seriously considered by chemists entering this field of work (Chapter 2);

(d) a wide variety of transformations is possible using enzymes and whole cells: some of these transformations cannot be emulated by currently available chemical reagents (Chapters 4 and 5).

1.3. THE FUTURE: PROSPECTS FOR BIOTRANSFORMATIONS

Obviously the number of applications using enzymes such as esterases, lipases and, to a lesser extent, dehydrogenases will continue to increase rapidly over the next few years. The use of enzymes with different catalytic activities will be the focus of much attention: the potential for enzymes which catalyse carbon–carbon bond formation (for example the aldolases) is very great.

The use of genetically manipulated micro-organisms to accomplish useful multistage processes will be increasingly in evidence. Two examples will have to suffice to demonstrate the possibilities. Glucose can be converted into ascorbic acid via 2,5-diketo-D-gluconate and 2-keto-L-gulonate. The conversion of glucose into 2,5-diketo-D-gluconate uses a mutant strain of *Erwinia*, while the next stage, reduction of the gluconate into the gulonate, is effected by *Corynebacterium*. The reductase enzyme responsible for the second stage of the process was identified, then the relevant gene was cloned and expressed in the organism *Erwinia herbicola*. The net result was a genetically engineered organism that was capable of converting D-glucose into 2,keto-L-gulonate. Although the yields and productivity obtained with the new organism were very low, the technical achievement was noteworthy and acts as a pointer for the future.

The second example concerns the microbiological conversion of tryptophan into indigo. Native *Escherichia coli* can convert tryptophan into indole. Incorporation of the enzyme naphthalene dioxygenase into the bacterium causes the indole to be converted into the diol (**10**) *en route* to indigo.

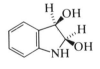

(**10**)

The modification of enzymes by site-directed mutagenesis in order to extend the range of the natural catalytic properties is well established.

Modification of yeast dehydrogenase by site-directed mutagenesis has given rise to a catalyst with different oxidation properties. For example, octanol is a good substrate for a mutant, but not the wild-type enzyme. The protease subtilisin has been modified so as to alter the pH optimum for the catalytic reaction.

Recently, a more dramatic alteration of the selectivity of an enzyme has been reported. Three amino acid substitutions have been made in the lactate dehydrogenase from *Bacillus stearothermophilus*. The wild-type enzyme has a catalytic specificity for pyruvate over oxaloacetate of *ca.* 1000, whereas the triple mutant has a specificity for oxaloacetate over pyruvate of 500. The malate dehydrogenase so produced exhibits a reasonable catalytic rate and is allosterically controlled by fructose-1,6-bisphosphate [10].

Perhaps one of the most fascinating concepts that has been explored recently is the use of antibodies as catalysts. The basic idea involves raising antibodies to a stable mimic of the transition state of the reaction under investigation. Given that the transition state mimic is a good one, the antibody should have the capacity to stabilize, to some extent, the appropriate high energy transition state [11].

The future holds many fascinating rewards for workers in the area of biotransformations. At the same time very many more non-specialist laboratories will find advantage in using enzymes or whole cells as catalysts for key reactions.

REFERENCES

1. We are grateful to Dr A. J. Willetts (Department of Biological Sciences, Exeter University) for help in compiling Appendix 1 and Dr J. Leaver (Glaxo Group Research, Greenford) for compiling Appendix 2.

2. M. Dixon and E. C. Webb, Enzyme classification, *Enzymes*. Academic Press, New York, 1979.

3. R. Aksen and S. Ernback, *Eur. J. Biochem.* **18**, 351 (1971); B. Solomon, R. Koppel, G. Pines and E. Katchalski-Katzir, *Biotechnol. Bioeng.* **28**, 1213 (1986); E. Guibé-Jampel and G. Rousseau, *Tetrahedron Lett.* **28**, 3563 (1987); A. Wiseman ed., *Principles of Biotechnology*. Blackie, University of Surrey, 1985; manufacturers of solid supports generally recommend procedures for immobilization.

4. M. D. Bednarski, H. K. Chenault, E. S. Simon and G. M. Whitesides, *J. Am. Chem. Soc.* **108**, 1283 (1987).

5. L. Duhamel and T. Herman, *Tetrahedron Lett.* **26**, 3099 (1985).

6. R. Noyori, T. Ohkuma, M. Kitamura, H. Takaya, N. Sayo, H. Kumobayashi and S. Akutagawa, *J. Am. Chem. Soc.* **109**, 5856 (1987); see also T. Kikukawa, Y. Iizuka, T. Sugimwa, T. Harada and A. Tai, *Chem. Lett.* 1267 (1987); see also M. Kitamura, T. Ohkuma, S. Inoue, N. Sayo, H. Kumobayashi, S. Akutagawa, T. Ohta, H. Takaya and R. Noyori, *J. Am. Chem. Soc.* **110**, 629 (1988).

7. D. H. R. Barton, J. Boivin, M. Gastiger, J. Morzycki, R. S. Hay-Motherwell, W. B. Motherwell, N. Ozbalik and K. M. Schwartentuber, *J. Chem. Soc., Perkin Trans. 1* 947 (1986); P. Leduc, P. Battioni, J. F. Bartoti and D. Mansuy, *Tetrahedron Lett.* **29**, 205 (1988).

8. C. H. Heathcock, in *Asymmetric Synthesis*, Vol. 3 (ed. J. D. Morrison), p. 111. Academic Press, New York, 1983; M. Braun, *Angew Chem., Int. Ed.* **26**, 24 (1987).

9. E. T. Kaiser and D. S. Lawrence, *Science* **226**, 505 (1984); C. Murali and E. H. Creaser, *Protein Eng.* **1**, 55 (1986); Y. Gao, R. M. Hanson, J. M. Klunder, S. Y. Ko, S. Masamune and K. B. Sharpless, *J. Am. Chem. Soc.* **109**, 5765 (1987).

10. A. R. Clarke, C. J. Smith, K. W. Hart, H. M. Wilks, W. N. Chia, T. V. Lee, J. J. Birktoft, L. J. Banaszak, B. A. Barstow, T. Atkinson and J. J. Holbrook, *Biochem. Biophys. Res. Commun.* **148**, 15 (1987).

11. A. D. Napper, S. J. Benkovic, A. Tramontano and R. A. Lerner, *Science* **237**, 1041 (1987); R. A. Lerner and A. Tramontano, *Sci. Am.* **42**, March (1988).

APPENDIX 1.1

Micro-organism	Ease of growth/ Ease of Handling*	Comments
Absidia spp.	1/2	
Achromobacter spp.	Reclassified as *Alcaligenes, Acinetobacter, Pseudomonas* or *Moraxella* spp.	
Acinetobacter lowfii	1/3	Human pathogen (ACDP Group 2)
Acremonium roseolum	1/1	
Arthrobacter sp.	1/2	
Alcaligenes spp.	1/2	
Arthrobacter simplex	2/2	
Aspergillus alliaceus	1/1	
A. foetidus	1/1	
A. fumigatus	1/3	Human pathogen (ACDP Group 2)
A. niger	1/1	Can produce mycotoxins
A. nigricans		
A. ochraceus		
A. oryzae	1/1	
A. parasiticus		
A. sojae		
A. tamarii		
Bacillus subtilis (subsp. niger)	1/2	
B. coagulans	1/2	
Bakers' yeast	see *Saccharomyces cerevisiae*	
Beauveria bassiana	1/1	
B. sulfurescens	see *Beauvaria bassiana*	
Botrytis allii	2/1	
Calonectria decora	1/1	
Candida cylindracea	see *Candida rugosa*	
C. guilliermondi	1/1	
C. lipolytica	1/1	
C. rugosa	1/1	
Clostridium acetobutylicum	2/2	
Cordyceps spp.	1/1	
Corynebacterium equi	1/1	Low pathogenicity for laboratory animals but does not represent a hazard for humans
C. glutamicum	2/2	
C. simplex	see *Arthrobacter simplex*	
Cunninghamella blakesleena		
C. echinulata	1/1	
C. elegans		

APPENDIX 1.1 (continued)

Micro-organism	Ease of growth/ Ease of Handling*	Comments
Curvularia lunata	1/1	
Cylindrocarpon radicola	2/1	
Daedalea spp.	2/1	
Diaporthe spp.	1/1	
Erwinia herbicola	2/2	
Escherichia coli	1/3	Human pathogen (ACDP Group 2)
Eubacterium lentrum	1/2	
Flavobacterium lutescens	1/2	
F. oxydans	1/2	
Fusarium equiseti	1/2	
Gibberella fujikuroi	1/1	
Gliocladium roseum	1/2	
Glomus nigricans	2/2	
Gluconobacter oxydans	1/2	
Gongronella lacrispora	1/1	
Helicostylum pyriforme	1/1	
Helminthosporum spp.	2/1	
Lactobacillus spp.	1/2	
Microsporum spp.	2/3	Human pathogens (ACDP Group 2)
Monosporium olivaceum	1/2	
Moraxella spp.	1/3	Human pathogens (ACDP Group 2)
Mortierella isabellina	1/1	
Mucor spp.	1/1	
Nocardia spp.	1/1	*N. asteroides* and *N. brasiliensis* are pathogenic
Ophiobolus herpotrichus	3/1	
Paecilomyces lilacinus	1/1	
Pellicularia filamentosa	1/2	
Penicillium crysogenum } *P. aurantiogriseum* } *P. griseofulvum* }	1/1	
P. lilancinium	see *Paecilomyces lilacinus*	

APPENDIX 1.1 (continued)

Micro-organism	Ease of growth/ Ease of Handling*	Comments
P. spinulosum	1/1	
P. urticae	see *P. griseofulvum*	
Proteus mirabilis	1/3	Human pathogen (ACDP Group 2)
Pseudomonas aeruginosa ⎫		*Human pathogens*
P. oleovorans ⎬	1/3	(ACDP Group 2)
P. putida ⎭		
Rhizophagus nigricans	see *Glomus nigricans*	
Rhizopus arrhizus	see *R. oryzae*	
R. circinans	1/1	
R. nigricans	see *R. stolonifer*	
R. stolonifer ⎫	1/1	
R. oryzae ⎭		
Rhodococcus spp.	1/2	
Saccharomyces cerevisiae	1/1	
Sepedonium chrysospermum	2/1	
Sporotrichum sulphurescens	see *Beauvaria bassiana*	
Stachylidium theobromae	see *Verticillium theobromae*	
Streptomyces fradiae ⎫		
S. griseus ⎪		
S. lavendulae ⎬	1/1	
S. platensis ⎪		
S. punipalus ⎪		
S. roseochromogenes ⎭		
Syncephalastrum racemosum	2/2	
Verticillium theobromae	1/1	
Wojnowicia graminis	2/2	
Xanthobacter autotrophicus	1/2	
Xanthomonas citri	2/2	

* Marked on a scale 1 → 3: 1, easy; 2, moderately easy; 3, difficult. Thus 1/3 signifies that the organism is easy to grow but difficult to handle, while 3/2 signifies that the organism is difficult to grow but is moderately easy to handle.

APPENDIX 1.2

Some commercially available enzymes and examples of the reactions that they catalyse

Enzyme (E.C. number)	Typical reaction	Notes
1. Oxidoreductases		
L-Alanine dehydrogenase (1.4.1.1)	L-Ala + H_2O + NAD → pyruvate + NH_3 + NADH	
Alcohol dehydrogenase (1.1.1.1)	Alcohol + NAD → aldehyde/ketone + NADH + glycine	
Alcohol oxidase (1.1.3.13)	RCH_2OH + O_2 → RCHO + H_2O_2	1
Aldehyde dehydrogenase (1.2.1.5)	RCHO + NAD + H_2O → RCO_2H + NADH	
D/L Amino acid oxidase (1.4.3.3/2)	Amino acid + H_2O + O_2 → 2-oxo-acid + NH_3 + H_2O_2	
Cholesterol oxidase (1.1.3.6)	Cholesterol + H_2O → 4-cholesten-3-one	
Choline oxidase (1.1.3.17)	Choline + H_2O → betaine aldehyde + H_2O_2	
Cytochrome C reductase (1.6.99.3)	NADH + acceptor → NAD + reduced acceptor	
Cytochrome C oxidase	4-Ferrocyt·c + O_2 → 4-Ferricyt·c + $2H_2O_2$	
Diamine oxidase (1.4.3.6)	RCH_2NH_2 + H_2O + O_2 → RCHO + NH_3 + H_2O_2	
Formate dehydrogenase (1.2.1.2)	HCOOH + NAD → CO_2 + NADH	
L-Fucose dehydrogenase (1.1.1.122)	L-Fucose + NAD → L-fucono-1,5-lactone + NADH	
Galactose dehydrogenase (1.1.1.48)	D-Galactose + NAD → D-galactono-γ-lactone + NADH	
Galactose oxidase (1.1.3.9)	D-Galactose + O_2 → D-galacto-hexodialdose + H_2O_2	
Glucose dehydrogenase (1.1.1.47)	β-D-Glucose + NAD(P) → D-gluconolactone + NAD(P)H	
Glucose oxidase (1.1.3.4)	β-D-Glucose + O_2 → D-gluconolactone + H_2O	
Glucose-6-P-dehydrogenase	D-G6P + NADP → 6-Phospho-D-gluconate + NADPH	
L-Glutamic dehydrogenase (1.4.1.3)	L-Glutamate + H_2O + NAD(P) → 2-oxoglutarate + NH_3 + NAD(P)H	
Glutathione reductase	NAD(P)H + oxidized glutathione → NAD(P) + glutathione	
Glyceraldehyde-3P-dehydrogenase (1.2.2.12)	3-Phosphoglycerate + Pi + NAD → D-glyceraldehyde-3-P + NADH	
Glycerol dehydrogenase (1.1.1.6)	Glycerol + NAD → dihydroxy-acetone + NADH	3

Glycerophosphate dehydrogenase (1.1.1.8)	Dihydroxyacetone-P + NADH → sn-glycerophosphate + NAD	
Glycollate oxidase (1.1.3.1)	Glycollate + O_2 → glyoxylate + H_2O_2	4
Glyoxylate reductase (1.1.1.26)	Glyoxylate + NADH → glycollate + NAD	5
β-Hydroxyacyl CoA reductase (1.1.1.35)	L-3-Hydroxyacyl-CoA + NAD → 3-oxyacyl-CoA + NADH	
β-Hydroxybutyrate dehydrogenase (1.1.1.30)	D-β-Hydroxybutyrate + NAD → acetoacetate + NADH	
3α-Hydroxysteroid dehydrogenase (1.1.1.50)	Androsterone + NAD(P) → 5α-androstane-3,17-dione + NAD(P)H	6
β-Hydroxysteroid dehydrogenase (1.1.1.51)	Testosterone + NAD(P) → 4-androstene-3,17-dione + NAD(P)H	
7α-Hydroxysteroid dehydrogenase (1.1.1.159)	3α,7α,12α-Trihydroxy-5β-cholanate + NAD → 3d,12α-dihydroxy-7-oxo-5β-cholanate + NADH	8
3α,20β-Hydroxysteroid dehydrogenase (1.1.1.53)	17,20β,21-Trihydroxypregn-4-ene-11-dione + NAD → cortisone + NADH	9
myo-Inositol dehydrogenase (1.1.1.18)	myo-Inositol + NAD → 2,4,6/3,5-pentahydroxycyclohexanone (scyllo inosose) + NADH	
Isocitrate dehydrogenase (1.1.1.42)	Isocitrate + NADP → α-keto-glutarate + NADPH + CO_2	
L-Lactate-2-monooxygenase (1.13.12.4)	L-Lactate + O_2 → acetate + CO_2 + H_2O	
Lactate dehydrogenase (1.1.1.27)	Pyruvate + NADH → lactate + NAD	10
Lipoamide dehydrogenase (1.6.4.3)	DL-Lipoamide + NADH → DL-Dihydrolipoamide + NAD	
Lipoxidase (1.13.11.12)	Linoleate + O_2 → 13-hydroper-oxyoctadeca-9,11-dienoate	11
Malic dehydrogenase (1.1.1.37)	Oxaloacetic acid + NADH → L-malate + NAD	12
Malic enzyme (1.1.1.37)	L-Malate + NADP → pyruvate + CO_2 + NADPH	13
Monoamine oxidase (1.4.3.4)	RCH_2NH_2 + H_2O + O_2 → RCHO + NH_3 + H_2O_2	14
Octopine dehydrogenase (1.5.1.11)	Pyruvate + L-arginine + NADH → octopine + NAD + H_2O	
6-Phosphogluconic hydrogenase (1.1.1.44)	6-Phospho-D-gluconate + NADP → D-ribulose-5-P + CO_2 + NADPH	
3-Phosphoglycerate dehydrogenase (1.1.1.95)	3-Phosphoglycerate + NAD → phosphohydroxypyruvate + NADH	
Polyol dehydrogenase (1.1.1.14)	Xylitol + NAD → D-xylose + NADH	
Saccharopine dehydrogenase (1.5.1.7)	L-Lysine + α-ketoglutarate + NADH → saccharopine + NAD + H_2O	

APPENDIX 1.2 (continued)

Enzyme (E.C. number)	Typical reaction	Notes
Sarcosine dehydrogenase (1.5.99.1)	Sarcosine + acceptor + H_2O → HCHO + glycine + reduced acceptor	
Sarcosine oxidase (1.5.3.1)	Sarcosine + H_2O + O_2 → HCHO + H_2O_2 +	
Sorbitol dehydrogenase (1.1.1.14)	D-Sorbitol + NAD → D-fructose + NADH	15
Uricase (1.7.3.3)	Urate + O_2 → unidentified products	
Xanthine oxidase (1.2.3.2)	Xanthine + H_2O + O_2 → uric acid + superoxide	16

2. Transferases

Enzyme (E.C. number)	Typical reaction	Notes
Acetate kinase (2.7.2.1)	ATP + acetate → acetylphosphate + ADP	17
Adenosine 5'-triphosphate sulphurase (2.7.7.4)	Adenosine-5'-phosphosulphate + Pi → ATP + Sulphate	
Adenylate kinase (Myokinase) (2.7.4.3)	2ADP → ATP + AMP	
Arginine kinase (2.7.3.3)	L-Arginine + ATP → N-phospho-L-arginine + ADP	
Carnitine acetyltransferase (2.3.1.7)	Ac-L-carnitine + CoA → L-carnitine + AcCoA	
Catechol-O-methyl transferase (2.1.1.6)	S-adenosyl-L-methionine + N-acetyl-serotonin → S-adenosyl-L-homocys-teine + N-acetyl-5-methoxytryptamine	
Chloramphenicol acetyltransferase (2.3.1.28)	Chloramphenicol + AcCoA → chloramphenicol-3-Ac + CoA	
Choline acetyltransferase (1.1.3.6)	Choline + AcCoA → Ac-choline + CoA	
Creatine phosphokinase (2.7.3.2)	Phosphocreatine + ADP → ATP + creatine	
Fructose-6-P kinase (2.7.1.11)	F6P + ATP → F-1,6-bisP + ADP	
Galactokinase (2.7.1.6)	D-Galactose + ATP → galactose-1P + ADP	
Gentamycin-3-acetyl transferase (2.3.1.60)	Gentamycin C + AcCoA → 3-N-acetylgentamycin C + CoA	18
Glucokinase (2.7.1.2)	D-Glucose + ATP → D-glucose-6-P + ADP	
Gluconate kinase (2.7.1.12)	D-Gluconate + ATP → 6-phospho-D-gluconate + ADP	
Glutamic oxaloacetic transaminase (2.6.1.1)	2-Oxoglutarate + L-aspartate → L-glutamate + oxaloacetate	
Glutamic pyruvic transaminase (2.6.1.2)	2-Oxoglutarate + L-alanine → L-glutamate + pyruvate	
Glutamyltranspeptidase (2.3.2.2)	(5-L-Glutamyl)-peptide + amino acid → peptide + 5-L-glutamyl amino acid	
Glutathione S-transferases (2.5.1.18)	RX + glutathione → HX + RSG	19

APPENDIX 1.2 (continued)

Enzyme (E.C. number)	Typical reaction	Notes
Glycerokinase (2.7.1.30)	Glycerol + ATP → L(α)-glycerophosphate + ADP	20
Hexokinase (2.7.1.1)	Glucose + ATP → glucose-6-P + ADP	21
NAD kinase (2.7.1.23)	NAD + ATP → NADP + ADP	
Ornithine carbamyl transferase (2.1.3.3)	L-Ornithine + carbamoyl-P → L-citrulline + Pi	
Phosphoglycerate mutase (2.7.5.3)	2,3-bisP-glycerate + 2-phospho-D-glycerate → 3-phosphoglycerate + 2,3-bisP-D-glycerate	
3-Phosphoglyceric phosphokinase (2.7.2.3)	1,3-bisP-glycerate + ADP → 3-P-glycerate + ATP	
Phosphoribulokinase (2.7.1.19)	D-Ribulose-5P + ATP → D-ribulose-1,5-bisP + ADP	22
Phosphotransacetylase (2.3.1.8)	Acetylphosphate + CoA → AcCoA + Pi	23
Pyruvate kinase (2.7.1.40)	PEP + ADP → pyruvate + ATP	24
Sucrose phosphorylase (2.4.1.7)	Sucrose + Pi → G-1-P + fructose	25
Transaldolase (2.2.1.2)	D-F6P + D-erythrose-4-P → D-glyceraldehyde-3-P + sedoheptulose-7-P	
Transketolase (2.2.1.1)	D-Xylulose-5-P + D-ribose-5-P → D-glyceraldehyde-3-P + sedoheptulose-7-P	26
Uridine-5′-bisphosphoglucose pyrophosphorylase (2.7.7.9)	UDP-Glucose + PPi → D-G-1-P + UTP	
Uridine-5′-bisphosphoglucuronyl transferase(2.4.1.17)	UDP-Glucuronate + acceptor → UDP + acceptor-β(D)-glucuronide	27

3. Hydrolases

Acetylcholinesterase (3.1.1.7)	Acetylcholine + H$_2$O → choline + acetate	28
Acid phosphatase (3.1.32)	α-Naphthyl phosphate + H$_2$O → α-naphthol + Pi	
Acylase 1 (3.5.1.14)	N-Acylamino acid + H$_2$O → fatty acid anion + amino acid	
Adenosine deaminase (3.5.4.4)	Adenosine + H$_2$O → inosine + NH$_3$	
Apyrase (3.6.1.5)	ATP + H$_2$O → ADP + Pi and ADP + H$_2$O → AMP + Pi	
5′-Adenylic acid deaminase (3.5.4.6)	5′-AMP + H$_2$O → 5′-IMP + NH$_3$	
Agarase (3.2.1.81)	Hydrolysis of 1,3β-D-galactosidic linkages in agarose giving tetramer as the predominant product	
Alkaline phosphatase (3.1.2.1)	Removal of 5′-phosphate groups from DNA, RNA, ribo- and deoxyribo-nucleoside triphosphates	

APPENDIX 1.2 (continued)

Enzyme (E.C. number)	Typical reaction	Notes
Aminoacylproline hydrolase (prolidase)(3.4.13.9)	Aminoacyl-L-proline + H_2O → amino acid + L-proline	
α-Amylase (3.2.1.1)	Endohydrolysis of 1,4α-D-glycosidic linkages in polysaccharides containing three or more 1,4α-linked D-Glc units	
Amyloglucosidase (3.2.1.3)	Hydrolysis of terminal 1,4-linked α-D-Glc residues successively from non-reduced ends of chains with release of β-D-glucose	
Arginase (3.5.3.1)	L-Arginine + H_2O → ornithine + urea	30
Asparaginase (3.5.1.1)	L-Asparagine + H_2O → L-aspartate + NH_3	
Carboxypeptidases (3.4.17)	Peptide hydrolysis	
Cathepsin C (3.4.14.1)	Dipeptidyl-polypeptide + H_2O → dipeptide + polypeptide	31
Cellulase (3.2.1.4)	Endohydrolysis of 1,4β-glycosidic linkages in cellulose, lichenin and cereal β-D-glucans	
Chitinase (3.2.1.14)	Chitin → N-Ac-D-glucosamine	
Cholinesterase, acetyl (3.1.1.7)	Ac-choline + H_2O → cholin + acetate	32
Cholinesterase, butyryl (3.1.1.8)	Butyrylcholine + H_2O → choline + butyrate	33
Choloylglycine hydrolase (3.5.1.24)	Glycocholic acid + H_2O → glycine + cholic acid	
Chymopapain (3.4.22.6)	N-α-Benzoyl-L-arginine ethyl ester + H_2O → N-α-benzoyl-L-arginine + EtOH	
α-Chymotrypsin (3.4.21.1)	Hydrolyses peptides, amides and esters preferably at the carboxyl groups of hydrophobic amino acids	
Collagenase (3.4.24.3)	Hydrolyses collagen. Preferential cleavage is at —Gly	
Creatinase (3.5.3.3)	Creatine + H_2O → urea + sarcosine	
Creatininase (3.5.2.10)	Creatinine + H_2O → creatine	
Deoxyribonuclease (3.1.21.1)	Endonucleolytic cleavage	
Dextranase (3.2.1.11)	Dextran → isomaltose	
Esterase (3.1.1.1)	Carboxylic ester + H_2O → alcohol + carboxylic acid	
D-Fructose-1,6-bisphosphatase (3.1.1.11)	D-F-1,6-bisP + H_2O → D-F6P + Pi	
α,β-Galactosidase (3.2.1.22.23)	Hydrolysis of terminal, non-reduced α-D-galactose residues in galactosides including galactose oligosaccharides, galactomannans and galactolipids	

APPENDIX 1.2 (continued)

Enzyme (E.C. number)	Typical reaction	Notes
Glucose 6-phosphatase (3.1.3.9)	D-G6P + H_2O → D-Glc + Pi	2
Glucosidase (3.2.1.20)	Maltose → glucose	
β-Glucuronidase (3.2.1.31)	β-D-Glucuronide + H_2O → alcohol + D-glucuronate	
Glutaminase (3.5.1.2)	L-Glutamine + H_2O → L-glutamate + NH_3	
Glyoxylase II (3.1.2.6)	(S)-2-Hydroxyacylglutathione + H_2O → glutathione + 2-hydroxy acid	
Hyaluronidase (3.2.1.3⁵⁾	Random hydrolysis of 1,4 linkages between 2-acetamido-2-deoxy-β-D-glc and D-glucuronate in hyaluronate	
Inorganic pyrophosphatase (3.6.1.1)	PPi + H_2O → 2Pi	
Isoamylase (3.2.1.68)	Hydrolysis of 1,6-α-D-glucosidic branch linkages in glycogen, amylopectin and their β limit dextrins	
Leucine aminopeptidase (3.4.11.1)	L-Leucinamide + H_2O → L-Leucine + NH_3	
Lipase (3.1.1.3)	Triacylglycerol + H_2O → glycerol + fatty acids	
Lysozyme (3.2.1.17)	Hydrolysis of 1,4-β linkages between N-acetylmuramic acid and 2-acetamido-2-deoxy-D-glc residues in a mucopolysaccharide or mucopeptide.	
NADase (3.2.2.5)	β-NAD + H_2O → nicotinamide + ADP-ribose	34
Pectinase (3.2.1.15)	Random hydrolysis of 1,4-α-D-galactosiduronic linkages in pectate and other galacturonans	
Pectinesterase (3.1.1.11)	Pectin + $n H_2O$ → pectate + methanol	
Penicillinase (3.5.2.6)	Benzylpenicillin + H_2O → benzylpenicilloic acid	
Phospholipase A2 (3.1.1.4)	Lecithin + H_2O → L-α-lysolecithin + fatty acid	35
Phospholipase C (3.1.4.3)	Lecithin + H_2O → choline phosphate + 1,2-diacylglyceride	36
Phospholipase D (3.1.4.4)	Lecithin + H_2O → choline + phosphatidate	37
Pyrophosphatase, nucleotide (3.6.1.9)	Dinucleotide + H_2O → 2 mononucleotides	
Sphingomyelinase (3.1.4.12)	Sphingomyelin + H_2O → N-acylsphingosine + choline + P	
Sulfatase (3.1.6.1)	Phenol sulphate + H_2O → phenol + sulphate	
Trypsin (3.4.21.4)	Protease, preferential cleavage at Arg-, Lys-	29
Urease (3.5.1.5)	Urea + H_2O → CO_2 + $2NH_3$	

APPENDIX 1.2 (continued)

Enzyme (E.C. number)	Typical reaction	Notes
4. Lyases		
Aconitase (4.2.1.3)	Citrate → *cis*-aconitate → *iso*-citrate	
Adenylsuccinate lyase (4.3.3.2)	Adenylsuccinic acid → 5'-AMP + fumaric acid	
Aldolase (4.1.2.13)	D-Fructose-1,6-bisphosphate → dihydroxyacetone-P + D-glyceraldehyde-3P	38
Amino acid decarboxylases (4.1.1)	Amino acid → amine + CO_2	
Aspartase (4.3.1.1)	L-Aspartate → fumarate + NH_3	
Citrate lyase (4.1.3.6)	Citrate → oxaloacetate + acetate	
Citrate synthase (4.1.3.7)	H_2O + oxaloacetate + AcCoA → citrate + CoA	
Crotonase (4.2.1.17)	CrotonoylCoA + H_2O → HydroxybutyrylCoA	
Enolase (4.2.1.11)	2-Phosphoglycerate → PEP + H_2O_2	
Fumarase (4.2.1.2)	L-Malate → fumarate + H_2O	
L-Glutamic decarboxylase	L-Glutamic acid → 4-aminobutyrate + CO_2	39
Glyoxylase I (4.4.1.5)	(*S*)-Lactoylglutathione → glutathione + methylglyoxal	
Histidase (4.3.1.3)	L-Histidine → urocanic acid + NH_3	
Mandelonitrile lyase (4.1.2.10)	Mandelonitrile → benzaldehyde + HCN	
Orotidine-5'-monophosphate decarboxylase (4.1.1)	Orotidine-5'-P → Uridine-5'-P + CO_2	
5. Isomerases		
Mutarotase (5.1.3.3)	α-D-Glucose → β-D-glucose	40
Phosphoglucose isomerase (5.3.1.9)	Fructose-6-P → glucose-6-P	41
Phosphomannose isomerase (5.3.1.8)	D-Mannose-6-P → D-fructose-6-P	
Phosphoriboisomerase (5.3.1.6)	D-Ribose-5-P → D-ribulose-5-P	42
Tautomerase (5.3.2.1)	Keto-phenylpyruvate → enol-phenylpyruvate	43
Uridine-5'-diphosphogalactose-4-epimerase (5.1.3.2)	UDP-galactose → UDP-glucose	
6. Ligases		
Aminoacyl RNA^t synthetase (6.1.1)	ATP + t-RNA + amino acid → AMP + PPi + aminoacyl RNA	
L-Glutamine synthetase (6.3.1.2)	ATP + L-glutamate + NH_3 → ADP + L-glutamine + Pi	

Notes:

(1) Acts on low molecular weight primary alcohols and unsaturated alcohols, but branched-chain and secondary alcohols are not attacked. (2) Also acts on D-galactose-1P. (3) Also acts on

1,2-propanediol. (4) Also oxidizes L-lactate and glyoxylate. (5) Reduces hydroxypyruvate to D-glycerate. (6) Also other 3α-hydroxysteroids. (7) Also other 3β- or 17β-hydroxysteroids. (8) Also oxidation of 7α-OH group of bile acids, alcohols or sulphates. (9) Also other 17,20,21-trihydroxy-steroids. (10) Also oxidizes other L-2-hydroxy-monocarboxylic acids. (11) Oxidizes other methylene interrupted polyunsaturated fatty acids. (12) Oxidizes some other 2-hydroxydicarboxylic acids. (13) Decarboxylates added OAA. (14) Acts on 1-ary, 2-ary and 3-ary amides. (15) Also other closely related sugars. (16) Oxidizes hypoxanthine, some other purines and pterins and aldehydes. (17) Propionate also acts as an acceptor. (18) Acetylates sisomian. (19) Where R = aliphatic, aromatic or heterocyclic radical and X = sulphate, nitrite or halide. Also catalyses the addition of aliphatic epoxides and arene oxides to glutathione, the reduction of polyol nitrate by glutathione to polyol + nitrite and certain isomerizations and disulphide interchange. (20) Dihydroxyacetone and L-glyceraldehyde can act as acceptors. UTP (and in yeast enzyme, ITP and GTP) can act as donors. (21) D-Glucose, mannose and glucosamine as acceptor, ITP and ATP as donor. (22) 2-Amidoadenosine can also act as acceptor. (23) Also acts with other short-chain AcCoAs. (24) Phosphorylates hydroxylamine and fluoride in the presence of CO_2. (25) Various ketoses and L-arabinose may replace fructose in the reverse reaction. (26) Broad specificity, e.g. converts hydroxypyruvate and RCHO into CO_2 and $RCHOHCOCH_2OH$. (27) A wide range of phenols, alcohols, amines and fatty acids can act as acceptor. (28) Acts on a wide variety of acetic esters; also catalyses transacetylations. (29) Hydrolyses peptides, amides and esters at bonds involving the carboxylic group of L-arg or L-lys. (30) Also hydrolyses α-N-substituted L-arginines and carnavine. (31) Also polymerizes dipeptide amides and transfers dipeptide residues. (32) Acts on a variety of acetic esters; also catalyses transacetylations. (33) Acts on a variety of choline esters and a few other compounds. (34) Also catalyses the transfer of ADP-ribose residues. (35) Also acts on PE, plasmalogen and phosphatides, removing the fatty acid attached to the 2 position. (36) Also on sphingomyelin. (37) Also with other phosphatidyl esters. (38) Also acts on $(3S,4R)$-ketose-1-P. (39) Brain enzyme also acts on cysteate, cysteine, sulphinate and aspartate. (40) Acts on L-arabinose, D-xylose, D-galactose, maltose and lactose. (41) Also catalyses the anomerization of D-glucose-6-P. (42) Also with D-ribose 5-bisP and D-ribose-5-triP. (43) Also acts with other acyl pyruvates.

—2—

Hydrolysis and Condensation Reactions

In this Section the enzyme catalysed hydrolysis reactions of esters, amides, epoxides and nitriles are reviewed. The formation of esters and amides using enzymes is also discussed.

2.1. CLEAVAGE AND FORMATION OF CARBOXYLIC ACID ESTER BONDS

2.1.1. Esterases and Lipases

Of the wide range of esterases available commercially, very few have been widely utilized in organic transformations. The most commonly used esterases have been pig liver esterase (E.C. 3.1.1.1), porcine pancreatic lipase (E.C. 3.1.1.3) and α-chymotrypsin (E.C. 3.4.21.1); others, such as the lipase from the yeast *Candida cylindracea*, are gaining popularity. Various microorganisms have been employed for certain hydrolyses and these are included in the following discussion where they perform similar reactions but in better yield, or where they afford better enantiomeric excesses than the reactions catalysed by isolated, partially purified, esterases or lipases.

Generally, esterases and lipases have been used for two basic transformations.

(i) Cleavage of a racemic ester to afford an optically active ester and an optically active acid. Thus by chemical hydrolysis of the resulting optically active ester both the (R) and the (S) acids may be obtained. This method has been utilized to provide starting materials for elegant syntheses of many natural products.

(ii) Removal of the acyl group from a racemic acylate to produce an optically active alcohol. Similarly, the recovered acylate may then be chemically hydrolysed to the chiral alcohol, thereby enabling both optically active alcohols to be available for further synthesis.

In addition, pro-chiral diesters have been hydrolysed to give high yields of optically active mono-esters. Increasingly, enzymes are becoming used as catalysts in esterification and transesterification. Examples of these different uses of esterases and lipases are described in the following Sections.

2.1.1.1. α-Chymotrypsin

One of the earliest enzymes to be investigated preparatively was α-chymotrypsin isolated from bovine pancreas. It is relatively inexpensive and is available commercially from sources such as Sigma Chemical Company Ltd., Boehringer, etc. The crystalline enzyme is stable indefinitely at $<5°C$, and aqueous solutions of low pH are stable for about two days. α-Chymotrypsin is extremely sensitive to temperature variations and aqueous reactions are best performed between 0°C and 35°C [1]. The addition of organic solvents causes a reduction in the rate of the hydrolyses; nevertheless, hydrolyses may be readily performed below 0°C. The optimum pH for hydrolyses with α-chymotrypsin is pH 7.8; any variation from this pH, however slight, will cause the rate of hydrolysis to decrease rapidly.

α-Chymotrypsin primarily catalyses the hydrolysis of amide bonds of proteins adjacent to the carbonyl groups of the aromatic amino acids, L-phenylalanine, L-tryptophan and L-tyrosine [2]. However, the catalytic activity is not restricted solely to amide hydrolysis; a wide range of ester and related functions is susceptible to hydrolysis with α-chymotrypsin [3,4]. A well-defined model for the active site of α-chymotrypsin has been proposed [5] in which the site is considered to consist of four pockets, each one of which uniquely corresponds to each of the four groups attached to the α-carbon atom of a specific substrate.

It was recognized [2,6] many years ago that α-chymotrypsin exhibits L stereospecificity towards its amino acid substrates. In all cases it can be assumed that D enantiomers of aromatic amino acid derivatives are non-substrates for α-chymotrypsin. Hence DL mixtures of amino acid esters may be readily resolved using α-chymotrypsin and there is a plethora of such resolutions documented in the literature [3]; such experiments can be carried out on substantial quantities of racemic material.

N-Acetyl-DL-phenylalanine methyl ester (**1**) has been resolved by treatment with α-chymotrypsin in aqueous methanol at 25°C for 2 h, when the optically pure L acid (**2**) was obtained in 44% yield [7].

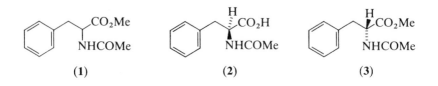

The unreacted D ester (3) can also be recovered from the reaction mixture in 34% yield. Similarly, both N-acetyl-DL-tyrosine ethyl ester [8] and N-acetyl-DL-tryptophan methyl ester [9] have been resolved to give the L acids in high yield and with good optical purity.

Berger et al. [10] have described a method for the synthesis of optically active α-amino acids via the hydrolysis of protected racemic esters using both α-chymotrypsin and subtilisin. In all cases the L-amino acid was obtained along with recovered D ester.

Aside from the resolution of amino acids various other types of molecules, such as *meso* diesters, have been successfully resolved with α-chymotrypsin. Diethyl β-acetamidoglutarate (4), lacking the structural features of typical substrates for this enzyme, has been hydrolysed, albeit slowly, to give (+)-ethylhydrogen β-acetamidoglutarate (5) [11]. The hydrolysis is followed at 27°C, pH 7.8, by the use of a pH meter during 5 h. When 83% hydrolysis has occurred, the reaction is interrupted and the solution worked up to give the optically pure monoester (5) in 57% isolated yield.

$$H \diagdown CH_2CO_2Et$$
$$MeCONH \diagup CH_2CO_2Et$$

$$MeCONH \diagdown CH_2CO_2H$$
$$H \diagup CH_2CO_2Et$$

(4)   (5)

The same workers [12] have also shown that the presence of an α- or β-acetamido group at a (developing) centre of asymmetry is sufficient to lead to stereospecificity in the hydrolysis of esters with α-chymotrypsin. Similarly, the presence of an hydroxyl group seems to be more effective in leading to stereoselective hydrolysis when in the β-position to the point of attack by the enzyme [13]. For instance, diethyl β-hydroxyglutarate (6) is slowly hydrolysed by α-chymotrypsin to give the (R) acid (7) with high stereoselectivity.

$$H \diagdown CH_2CO_2Et$$
$$HO \diagup CH_2CO_2Et$$

$$HO \diagdown CH_2CO_2H$$
$$H \diagup CH_2CO_2Et$$

(6)   (7)

The hydrolysis was carried out at pH 7.8 in aqueous solution during 15 h when an 85% yield of the required acid was obtained. The analogous dimethyl ester of β-hydroxyglutaric acid behaved in an identical fashion leading to the (R) monoester in 88% yield. Unfortunately, the enantiomeric excess of the product was very dependent upon the reaction conditions (*vide infra*) and it is advisable to convert the alcohol unit to the methoxymethyl ether prior to hydrolysis in the presence of α-chymotrypsin. With this modification, the (R) monoester (8) was obtained in ca. 95% yield (e.e. = 93%) [14].

MeOCH₂O̲ CH₂CO₂H
H CH₂CO₂Me

(8)

The hydrolysis of the above *meso* diesters has also been investigated using micro-organisms [15]. It has been shown that the *meso* diester **(6)** can be selectively cleaved with *Acinetobacter lowfii* to give the (*R*) acid **(7)**, whereas both *Corynebacterium equi* and *Arthrobacter* sp. exhibit the opposite enantioselectivity and produce the (*S*) acid **(9)**.

H̲ CH₂CO₂H
HO CH₂CO₂Et

(9)

The β-hydroxybutyrate **(10)** was hydrolysed much more slowly and without selectivity using α-chymotrypsin; only 25% hydrolysis was observed after 40 h and the isolated acid was racemic. Similarly ethyl DL-β-acetamido-butyrate **(11)** was not hydrolysed by α-chymotrypsin under the same

H CH₂CO₂Et   H CH₂CO₂Et
HO CH₃   MeCONH CH₃

(10)   **(11)**

conditions [12]. It is obvious that the presence of the second carboethoxy group in the β-position to the developing centre of asymmetry is essential for a rapid rate of hydrolysis and high stereoselectivity.

The effect of α-chymotrypsin catalysed hydrolysis of DL-diethyl *N*-acetyl-aspartate **(12)** was examined [16]; again, α-chymotrypsin displayed L stereo-specificity and afforded a 37% isolated yield of optically pure (*S*)-(+)-β-ethyl α-hydrogen-*N*-acetyl-aspartate **(13)** within a very short time. The

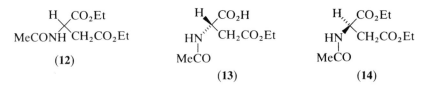

H CO₂Et   H CO₂H   H CO₂Et
MeCONH CH₂CO₂Et   HN CH₂CO₂Et   HN CH₂CO₂Et
                     | |
                     MeCO   MeCO

(12)   **(13)**   **(14)**

unreacted (*R*) diester **(14)** was recovered from the reaction mixture in 34% yield. This is, therefore, a convenient process for the preparation of deriva-tives of D- and L-*N*-acetylaspartic acid.

Racemic diethyl α-benzylsuccinate **(15)** is hydrolysed stereospecifically by α-chymotrypsin during 18 h at pH 7.2 to give (*R*)-(+)-β-ethyl α-benzyl-succinate **(16)** in 45% yield along with a 44% yield of (*S*)-(−)-diethyl α-benzylsuccinate (17) [17]. However, racemic dimethyl α-methyl succinate is only hydrolysed with partial stereoselectivity, the (S) enantiomer hydrolysing about six times more rapidly than the (R) enantiomer.

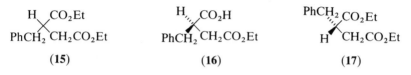

(15) (16) (17)

Similarly, DL-ethyl α-methyl-β-phenylpropionate (18) is hydrolysed stereospecifically leading to optically pure (S)-α-methyl-β-phenylpropionic acid in 42% yield, along with a 40% recovery of optically active unhydrolysed ester [17].

$$\begin{array}{c} H \diagdown CO_2Et \\ \diagup \diagdown \\ PhCH_2 \diagup CH_3 \end{array}$$

(18)

From some of the examples cited above, it can be seen that the presence of an acetamido group α or β to the developing centre of asymmetry results in the formation of carboxylic acids having the (S) configuration. The presence of an acetoxy group in either position has been shown to cause a change in selectivity [18]. Comparison of the rates of hydrolysis of both (R)-(+)- and (S)-(−)-ethyl α-acetoxypropionates showed that the (R) enantiomer is hydrolysed almost three times faster than the (S) enantiomer.

More recently, α-chymotrypsin has been used to distinguish between two different ester functionalities. Whereas acetate and dihydrocinnamate esters are hydrolysed at almost identical rates with hydroxide ion [19], α-chymotrypsin has been shown to be completely specific in removing dihydrocinnamate protection from simple mixed acetate-dihydrocinnamate diesters [20].

In order to evaluate the application of this method more fully, a range of such diesters, e.g. (19), (21) and (23), was prepared and their reactivities towards α-chymotrypsin were investigated (Scheme 2.1) [21].

Literature data [22] suggest that in such systems as these [i.e. (19), (21) and (23)], the 6β-functionality should be preferentially hydrolysed with hydroxide ion, and indeed, treatment of (19) with an equivalent of sodium hydroxide at 25°C gave a 53% yield of the 1β-acetate (20). However, 35% of the diol was also isolated. In contrast, α-chymotrypsin-catalysed hydrolysis of (19) afforded the 1β-acetate (20) in quantitative yield by specific removal of the dihydrocinnamoyl group. Similarly, the isomeric 6α-dihydrocinnamate (21) was converted into the 1β-acetate (22) in 96% yield with α-chymotrypsin although the rate of hydrolysis was significantly slower. In the case of the 6β-acetate (23), the enzyme mediated hydrolysis was appreciably slower and, therefore, not appropriate for large-scale work. This is due, to a large extent, to the marked insolubility of this compound in water.

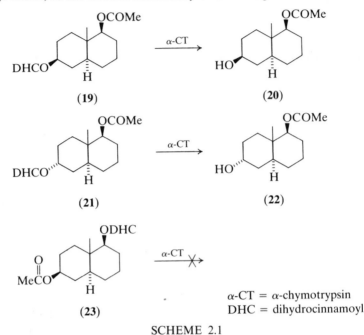

α-CT = α-chymotrypsin
DHC = dihydrocinnamoyl

SCHEME 2.1

The hydrolysis of a 3α,6β-protected tropanediol (24) was also examined and it was found that the two ester groups could be differentiated (Scheme 2.2). Hydrolysis with hydroxide ion effected removal of the 6β-acetate to give the alcohol (25) in 65% yield. Enzyme mediated hydrolysis, on the other hand, selectively removed the dihydrocinnamoyl group to give the alcohol (26) in quantitative yield.

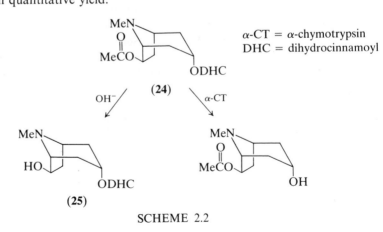

α-CT = α-chymotrypsin
DHC = dihydrocinnamoyl

SCHEME 2.2

All these enzyme mediated hydrolyses were carried out at pH 7.8 at 25°C. Due to the low aqueous solubility of the substrates, acetonitrile was used as a cosolvent. However, the amount of organic solvent was kept below 5% due to the known inactivation of α-chymotrypsin by such additives [23].

These results demonstrate the usefulness in preparative organic chemistry of α-chymotrypsin for selective ester cleavage and, due to the mildness of the conditions employed, racemization and other such problems are minimized.

α-Chymotrypsin has also been shown to hydrolyse a series of 5-(4*H*)-oxazolones bearing various substituents in the 2 and 4 positions (Scheme 2.3) [24]. Some enantiomeric enrichment occurred when both substituents (R– and R^1–) were relatively bulky. Similar results were obtained when subtilisin was employed as the hydrolysing agent.

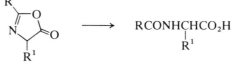

SCHEME 2.3

Workers in Switzerland [25] have employed α-chymotrypsin catalysed hydrolysis as one of the steps in a 15-stage synthesis of (+)-conglobatin. They showed that, on treatment of the *meso*-diester (27) with α-chymotrypsin in aqueous solution for 12 days, the (2*S*,4*R*) enantiomer of 2,4-dimethylglutamic acid (28) was obtained in 50% yield (77% enantiomeric excess).

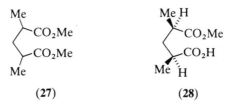

(27)   (28)

Although it had generally been considered that α-chymotrypsin lost a certain amount of its catalytic activity in solutions of high concentrations of water-miscible organic solvents [26], it has recently been shown by Japanese workers that aromatic amino acids may be converted into their ethyl esters by incubation with α-chymotrypsin in ethanol containing only very small amounts of water [27]. Although the activity of α-chymotrypsin was only marginally decreased in solutions of high ethanol concentration, a trace of water was found to be essential for esterification to occur; total inhibition of the enzyme occurred in the absence of water. The amino acids investigated in this study were *N*-acetyl-L-tryptophan and *N*-acetyl-L-tyrosine. Interestingly, however, under the conditions required for esterification in these cases, it was found that *N*-acetyl-D-tryptophan could not be esterified, indicating that α-chymotrypsin maintains its native conformation under the reaction conditions.

French workers have also shown that immobilized α-chymotrypsin served as an excellent catalyst for the esterification of *N*-acetyl-L-tyrosine with ethanol [28]. The α-chymotrypsin was immobilized on silica and the reaction was optimized such that esterification proceeded giving a 40% yield.

In summary, α-chymotrypsin is one of the most versatile proteolytic enzymes for large-scale resolutions of carboxylic acids. Not only is it commercially available from a wide variety of sources, but it is also relatively cheap, and by using the active site model so excellently described by Cohen [5], its stereoselectivity can be predicted with some confidence.

2.1.1.2. Pig Liver Esterase

Another hydrolytic enzyme that has been extensively studied in recent years is pig liver esterase (E.C. 3.1.1.1). The first documented use was for the hydrolysis of a dimethyl glutarate. It had been shown in 1961 [13] that α-chymotrypsin could be employed to hydrolyse selectively the pro-(S) ester group of dimethyl β-hydroxyglutarate (**29**) to give the mono-acid (**30**). By

$$\underset{\textbf{(29)}}{\overset{\displaystyle H \diagup CH_2CO_2Me}{\underset{\displaystyle HO \diagdown CH_2CO_2Me}{\diagdown}}} \qquad \underset{\textbf{(30)}}{\overset{\displaystyle HO \diagup CH_2CO_2H}{\underset{\displaystyle H \diagdown CH_2CO_2Me}{\diagdown}}}$$

duplication of these hydrolytic conditions, Huang *et al.* [29] successfully hydrolysed dimethyl β-hydroxy-β-methylglutarate (**31**) to give the mono-acid with high optical purity. However, the method had significant disadvantages in that the rate of hydrolysis was slow and virtually a stoichiometric amount of α-chymotrypsin was required. In contrast, pig liver esterase (PLE) was found to hydrolyse the compound (**31**) much more efficiently to give the half-ester (**32**) in 62% yield. Conversion of this mono-acid into mevalonolactone (**33**) afforded material whose specific rotation suggested that the optical purity was high and that the esterase had cleaved the pro-(R) methyl ester group of the compound (**31**).

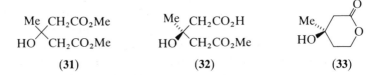

duplication of these hydrolytic conditions — *(structures 31, 32, 33)*

Wilson *et al.* [30] have modified this procedure by increasing the buffering capacity of the reaction mixture coupled with periodic additions of base and substrate. By this method the reaction may be scaled up to allow access to 10–15 g of the half-ester (**32**).

Mohr *et al.* [31] have recently re-investigated the hydrolysis of 3-hydroxy-glutarates with both pig liver esterase and α-chymotrypsin and have shown that, in contrast to earlier reports [13,32], hydrolyses with both of these enzymes are not highly enantioselective. It may be that in the earlier work too much reliance was placed on the observed value of the optical rotation of the product, e.g. (**30**). (The specific rotation is ca. $-1.7°$; because this value is very low and the material exists as a viscous oil, it is obviously dangerous to rely on the optical rotation as a method for evaluating enantiomeric excess.) Moreover, Rosen *et al.* [33] observed that the optical rotation of compounds such as (**30**) is critically dependent on the concentration of the solution employed. A totally different explanation is that the enzymes employed were not quite the same; they were in fact obtained from different sources.

In order to circumvent the problems involved in the measurements of the specific rotation and in order to assess accurately the enantiotopic distribution of the product, the monoester (**30**) was transformed into a diastereomeric triester (**34**) in two steps, avoiding crystalline intermediates.

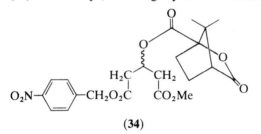

(**34**)

This triester could then be assayed accurately by HPLC and the enantiomeric excesses of the products determined. Using this method, and by carefully selecting the pH for the hydrolysis, the best enantiomeric excess that could be obtained was 68% in favour of the (*R*) isomer by using α-chymotrypsin at pH 7.0. Pig liver esterase preferentially afforded the other enantiomer as was previously claimed [32], but the enantiomeric excess was very poor (e.e. = 22%).

The enzymic hydrolysis of a variety of β,γ-epoxy esters (synthetic precursors of β-hydroxy esters) using pig liver esterase has been investigated. The epoxyhexanedioate (**35**), on incubation with PLE at pH 7.0, was found to undergo rapid hydrolysis and both the resulting acid (**36**) and recovered diester were shown, by chemical methods, to have been obtained in >95% enantiomeric excess. Similarly, the epimeric epoxydiester (**37**) also underwent hydrolysis with PLE to give the mono-acid (**38**) in 90% yield; this was shown to have been obtained in >95% enantiomeric excess.

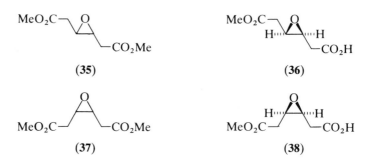

(35)   (36)

(37)   (38)

Thus, with respect to the prochiral C (β) atom bonded to an O atom, PLE cleaves the same ester group as in the case of 3-hydroxyglutarate but with much higher specificity.

Since Mohr *et al.* had proposed elsewhere [34] (see later) that only substituents in the α and/or β position (with respect to the ester group which is cleaved) are essential for selectivity during PLE hydrolysis, the epoxyester (**39**) was investigated. This was shown to be an excellent substrate for PLE. Although hydrolysis did not stop completely after consumption of 0.5 equivalents of base (indicating the reaction was not totally enantioselective), reasonable yields of both the acid (**40**) and the recovered ester were obtained. In fact, by simply stopping the enzymic reaction after 40% or 60% hydrolysis, both ester and acid, respectively, are obtainable in extremely high optical yield.

(39)   (40)

The hydrolysis of a variety of other 3-substituted glutarates has also been studied by other workers. The diethyl glutarate (**41**) has been shown to undergo hydrolysis with PLE at pH 8 during 15 h to afford the monoester (**42**) in 69% enantiomeric excess [35]. In this case the enzyme is pro-(*R*) selective and the ester is obtained in 80% yield. The optical purity was determined by chemical conversion into the valerolactone (**43**). Interestingly,

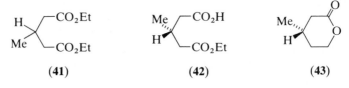

(41)   (42)   (43)

the dimethyl ester (**44**) was hydrolysed with significantly higher selectivity to give the monoester (**45**) in 86% yield and with 90% enantiomeric excess [36].

The synthetic value of this process has been demonstrated by conversion of the optically active ester into verrucarinic acid (46) and verrucarin A, a naturally occurring macrocyclic trichothecane ester [37]. Note that this bioconversion [(44)→(45)] is best performed at −10°C in methanol/water (1:4) (see later).

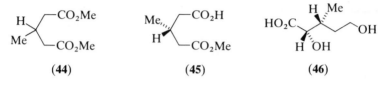

(44)   (45)   (46)

A strategy for altering the enantioselectivity of the PLE-catalysed hydrolysis of a β-aminoglutarate has been demonstrated by Ohno and co-workers [38]. The β-aminoglutarate (47) (which was only slowly hydrolysed with α-chymotrypsin) was efficiently hydrolysed with PLE, resulting in a 94% yield of the monoester (48) (Scheme 2.4).

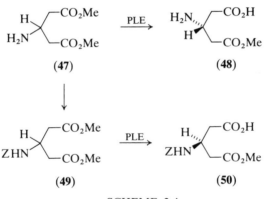

SCHEME 2.4

The product proved to be the (3R) monoester but the material was only obtained in 42% enantiomeric excess; this poor enantioselectivity was due to the fact that the substrate was partially hydrolysed under the reaction conditions in the absence of the enzyme. In order to circumvent this problem, the hydrolysis was investigated using the N-benzyloxycarbonyl derivative (49), readily obtained from (47) in 90% yield. Protection of the amino group was thought to be essential to prevent it participating in the chemical hydrolysis through hydrogen bonding (thus causing the deleterious effect on the optical yield). On incubation of the protected ester (49) with PLE under the usual conditions (pH 8.0 phosphate buffer at 25°C for 1.5 h) it was found that hydrolysis proceeded in 93% yield. Surprisingly, the product was the (3S) monoester (50) which was obtained in 93% enantiomeric excess. The esterase showed an altered pattern of stereoselectivity because the pro-(R) methyl ester had been cleaved. Hence, both enantiomers of the monoester are readily available by enzymic methods. As an extension of this work, various N-substituted dimethylglutarates were subjected to PLE hydrolysis and it was shown that by careful selection of the nitrogen substituent both the enantioselectivity of the enzyme and the optical purity of the product could be controlled [39]. Generally speaking, the optical purity of the products decreases considerably with increasing chain length of the N substituent and, in certain cases, the absolute configuration of the product may be reversed. Bulky acyl groups such as pivaloyl or cyclohexanoyl also cause a reversal in enantioselectivity. Thus by careful selection of the N protecting group, the required enantiomer may be obtained in good yield. Some of the more useful results are shown in Table 2.1.

Table 2.1

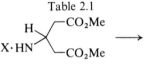

X	Yield (%)	Stereochemistry of mono-ester	e.e. (%)
CH₃CO–	81	R	93
(CH₃)₃C·CO–	50	S	93
CH₃CH=CH·CO–	60	S	100
PhCH₂OCO–	93	S	93

Preparation 2.1. Preparation of methyl hydrogen (3S)-3-[(benzyloxycarbonyl)-amino]glutarate (**50**)

To a solution of dimethyl 3-[(benzyloxycarbonyl)amino]glutarate (**49**) (465 mg, 1.5 mmol) in acetone (1.5 ml) and 0.05 M phosphate buffer (45 ml, pH 8) was added pig liver esterase (Sigma, NoE-3128 Type I; 0.22 ml, 300 units). The mixture was incubated for 7 h at 25°C and then acidified to pH 3.0 with hydrochloric acid and extracted with dichloromethane. The organic layer was dried with sodium sulphate, concentrated *in vacuo*, and purified by chromatography on silica gel (eluted with ether) to give the monoester (**50**) (410 mg) $[\alpha]_D^{25} + 0.69°$ (c = 7.50, chloroform). Optically pure (**50**) was obtained by recrystallization (dichloromethane–n-hexane); m.p. 97.0–97.5°C; $[\alpha]_D^{25} + 0.72°$ (c = 7.5, chloroform); IR (KBr); 3330, 1740, 1725, 1705 cm⁻¹; ¹H-NMR (CDCl₃) δ: 2.70 (d, J = 5.8 Hz, 4H), 3.64 (s, 3H), 4.37 (m, 1H), 5.08 (s, 2H), 5.66 (bd, 1H) 7.35 (s, 5H) 8.30 (s, 1H). *m/z* 295.

The synthetic potential of these results was demonstrated [38] by conversion of the two compounds (**48**) and (**50**) into the enantiomeric azetidin-

2-ones (**51**) and (**52**) which have themselves been transformed into the carbapenems, ($-$)-asparenomycin C [40] and ($-$)-carpetimycin A [41] and the antibacterial compound, ($+$)-negamycin [42]: moreover, the structure–stereospecificity relationships revealed by such studies gave information about the topography of the active site of PLE thus allowing an "active site model" to be constructed [38].

(**51**)   (**52**)

In view of the fact that PLE is of limited use due to the cost and the impracticality of using the enzyme on a very large scale, Ohno assayed ca. 500 species of micro-organisms as potentially useful agents for the hydrolysis of the amino-glutarate (**49**) [43]. All the tested strains preferentially hydrolysed the pro-(R) ester group to yield the acid (**50**); some micro-organisms (e.g. *Achromobacter lyticus* IFO 12725 and IFO 12726 and *Gluconobacter dioxyacetonicus* IFO 3271) produced the acid in optical yields equal to or higher than PLE (e.e. = 93%). Indeed, *Flavobacterium lutescens* (IFO 3084 and IFO 3085) hydrolysed the compound (**49**) more selectively (e.e. = 97%) and efficiently than PLE.

The synthetic potential of PLE hydrolysis of glutarate diesters has also been investigated by Francis and Jones [44]. A range of diesters, substituted at C-3 with a variety of alkyl groups (**53**), was studied and it was demonstrated that not only was the enzyme remarkably tolerant to the size of the substituent at C-3, but under controlled conditions (pH 7) the hydrolyses proceeded in good yield with complete [pro-(S)] enantiotopic specificity (Scheme 2.5).

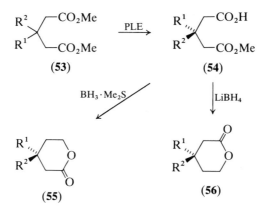

SCHEME 2.5

The resulting monoesters (54) were converted into the lactones (55) and (56) and the enantiomeric excesses were determined by GLC analyses of the *ortho*-esters produced from the lactones using (2R,3R)-butanediol. In all cases the enantiomeric excesses were found to be >99%. This is in marked contrast to the enantiomeric excess values for products from similar reactions reported elsewhere in the literature [29,35] but this is due, in part, to competing chemical hydrolysis at pH 8 in the earlier work; at pH 7, only enzymic hydrolysis is observed. On a salutary note, Jones has shown that with a different batch of the enzyme, the same high levels of enantiomeric excess could not be obtained [45]. Nevertheless, the optical yields observed were still in the order of 70% and the hydrolysis was shown to be pro-(S) selective for diesters with small C-3 substituents but reversing to a pro-(R) preference for larger groups. A detailed study of the hydrolysis of dimethyl 3-methylglutarate (44) was carried out in order to optimize the reaction conditions. It was found that lowering the pH from 7 to 6 caused a slight decrease in the enantiomeric excess of the product and similarly addition of acetone or acetonitrile had a retrograde effect. On the other hand, the presence of either 5% DMSO or 5% methanol was beneficial. In fact, 5% methanol appeared to overcome the negative effect obtained by lowering the pH. It was eventually demonstrated that the best conditions for the hydrolysis was to perform the reaction at $-10°C$ in 10–20% aqueous methanol when the monoester was obtained in 97% enantiomeric excess. Using these conditions the 3-benzyl derivative, for which the hydrolysis is of opposite enantiotopic selectivity, was transformed into its monoester with 81% enantiomeric excess.

Preparation 2.2.   Preparation of methyl hydrogen (3R)-3-methylglutarate (45)

A rapidly stirred solution of dimethyl 3-methylglutarate (44) (1.0 g, 5.75 mmol) in 0.03 M KH_2PO_4 buffer/methanol (30 ml; ratio 4:1) at $-10°C$ was neutralized to pH 7 with 0.5 M aqueous NaOH by using a pH-stat. PLE (1000 units) was added and the pH of the solution was maintained at 7. The hydrolysis proceeded until 1 equivalent of base had been consumed (4 h) at which point it stopped. The reaction mixture was frozen at $-78°C$ and then just thawed and a saturating amount of sodium chloride was added. The resulting cold solution was washed with ether (2 × 30 ml) to remove unreacted diester then acidified to pH < 2.5 with concentrated HCl. Extraction with ether (3 × 70 ml) followed by evaporation of the dried ($MgSO_4$) ether extracts and Kugelrohr distillation yielded the acid ester (45) (860 mg, 94% yield, e.e. = 97%; b.p. 65–68°C/0.05 mmHg; IR v_{max} 1713, 1735, 2410–3712 cm^{-1}; δ 1.02 (3H, d, $J = 6$ Hz), 1.93–2.70 (5H, m), 3.67 (3H, s) and 9.45–10.33 (1H, bs).

Sih and co-workers [46] have studied the hydrolysis of the di-substituted glutarate (27) in an attempt to explore further the enantiotopic specificity of PLE. Incubation of the diester (27) with PLE at pH 8 for 1 h gave the monoester (28) in 85% yield. This material was converted into the lactone (57) which was obtained in 64% enantiomeric excess. In this case, PLE preferentially cleaved the pro-(S) ester group of the compound (**27**), thus displaying the same catalytic profile as α-chymotrypsin for this reaction.

The same diester (**27**) was also subjected to incubation with various micro-organisms and, although most of these preferentially attacked the pro-(R) ester group, the monoesters were obtained in varying degrees of optical purity. Only one organism, *Gliocladium roseum*, was found to effect the cleavage in high yield (95%) and by a process of recycling, the product (**58**), showing 98% enantiomeric excess, could be obtained.

(**57**)   (**58**)

Ramos Tombo *et al.* have also investigated the hydrolysis of the corresponding (SS,RR) diethyl ester (**59**) using a range of micro-organisms [47]. By variation of the organism used as catalyst for the hydrolysis, the stereochemistry of the product could be controlled. Thus, incubation of the compound (**59**) with Isolate K1 afforded the (S,S) monoester (**60**) in 50% yield (e.e. = 50%), whereas incubation with *Streptomyces* sp. R1186 gave the (R,R) monoester (**61**) in 48% yield (e.e. = 66%).

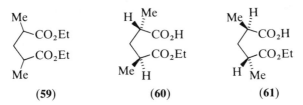

(**59**)   (**60**)   (**61**)

The PLE hydrolysis of a range of dimethyl esters of symmetrical dicarboxylic acids has also been the subject of a study by Mohr *et al.* [34] who investigated both acyclic and cyclic *cis* diesters. Included in this investigation were mono- and di-substituted glutarates, malonates and succinates and the cyclic esters shown in Scheme 2.6.

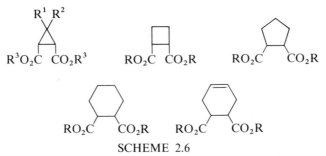

SCHEME 2.6

From the results obtained the following conclusions were proposed:

(1) In order to obtain high stereoselectivity, the ester group must be no further away from the induced prochiral centre than the β-position.

(2) Diesters bearing a higher substitution pattern are hydrolysed with higher stereoselectivity than their less-substituted counterparts.

(3) Cyclic diesters show higher stereoselectivity than the corresponding acyclic analogues.

(4) Substituents of different polarity and different size show opposite effects on selectivity of the enzymic hydrolysis. For example, while the compounds (27) and (44) are hydrolysed with pro-(S) specificity (*vide supra*), for the compounds (29) and (62) the pro-(R) group is preferentially attacked.

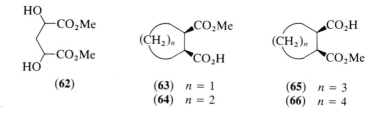

Cyclic *meso*-diesters have been shown [48] to be excellent intermediates for the enzymic syntheses of useful chiral bicyclic lactones. (Such lactones are also available by oxidation of *cis* diols with horse liver alcohol dehydrogenase, but this strategy requires the use of expensive and unstable cofactors [49] [see Section 4.1].) In the case of these cyclic *meso*-diesters [48], PLE at pH 7 exhibits interesting enantioselectivity in that for the small ring diesters, the pro-(S) ester group is selectively cleaved to yield the monoesters such as (63) and (64) in high yield and excellent enantiomeric excess. The enantiotopic specificity undergoes reversal for the cyclopentyl diester in that it is the pro-(R) ester group that is preferentially cleaved to give the half-ester (65). Although in this case the half-ester (65) was only obtained in 17% enantiomeric excess, the reversal of specificity is complete for the cyclohexyl diester and the resulting product (66) was obtained in 97% enantiomeric excess.

Concurrently, Schneider *et al.* reported their results on the hydrolysis of almost identical cyclic *cis,meso*-diesters [50]. In most cases the reported yields and enantiomeric excesses were the same within the bounds of experimental error. In addition, the corresponding cyclohexenyl compound behaved in an identical manner, a result also described by both Gais and Lukas [51] and Kobayashi *et al.* [52]. Thus, incubation of this compound with PLE produced enantiomerically pure (1S,2R) half-ester (67) in 99% yield. The reaction could be conveniently carried out on a molar scale.

Preparation 2.3. Preparation of methyl hydrogen (1*S*,2*R*)-cyclohex-4-ene-1,2-dicarboxylate (**67**)

To dimethyl *cis*-cyclohex-4-ene-1,2-dicarboxylate (10 mmol) suspended in 0.1 M phosphate buffer (50 ml, pH 8) was added pig liver esterase (500 units) with vigorous stirring. The pH value was kept within the 7.5–8.0 range by addition of 1 N sodium hydroxide. After consumption of 1-mol equivalent of base the mixture was homogeneous. The pH value was adjusted to 9 and the aqueous phase was extracted with ether. The organic layer was washed with water and the combined aqueous solution was acidified to pH 2.5. The aqueous layer was extracted with ether, dried and evaporated *in vacuo* to yield the monoester (**67**) (95%); $[\alpha]_D^{20}$ + 14.6° (*c* = 0.2, ethanol); ^{1}H-NMR (CDCl$_3$) δ: 2.0–2.8 (4H, m), 2.8–3.2 (2H, m), 3.7 (3H, s), 5.65 (2H, bs), 10.1 (1H, bs).

One dichotomy between Jones' [49] and Schneider's [50] work was the stereochemistry assigned to the product obtained from hydrolysis of the cyclopropyl diester. Whereas Jones reported this diester to be enantioselectively cleaved to give the (−)-(1*R*,2*S*) half-ester (**63**) in high yield and excellent enantiomeric excess, Schneider reported that the opposite enantiotopicity was observed for the reaction and the product was the (−)-(1*S*,2*R*) half-ester.

A recent publication clarifies the issue: the product is indeed the (−)-(1*R*,2*S*) compound (**63**) [53]. Schneider *et al.* [50] transformed the appropriate dimethyl ester (**68**) into the (−)-(1*S*,2*R*) half-ester (**69**) in good yield and with high enantioselectivity.

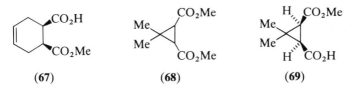

        (**67**)   (**68**)   (**69**)

Björkling *et al.* [54] have also shown that the appropriate *cis*-cyclohexyl diester is cleaved with pro-(*R*) specificity to give the (1*S*,2*R*) monoester (**66**) in an optically pure form. Interestingly, the analogous diethyl ester, although being readily hydrolysed with PLE at pH 7.5, gave racemic monoester. These workers also showed that subtilisin (E.C. 3.4.21.14) did not hydrolyse either of these diesters. In contrast, the isomeric *trans* diesters were hydrolysed by both PLE and subtilisin to give the corresponding monoesters (**70**). In no case was the enantiomeric excess >45%, although optically pure material could be obtained by repeated crystallization.

Incubation of the *cis* diester (**71**) with PLE at pH 7 for 4 h afforded the monoester (**75**) in 82% yield, but in only 34% enantiomeric excess [55]. Similar hydrolyses were carried out on the two related diesters (**72**) and (**73**), but in both cases, although hydrolysis was chemically efficient, the enantio-

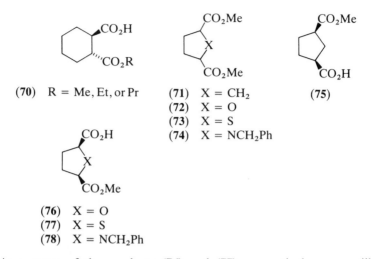

(70) R = Me, Et, or Pr   **(71)** X = CH$_2$ **(75)**
                              (72) X = O
                              (73) X = S
                              (74) X = NCH$_2$Ph

(76) X = O
(77) X = S
(78) X = NCH$_2$Ph

meric excesses of the products **(76)** and **(77)**, respectively, were still disappointingly low (e.e. $\simeq 45\%$). PLE preferentially hydrolysed the pro-(S) ester group in all three cases.

As an addition to this range of heterocyclic diesters, Kurihara *et al.* have investigated the hydrolysis of the pyrrolidine **(74)** in an approach to the synthesis of various β-lactams [56]. Again, PLE preferentially hydrolysed the pro-(S) ester group and the half-ester **(78)** was obtained in 85% yield with 80% enantiomeric excess.

Preparation 2.4. Preparation of (1S,4R,5R,6S)-5,6-dimethylmethylenedioxy-3-methoxycarbonyl-7-oxabicyclo[2.2.1]hept-2-ene-2-carboxylic acid (**80**)

To a solution of dimethyl 5,6-dimethylmethylenedioxy-7-oxabicyclo[2.2.1]hept-2-ene-2,3-dicarboxylate **(79)** (3.0 g) in acetone (30 ml) and 0.1 M phosphate buffer (300 ml, pH 8) was added pig liver esterase (3 ml, 4140 units). The mixture was incubated for 4 h at 32°C, and then acidified to pH 4 with 2 M hydrochloric acid and extracted with ethyl acetate. The organic layer was washed with water, dried over sodium sulphate and concentrated *in vacuo* to afford 2.73 g (96%) of monoester as a white solid; $[\alpha]_D^{20}$ $-37.1°$ ($c = 1.0$, chloroform). The product was dissolved in hot carbon tetrachloride; after standing at room temperature a small amount of material of very low optical purity was removed by filtration. The filtrate was concentrated and the solid was recrystallized twice from carbon tetrachloride–n-hexane to give optically pure **(80)**: m.p. 115.5–117.5°C; $[\alpha]_D^{20}$ $-49°$ ($c = 1.0$, chloroform); v_{max} (KBr) 1725, 1650 and 1622 cm^{-1}; δ (CDCl$_3$) 1.37 (3H, s), 1.52 (3H, s), 4.0 (3H, s), 4.54 (2H, s), 5.20 (2H, s), 5.24 (1H, s).

The tricyclic compound **(79)** (Scheme 2.7) also undergoes enantioselective hydrolysis with PLE [57] at pH 8 in the presence of 10% acetone and the half-ester **(80)** is obtained in 96% yield. Although the enantiomeric excess of this material is only ca. 77%, the optical purity can be improved to >95% by crystallization. The synthetic utility of the half-ester (**80**) was demonstrated by its conversion into both D- and L-ribosides and into the nucleosides, (+)-showdomycin and (−)-6-azapseudouridine.

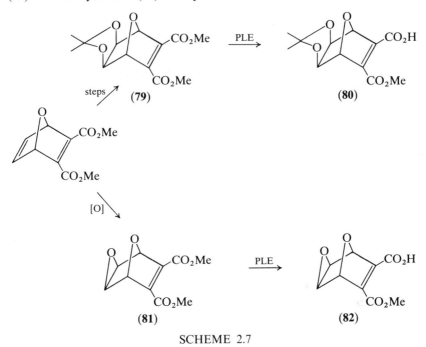

SCHEME 2.7

Similarly, the epoxide (**81**) is enantiotopically hydrolysed with PLE to give the monoester (**82**) in quantitative yield. This material has been transformed into the known inhibitor of RNA synthesis, cordycepin. Interestingly, the saturated system (**83**) was hydrolysed very slowly with PLE. After 24 h, only a 10% yield of the optically active half-ester (**84**) was obtained.

(83) (84)

The norbornene system (**85**) has also been subjected to PLE hydrolysis, whereupon the half-ester (**86**) was obtained in quantitative yield and in ca. 80% enantiomeric excess. This material has been converted into the carbocyclic nucleosides (−)-aristeromycin and (−)-neplanocin A [58].

(85)   (86)

Bloch *et al.* [59] have studied the hydrolysis of another range of diesters: (87)–(89), (91) and (92). The enzymic hydrolyses were carried out in pH 8 phosphate buffer at 30°C and only the diester (92) failed to react. The

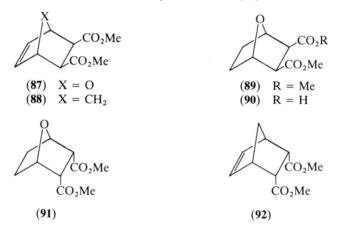

(87) X = O   (89) R = Me
(88) X = CH₂   (90) R = H

(91)   (92)

experiments showed that compounds with equatorial ester groups were good substrates for PLE and that the presence of the oxygen bridge was essential for selectivity. The presence of an endocyclic double bond had a deleterious effect on the reaction. For example, the diesters (87) and (89) were easily hydrolysed in 3 h to provide the corresponding half-esters [e.g. (90)] in excellent enantiomeric excess, whereas the hydrolysis of compound (91) took almost 50 h and the half-ester was obtained in 64% enantiomeric excess. This material could, however, be obtained in an optically pure state by crystallization. In contrast, the diester (92) was only slowly hydrolysed with very little selectivity.

Enantioselective hydrolysis of another *meso*-diester has afforded an intermediate for the synthesis of (+)-biotin [60]. Initial experiments on the methyl ester (93) showed that, as expected, the pro-(S) ester group was selectively cleaved but the half-ester (95) was only obtained in 38% enantiomeric excess. In contrast, the bis-propyl ester (94) proved to be a much better substrate and the monoester (96) was obtained in 85% yield and with 75% enantiomeric excess.

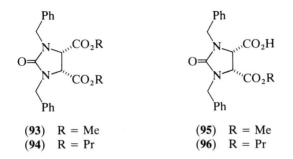

(**93**) R = Me
(**94**) R = Pr

(**95**) R = Me
(**96**) R = Pr

Disubstituted malonates are also good substrates for PLE; the half-ester (**98**) is obtained in 86% yield and 59% enantiomeric excess on hydrolysis of the diester (**97**) for 24 h at pH 7 [60].

(**97**)

(**98**)

Similar hydrolyses of aralkyl malonates have been reported by Björkling *et al.* [61]. A selection of these results is shown in Table 2.2. The enantiomeric excesses obtained from the hydrolyses were much enhanced by the incorporation of 50% dimethylsulphoxide in the incubation mixture and the reactions could be readily performed on a multigram scale. The use of α-chymotrypsin as catalyst was also investigated, but, although enantiomerically pure half-esters were obtained, the extremely low reaction rates preclude the use of this enzyme for large-scale synthesis.

Table 2.2

R^1	R^2	Yield (%)	e.e. (%)
H	H	>85	45
MeO	H	>85	82
MeO	MeO	>85	93

Simple alkyl malonates have been the subject of another study by Björkling *et al.* [62]: series of methyl and ethyl diesters (**99**) were investigated. They showed that, in the case of the dimethyl esters, and for

(99)   R = Me or Et   (100)   n = 2–4
       n = 2–7

$n = 2$ to 4, up to 73% enantiomeric excess of the (S) enantiomer of the half-ester (100) can be obtained; for the dimethyl esters, with $n = 5$ to 7, an inversion of enantioselectivity occurred and the (R) enantiomer was obtained as the major product. All hydrolyses were carried out at pH 7.5 in 25% dimethylsulphoxide at 25°C.

A similar enantioselectivity was observed for the corresponding diethyl esters, although the enantiomeric excesses obtained were much lower. This is thought to be due to the greater tendency of the ethyl ester to bind to a critical hydrophobic site on the enzyme. None of these simple alkyl malonates was observed to undergo hydrolysis with α-chymotrypsin.

Schneider et al. [63] observed slightly different enantiomeric excesses with similar substrates due, no doubt, to the different reaction conditions employed (i.e. pH 8 in aqueous phosphate buffer). Essentially the results of both groups of workers were entirely complementary.

A further range of malonates was investigated by Luyten et al. [64] who concluded that for the hydrolysis of 2,2-dialkylated dimethyl malonates with PLE the enzyme promotes preferential attack at the [pro-(S)]-methoxycarbonyl group. Although some malonates were found to react only sluggishly, the t-butoxymethyl compound (101) was shown to yield the optically active (e.e. = 96%) monoester (+)-(R)- (102) in 90% yield after incubation in pH 7 phosphate buffer for 6.25 h. The employment of dimethylsulphoxide (10% v/v) as an additive caused a decrease in the rate of reaction although the optical purity remained high, whilst methanol (10% v/v) gave a better yield but a lower optical purity (e.e. = 86%).

(101)   (102)

A range of 3-hydroxy esters (103) has been studied by Wilson et al. [30] and, although enantiomeric excesses were quite variable, in all cases PLE was shown to hydrolyse preferentially the (R) enantiomer. The optical purity of the product (104) was dependent on the group R^1: when R^1 is ethyl, allyl or dimethoxyethyl, enantiomeric excesses of >94% can be obtained, whereas when R^1 is hexyl only a 26% enantiomeric excess was observed. Also, by allowing the starting ester to undergo extensive hydrolysis, the (S) enantiomer of the starting racemate could be recovered in excellent

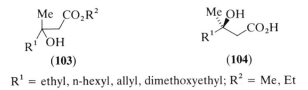

(**103**)   (**104**)

R^1 = ethyl, n-hexyl, allyl, dimethoxyethyl; R^2 = Me, Et

optical purity. Although both methyl and ethyl esters were investigated, no general trend as regards enantioselectivity was observed for these two series of compounds.

It has been noted that, for reactions involving PLE, hydrolyses of racemic allenic esters can be predicted and that the likely enantiomeric selectivity can be estimated [65]. The hydrolysis is consistently (S)-ester selective when the C-4 substituents are small or acyclic (Scheme 2.8). When these substituents are large or cyclic then the reaction is (R)-ester selective. The enantiomeric excesses of the products are highest when C-2 and C-4 are substituted with bulky groups. In general, the more highly substituted the allene, the slower the rate of hydrolysis. Ethyl esters in this series are hydrolysed more slowly than are the corresponding methyl esters.

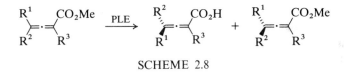

SCHEME 2.8

The enzymic hydrolysis of a dicarboxylic ester stops after removal of one of the ester groups [66], whereas the hydrolysis of diacetates does not terminate in the same fashion and the reaction product is often contaminated with the resulting diol. However, in some cases, the intermediate alcohol can be isolated in high yield and with excellent optical purity. For example, Wang and Sih [67] have utilized PLE in an alternative route to (+)-biotin (cf. ref. 60). Thus, incubation of the diacetate (**105**) with PLE in 10% aqueous methanol at pH 7 afforded the alcohol (**106**) in 70% yield and with an enantiomeric excess of 92%, i.e. the enzyme preferentially cleaved the pro-(R) acetoxy group of the diester (**105**). This alcohol can be readily converted into a known intermediate of (+)-biotin.

Further to their work on cyclic *meso*-dimethyl esters, Schneider and co-workers [68] have recently investigated the hydrolysis of the corresponding diacetates with PLE to furnish the alcohols (**107**). The hydrolyses were carried out in pH 7 phosphate buffer; unfortunately both the chemical yields and the enantiomeric selectivities were poor. However, the same hydrolyses were investigated with porcine pancreatic lipase (PPL) as catalyst and the results obtained were much more promising (see Section 2.1.1.3).

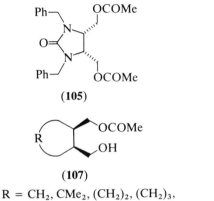

(105)   (106)

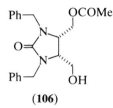

(107)

R = CH$_2$, CMe$_2$, (CH$_2$)$_2$, (CH$_2$)$_3$,
(CH$_2$)$_4$, (CH$_2$CH=)$_2$

It is of interest to note here that, in most cases, PPL and PLE showed opposite enantioselectivity. For example, the cyclohexenyl derivative (108) was hydrolysed with PPL to give the optically pure (−)-enantiomer in 96% yield, whereas PLE gave the (+)-enantiomer [(107), R = (CH$_2$CH=)$_2$] in 43% yield but in only 40% enantiomeric excess.

In an effort to improve the enantiomeric excess obtained from the diacetate (108) on hydrolysis with PLE, Guanti et al. [69] have investigated the effect of various additives on the rate and enantiomeric purity of the product. It was found that the addition of organic solvents decreased the rate of hydrolysis but had a beneficial effect on both the chemical yield and the enantiotopic selectivity. In the presence of 10% t-butanol, the alcohol [(107), R = (CH$_2$CH=)$_2$] could be obtained in 75% yield and with an

(108)   (109)

enantiomeric excess of 96%. The reaction could be performed readily on a multigram scale. Other additives also affect the reaction: lithium chloride marginally increases the enantiomeric excess of the product, whilst albumin is shown to be responsible for an increase in both the rate of hydrolysis and the asymmetric induction. In view of these results, the enzyme catalysed hydrolysis of a range of other substrates, e.g. compounds (27), (87), (107), [R = (CH$_2$)$_4$] and (109), was investigated in the presence of 10% t-butanol. In each case, a decrease in reactivity was observed, but no general trend regarding the enantiomeric excesses of the product could be established and none of the products showed a large improvement in their enantiomer ratios.

Schneider and Laumen [70] studied the propensity of a variety of esterases and lipases to hydrolyse the diacetates (110)–(112). A series of experiments

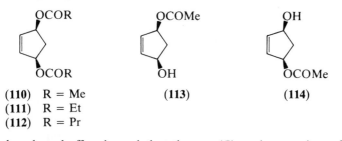

(110) R = Me	(113)	(114)
(111) R = Et		
(112) R = Pr		

in pH 7 phosphate buffer showed that the pro-(R) acyl group is preferentially hydrolysed by α-chymotrypsin, PLE and bakers' yeast, whereas the lipases (from *Rhizopus* sp. and *Candida cylindracea*) exhibited pro-(S) selectivity. However, the rate of hydrolysis with α-chymotrypsin, bakers' yeast and *Rhizopus* was so slow that it militated against their use in organic synthesis. In addition the chemical yield obtained was much higher for the acetate (110) than the propionate (111) and the butanoate (112). Thus, the acetate (110) was hydrolysed in the presence of PLE to give (−)-(1S,4R) alcohol (113) in 86% yield with 86% enantiomeric excess, whereas the lipase from *Candida cylindracea* afforded the (+)-(1R,4S) alcohol (114) in 82% yield. Although the optical yield of this latter process was low (50% enantiomeric excess), the purity of the product may be improved by fractional crystallization. Schneider and co-workers have described a simple method for the effective immobilization of PLE and the use of this material in the above hydrolysis [71]. Although there was no improvement in either chemical yield or enantiomeric excess of the product, with this modification the reaction was easily carried out on a 500-mmol scale. The complex is prepared by dialysing PLE against phosphate buffer and then mixing the enzyme with oxirane activated acrylic beads to provide a covalent bond between the polymer matrix and the enzyme. This catalyst may be stored at 0–5°C for several months. Furthermore, the immobilized enzyme has been used successfully for the preparation of compounds (32), (66), (113) and (115) on a 50 to 500-mmol scale.

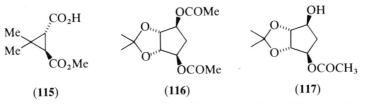

(115)	(116)	(117)

This work has since been complemented by the studies of Deardorff *et al.* [72] who investigated the hydrolysis of the diacetoxycyclopentene (110) and showed that, on incubation with electric eel acetylcholinesterase, the optically active alcohol (114) could be obtained in 94% yield (e.e. >99%). Thus both stereoisomers of this mono-alcohol can be obtained in high optical purity by careful selection of the enzymic catalyst (cf. ref. 70). Interestingly, a problem normally encountered in these types of acetate hydrolysis, i.e. concomitant removal of both ester functions, was not experienced with acetylcholinesterase. The relevant diol was only observed after prolonged exposure to the enzyme.

Electric eel acetylcholinesterase has also been utilized by Johnson and Penning in their synthesis of (−)-prostaglandin E$_2$ [73]. The protected diacetate (116) on incubation with electric eel acetylcholinesterase provided the required (S) alcohol (117) in 80% yield (e.e. = 98%). This material was transformed to the required prostaglandin in an overall yield of 51%.

Similarly, Pearson et al. [74] have shown that the racemic diacetoxycycloheptene (118) can be smoothly transformed into the alcohol (119) (39% yield, e.e. = 100%) on treatment with electric eel acetylcholinesterase. In contrast, the lipase from Candida cylindracea, although hydrolysing the

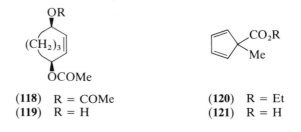

(118)   R = COMe   (120)   R = Et
(119)   R = H   (121)   R = H

diester (118) in similar yield, exhibited much lower enantioselectivity, the alcohol (119) being obtained in 44% enantiomeric excess.

Recently, another advantage of using PLE in organic synthesis has been demonstrated by Burger et al. [75]. Base hydrolysis of the cyclopentadiene derivative (120) gave only potassium ethyl carbonate and a mixture of methylcyclopentadienes. However, by using PLE in pH 8 phosphate buffer, the required acid (121) could be isolated in 66% yield; this is the first example of the isolation of a simple cyclopentadiene carboxylic acid. Similarly, the mild reaction conditions that can be employed to hydrolyse cyclopropylacetate with PLE led to the isolation of a good yield of cyclopropanol [76].

The use of PLE in the synthesis of an intermediate for carbocyclic analogues of ribonucleosides and nicotinamide has been demonstrated by Sicsic et al. [77]. The racemic cyclopentene derivative (122), on incubation with PLE in aqueous methanol at pH 7.5, allowed the isolation of the optically active acid (−)-(123) in 47% yield (e.e. = 97%) along with the recovered, less reactive, enantiomer of the starting ester (+)-(**122**) in 43% yield (e.e. = 87%).

(**122**)   (**123**)

The reaction was easy to perform on a 3-g scale.

In contrast to the popular use of PLE as a hydrolysing enzyme, Camboie and Klibanov [78] have developed an excellent method for the formation of optically active alcohols and esters from racemates via transesterification. Since most enzyme reactions are carried out in aqueous solution where the concentration of water is high and that of competing nucleophiles (e.g. alcohols) is low, then obviously the yield of transesterification products will be small. However, by utilizing a two-phase system, in which the enzyme is present in the aqueous phase and the ester and the alcohol are dissolved in the organic phase, Camboie and Klibanov have succeeded in preparing a number of optically active alcohols and their propionic esters. In order to avoid problems with emulsions and to simplify isolation, the enzyme was entrapped in either Sepharose or Chromosorb before use. Thus, the enzyme could be readily recovered and used repeatedly as the catalyst for the process. By using the above conditions, esters of 3-methoxy-1-butanol, 3-methyl-1-pentanol, 3,7-dimethyl-1-octanol and β-citronellol were prepared in excellent optical yield (e.e. >90%). A major disadvantage in using PLE in this process is its narrow substrate specificity, i.e. it will only work successfully with primary alcohols having no substituents closer to the hydroxyl group than the γ-position. In order to overcome this problem and hence broaden the scope of the process, the lipase from *Candida cylindracea* was investigated. This was shown to have a much wider substrate specificity in that it will accept secondary as well as primary alcohols. In addition, it shows a marked stereoselectivity in that, in the case of chiral alcohols, it will only accept one enantiomer as a nucleophile. As with PLE, the enantiomeric excesses of both the products and the recovered unreacted alcohols were extremely high. The efficacy of the yeast process was demonstrated by the reaction of tributyrin with a number of racemic alcohols and in each case only the (*R*) enantiomer reacted. (For other esterification and transesterification reactions catalysed by lipases see Section 2.1.1.3).

2.1.1.3. Lipases

In naturally occurring processes, enzymes always function in the presence of water. Equally, the majority of chemical uses of enzymes have employed mainly aqueous systems containing only small quantities of organic solvents. However, it has been shown that some lipases can act as excellent catalysts in almost anhydrous organic solution and the advantages of using enzymic catalysis in organic media are numerous [79].

Both porcine pancreatic lipase (PPL) [80] and yeast lipase are excellent catalysts for both esterification and transesterification in almost anhydrous solution and the processes are often highly regioselective and/or stereoselective [81]. Using yeast lipase (*Candida cylindracea*), a wide variety of α-halogenated acids was found to react with alcohols to yield both the optically active ester (**125**) and the unreacted enantiomer of the acid (**124**) (Scheme 2.9). On recovery of the enzyme, it was noted that only minor

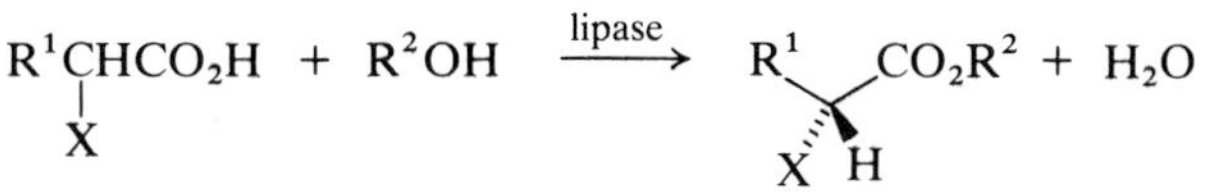

$$R^1CHCO_2H + R^2OH \xrightarrow{\text{lipase}} R^1 \diagdown CO_2R^2 + H_2O$$
$$\underset{X}{\mid} \qquad\qquad\qquad\qquad X\ H$$

(**124**)   (**125**)

SCHEME 2.9

inactivation of the lipase had occurred. In short, this lipase showed an overwhelming preference for esterification of the (*R*) enantiomer of some racemic carboxylic acids.

Although PPL displayed low catalytic activity in similar esterification reactions in organic media, it was found to be active and extremely stereoselective for *trans* esterification reactions in anhydrous organic systems (Scheme 2.10). High chemical and optical yields were obtained.

$$R^1CO_2CH_2CCl_3 + (xs)R^2OH \xrightarrow{\text{PPL}} R^1CO_2R^2 + R^2OH$$
$$\text{(optically active)} \quad \text{(optically active)}$$
$$+ CCl_3CH_2OH$$

NB: R^2 contains an asymmetric centre

SCHEME 2.10

The stereoselectivity of the process decreased when the difference in size between the substituents at the asymmetric carbon atom in R^2 is small. Most of the experiments were carried out in ether or heptane but many other water-miscible and water-immiscible solvents were investigated. In all cases, the presence of water was observed to have a detrimental effect in that the enzymic transesterification was overshadowed by hydrolysis. Once again the enzyme showed an overwhelming preference for esterification of the (*R*) enantiomer of the racemic alcohol. Using these methods, a number of optically active alcohols, acids and esters were prepared on a gram scale.

Preparation 2.5.   Preparation of (2R)-n-butyl 2-bromopropanoate (**125**, R^1 = Me, R^2 = n-Bu, X = Br) and (2S)-2-bromopropanoic acid

To a solution of 2-bromopropanoic acid (6.1 g) and n-butanol (11 ml) in dry hexane (400 ml) was added *Candida cylindracea* lipase (2.0 g *ex* Sigma Chemical Co.). The suspension was placed in an Erlenmeyer flask and shaken on an orbit shaker at 250 rpm and 30°C for 6 h. The enzyme was removed by filtration and the filtrate was washed with 0.5 M aqueous sodium hydrogencarbonate (3 × 80 ml). The organic phase was dried (Na_2SO_4) and the solvent evaporated. The residue was distilled to give the title ester (3.3 g, e.e. = 96%).
    To provide optically active acid the reaction was run for 14.5 h before work up as above. The aqueous phase was acidified to pH 1 (6 N HCl) and extracted with dichloromethane (3 × 80 ml). The combined organic fractions were dried (Na_2SO_4) and evaporated to give (2S)-2-bromopropanoic acid (1.3 g, e.e. > 99%).

It is also noteworthy that PPL is extremely thermostable. Not only can the dry lipase withstand heating at 100°C for many hours, but it also exhibits high catalytic activity at this temperature [82]. The lipase-catalysed transesterification between tributyrin and heptanol at 100°C was shown to be much faster than the same process at 20°C. This was shown to be a general phenomenon by replacement of the heptanol with a number of other primary and secondary alcohols when, in all cases, the transesterification was shown to be much faster at 100°C than at 20°C.

A practical use for the transesterification properties of PPL has been demonstrated by Veschambre who has successfully resolved (±)-sulcatol (**126**) using the enzyme in anhydrous ether [83]. After 4 days reaction the (S)-(+)-sulcatol obtained had an enantiomeric excess >99% and the (R)-(−)-sulcatol [produced via chemical hydrolysis of the recovered butyrate (**127**)] was ca. 80% optically pure. The advantages gained by using dehydrated PPL and 2,2,2-trifluoroethyl laurate for this process have been described [84].

MeCH(OH)CH$_2$CH$_2$CH=CMe$_2$

(**126**)

$$H_7C_3\overset{\overset{\displaystyle O}{\|}}{C}O$$

H$\quad$(CH$_2$)$_2$CH=CMe$_2$

Me

(**127**)

Ester interchange, catalysed by PPL in organic solvents and involving vinyl esters, such as vinyl butyrate, and primary or secondary alcohols, shows promise because the reverse reaction is virtually impossible [85]. The regio- and enantio-selectivities of the reaction remain to be investigated. The lipase preparation derived from the fungus *Mucor miehei* is available and has been used for the esterification of some carboxylic acids (e.g. octanoic acid) with some secondary alcohols (e.g. 2-hexanol). The stereoselectivities observed for the reactions are good (the above coupling gave a 48% yield of the (R) octanoate ester (e.e. = 83%) and recovered (*S*) alcohol (35% yield, e.e. = 87%)), but the amount of enzyme required for the conversions and the long reaction times needed (3–5 weeks) suggest further work is required to optimize the process [86].

The lipase from *Candida cylindracea* has been shown to hydrolyse octyl 2-chloropropionate (**128**) enantioselectively and by the use of this method a convenient process for the resolution of 2-chloropropionic acid has been demonstrated [87]. In contrast, no appreciable enantioselectivity was observed for the hydrolysis of the corresponding methyl ester. At the concentrations used in this work, the methyl ester was completely soluble in water, whereas the octyl ester formed an emulsion; it was postulated that in order to exhibit stereoselectivity the lipase needs to act at the substrate–water interface. This resolution can be performed successfully on ca. 14 g of racemic substrate when only the (*R*) stereoisomer is hydrolysed, affording the (*R*) acid in 92% yield and 96% enantiomeric excess. The (*S*) ester can be readily recovered from the reaction mixture in ca. 89% yield.

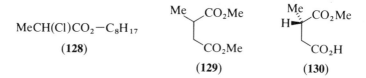

$$\text{MeCH(Cl)CO}_2-\text{C}_8\text{H}_{17}$$

(**128**)   (**129**)   (**130**)

From the range of lipases investigated, PPL was shown to be the enzyme of choice for the hydrolysis of 3-chloro-2-methyl propanoyl propionate [88]. By terminating the reaction after 38% hydrolysis, it is possible to prepare 3-chloro-2-methylpropanol containing 87% of the *l* form (e.e. = 74%). The non-hydrolysed ester was recovered in high yield and recycled. By carrying the degree of hydrolysis to 81% the non-hydrolysed ester can be recovered, in a highly optically enriched form, containing 90% of the *d* isomer (e.e. = 80%). Immobilized enzyme was also used successfully for this resolution and this has the distinct advantage that the enzyme can be recovered, with only minimal loss in activity, and recycled.

Racemic derivatives of dimethyl succinate can also be resolved successfully using PPL [89]. For example, the racemic diester (**129**), on incubation with PPL in phosphate buffer at pH 7.2, gave the unhydrolysed (*R*) ester (**129**) in 46% yield (e.e. = 95%) along with the hydrolysed half-ester (**130**) (38%; e.e. = 73%). Although the optical purity of the half-ester was low, it could be improved to >96% enantiomeric excess by simply recycling the material through a second enzyme catalysed hydrolysis after esterification with diazomethane.

Unfortunately, the ethyl and butyl diesters were hydrolysed much more slowly with a lot less stereoselectivity. Similarly, the choice of the α-substituent was critical; whilst racemic dimethyl 2-benzylsuccinate allowed recovery of the unreacted diester in 90% yield (e.e. >98%), reactions on the allyl and butyl analogues exhibited extremely low selectivity.

Lipases have been used for the hydrolysis of esters in prostaglandin chemistry. PPL has been proposed as the reagent of choice for the mild ester hydrolysis of sensitive prostaglandin derivatives [90]. The lipase was success-fully used in the hydrolysis of the ester (131) when the acid (132) was

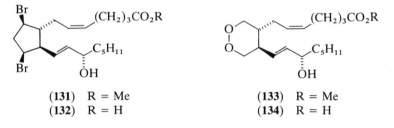

(131) R = Me
(132) R = H

(133) R = Me
(134) R = H

obtained in 92% yield. Various chemical methods were investigated for this hydrolysis but yields were lower (ca. 80%), reaction times were long and some side reactions accompanied hydrolysis.

The cyclic peroxide (133) has been converted into 10-nor-9,11-seco-PGH$_2$ (134) by lipase catalysed hydrolysis [91]. The reaction was performed in pH 8 buffer at 37°C and the prostaglandin analogue (134) was obtained in 87% yield.

PPL in aqueous solution has proved to be an excellent medium for the selective hydrolysis of meso-diesters resulting in the formation of optically active alcohols [92]. Thus, Kasel et al. [92] have investigated the hydrolysis of cyclic diesters (Scheme 2.11) and have demonstrated excellent chemical and

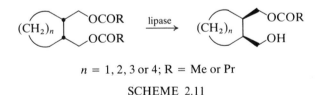

$n = 1, 2, 3$ or 4; R = Me or Pr

SCHEME 2.11

optical yields for reactions involving 2 g of substrate. Furthermore, in all these cases PPL proved to be consistently enantiotopically specific in pro-ducing the alcohol described in Scheme 2.11. The more lipophilic alcohols ($n = 3, 4$; R = Pr) were obtained in lower yields due to further hydrolysis to the corresponding diols. Schneider and co-workers subjected a similar range of cyclic diesters to hydrolysis with both PPL and PLE [68]. With PLE, no distinct trend was observed and low optical and chemical yields were obtained (see Section 2.1.1.2). In contrast, PPL showed remarkable enantiotopic selectivity in that the enzyme only hydrolysed one of the two ester groups giving rise, in each case, to the formation of only the alcohol with the general formula depicted in Scheme 2.11. In general, the optical yields were very high. In addition, hydrolyses with PLE had to be terminated by removal of the enzyme, whereas with PPL essentially only one ester group was hydrolysed and the reaction self-terminated.

More recently it has been reported [93] that *Pseudomonas fluorescens* lipase catalyses the hydrolysis of a range of cyclic acetates [for example the compounds (135)–(137)] to produce the corresponding optically active alcohols.

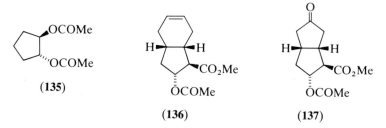

(135)   (136)   (137)

The observed optical purities of the products were generally extremely high. For the above trio the enzyme preferentially hydrolysed the (*R*) acetate.

The lipase from *Candida cylindracea* (CCL) has been used as the catalyst for the synthesis of both enantiomers of norbornenol [94]. Incubation of the lipase with the racemic acetate (138) to the point where 40% conversion had occurred allowed the isolation of the alcohol (+)-(139) in 36% yield (e.e. = 90%), along with the recovery of the partially resolved acetate. This acetate was then subjected to further hydrolysis until another 20% conversion was obtained. Work up (extraction and column chromatography) furnished unreacted acetate (−)-(138) in 38% yield which was shown to be 96%

(138)   (139)

enantiomerically pure after chemical hydrolysis to optically active norbornenol. Using this method 2.6 g of dextrorotatory alcohol and 3.8 g of laevorotatory alcohol were obtained from 10 g of racemic ester.

The *meso*-diester (140) has been shown to undergo hydrolysis with PPL in pH 7 phosphate buffer; the pro-(S) acetoxy group was preferentially cleaved [95], affording the alcohol (141). Interestingly, PLE exhibited the opposite enantiotopic specificity and hydrolysed the pro-(R) acetoxy group to give the enantiomer (142).

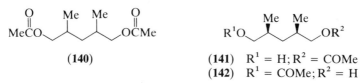

(140)

(141) R¹ = H; R² = COMe
(142) R¹ = COMe; R² = H

PPL has been shown to have a stereochemical preference for the pro-(R) acetoxy group of diester (143, R = CHCH$_2$) furnishing the monoester (144) [67]. In order to obtain an enantiomeric excess of ca. 95%, the optimum

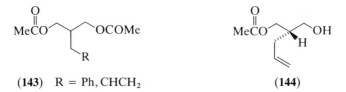

(143) R = Ph, CHCH$_2$ (144)

isolated yield of the monoacetate (144) was ca. 34%. The acetate has been used as an intermediate in the synthesis of (−)-A factor, a compound involved in microbial growth. PPL hydrolysis of the diacetate (143, R = Ph) gave (2R)-3-hydroxy-2-benzyl-n-propylacetate.

Preparation 2.6. Preparation of (2R)-3-hydroxy-2-benzyl-n-propyl acetate

A solution of (143; R = Ph) (35 g) in acetone (600 ml) was added to a solution of Triton X-100 (3 g) in deionized water (1400 ml). The suspension was emulsified by ultrasonification for 1 h. To the stirred and cooled emulsion of (143; R = Ph) was added PPL (Sigma Chemical Co., L-3126, 70 g) at −5°C. The reaction mixture was stirred vigorously with automatic neutralization of the liberated acetic acid with 1 M sodium hydroxide to keep the mixture at pH 7.0. After 2 h at −5°C, 70 ml of 1 M sodium hydroxide was consumed. At this point, the reaction was quenched by the addition of sodium chloride, ethyl acetate and Celite. The mixture was filtered and the organic layer was separated. The aqueous layer was then extracted three times with ethyl acetate. The ethyl acetate solution was dried (MgSO$_4$) and concentrated *in vacuo* to give an oil (34.5 g). This was chromatographed over silica (350 g). Elution with n-hexane–EtOAc (8:1) gave starting material (143, R = Ph) (13.5 g, 39%). Further elution with n-hexane–EtOAc (3:1) gave the title compound (16.2 g, 56%, e.e. = 86%), n_D^{24} 1.5046, $[\alpha]_D^{24}$ + 24.2° (e.e. = 86%) (c = 1.03, CHCl$_3$); v_{max} (film) 3460 (s), 1735 (s), 1605 (m) cm^{-1} [M. Ohno, personal communication].

PPL has also been used for the enantioselective hydrolysis of esters of epoxy alcohols (Scheme 2.12), thus providing an enzymic alternative to the Sharpless reactions as a route to useful chiral epoxides [96]. Since the enzyme

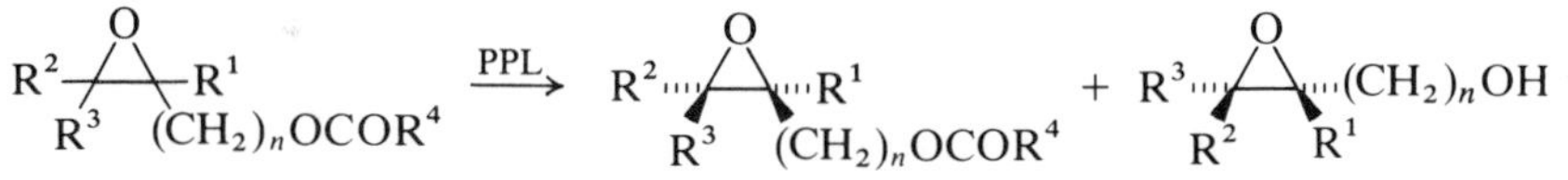

$$R^2 \underset{R^3 \quad (CH_2)_n OCOR^4}{\overset{O}{\triangle}} R^1 \quad \xrightarrow{PPL} \quad R^2 \underset{R^3 \quad (CH_2)_n OCOR^4}{\overset{O}{\triangle}} R^1 \quad + \quad R^3 \underset{R^2 \quad R^1}{\overset{O}{\triangle}} (CH_2)_n OH$$

SCHEME 2.12

is active at water/organic solvent interfaces, the solubility of the substrate in water is not essential and the hydrolyses are carried out at pH 7.8 by controlled addition of 7 M NaOH. Generally, the longer the ester group (R^4) the better the enantioselectivity of the process, although foaming and emulsification tend to be practical problems if R^4 is longer than C_5H_{11}. The reaction can be readily performed on a 300-g scale and, by careful monitoring of the reaction and termination after the optimum conversion has occurred, the products can be obtained with good optical purity (e.e. $\geqslant 90\%$). The major advantage of this method over transition-metal catalysis for chiral epoxide formation is that it is a much simpler process to perform. However, the method has the distinct disadvantage that only a maximum of 50% yield of one enantiomer can be attained when employing racemic substrates.

Preparation 2.7. Preparation of optically active glycidyl esters

A mixture of 300 g of epoxyester and 300 ml of water was placed in a 1-l three-necked flask equipped with a pH electrode, and the two-phase mixture was stirred vigorously with a magnetic stirring bar. Addition of 7.5 g of crude porcine pancreatic lipase (Sigma, Type II) initiated the hydrolysis. The pH was kept at 7.8 by addition of 7 M sodium hydroxide using a pH controller. When 60% of the theoretical amount of base required for complete hydrolysis of the ester had been added (ca. 6 h reaction time) the mixture was poured into 1 l of dichloromethane. The phases were separated and the aqueous phase re-extracted with two 200 ml portions of dichloromethane. The combined organic extracts were washed once with 300 ml of 10% sodium hydrogen carbonate and twice with 200 ml portions of water, dried ($MgSO_4$ containing a small amount of Na_2CO_3), and concentrated on a rotary evaporator. Distillation yielded > 100 g of (R)-ester (0.74 mol, 89% based on the theoretical yield of one enantiomer), b.p. 81–82°C (12 torr), with e.e. $> 92\%$.

Many other lipases from a variety of sources have been employed in organic syntheses. For instance, both pancreatin (hog pancreas extract) and steapsin (lipase from hog pancreas extract) have been utilized for the resolution of the acetate ($\pm$)-(**145**) affording (S) acetate and (R) alcohol (**146**), showing that the hydrolysis takes place with appreciable enantioselectivity [97]. Approximately 75% hydrolysis of ($\pm$)-(**145**) with pancreatin or steapsin gave the optically active acetate [(S)-**145**] in 20% yield and with 90% enantiomeric excess. Many micro-organisms were assayed for the asymmetric hydrolysis of the ester ($\pm$)-(**145**) and, although many *Mucor*

ClCH₂CH(Cl)CH₂OCOMe

(145)

(146)

species were found to hydrolyse the acetate stereoselectivity, their selectivity was much lower than that of pancreatin or steapsin.

An enzyme obtained from the yeast *Candida cylindracea* is readily available commercially in large quantities and is known as Lipase MY. This system has been used for the resolution of α-substituted cyclohexanols via esterification with lauric acid [98]. A range of α-substituted cyclohexanols was investigated and in all cases the lipase was shown to be specific for reactions involving (R) alcohols. In addition, it was shown that α-*trans* compounds were esterified faster than α-*cis* isomers. By this method both optically active alcohols are available from a racemic mixture with excellent enantiomeric excesses (> 80% in most cases). The process is readily amenable to scale up; for example, (±)-menthol has been successfully resolved on a 1-mol scale. Further work on the resolution of menthol using Lipase MY for asymmetric hydrolysis, ester interchange and ester formation has shown that almost optically pure (−)-menthol is readily obtainable in ester interchange between (±)-menthyl laurate and isobutanol [99]. The reaction is best carried out in an organic solvent using either the powdered lipase, or with the enzyme supported on glass beads [100].

Lipase MY has also been used for the enantioselective hydrolysis of a thio-ester [101]. (±)-3-Acetylthiocycloheptene (147) was shown to be hydrolysed more stereoselectively than the corresponding (±)-ester (148). In both cases the (+)-enantiomer was hydrolysed preferentially, although low chemical and optical yields were obtained.

(147) X = S
(148) X = O

(149)

(150)

Fluoromalonates have also been shown to be suitable substrates for resolution with Lipase MY [102]. For example, the laevorotatory malonate (150) was obtained in 91% enantiomeric excess on treatment of the diester (149) with Lipase MY in pH 7.3 phosphate buffer during 6 h. The use of various other enzyme systems, including PPL and a cellulase (*Trichoderma viride*) was investigated, but none were found to give as encouraging results as the lipase.

Lipoprotein lipase Amano 40 [103] *ex Pseudomonas aeruginosa* preferentially hydrolysed the (R) isomer of ($\pm$)-1,2-diacetoxy-3-chloropropane to leave the (S) isomer (151), the optical purity of which is estimated to be ca. 90% enantiomeric excess [104]. This material can be readily converted into the β-adrenergic blocking agent (S)-propranolol which can be obtained optically pure in this way.

(151)   (152)

Racemic 2-oxazolidinone esters have been resolved using both Lipase Amano 3 and Lipase PL266 from *Alcaligenes* species [105]: for example, treatment of the racemic acetate ($\pm$)-(152) with either of the above enzymes effected selective hydrolysis of the (R) acetate, allowing recovery of the (S) acetate [(S)-152] in good optical yield. In both cases the optical purities were better than 93% enantiomeric excess. Other lipases were investigated, but these gave only low selectivity. Certain micro-organisms gave enantiomeric excesses of > 91%, but they suffer the disadvantage that they are not as readily available.

The lipase from *Candida cylindracea* has been shown to effect regioselective deacylation of suitably protected sugars (Scheme 2.13) [106]. A variety of

SCHEME 2.13

different acyl sugars was investigated and although the octanoyl derivatives were found to be the best substrates in terms of enantioselectivity they caused practical problems during work-up. All things considered, the pentanoyl derivatives were found to be the best to work with from the preparative point of view. Incidentally, the acetyl derivatives are not substrates for the enzyme. The reaction can be successfully performed on a 50-mmol scale and the results are essentially the same as those from a small-scale preparation.

Recently, Japanese workers have shown that, as well as intermolecular transesterification, lipases can also be used for intramolecular transesterification, resulting in the formation of macrocyclic lactones (Scheme 2.14) [107]. A range of lipases was investigated but only PPL and Lipase P from *Pseudomonas* sp. were found to catalyse the lactonization reaction efficiently. In addition, polar solvents such as ethyl acetate or dimethylsulphoxide gave extremely poor results. The reaction was shown to be best performed in an organic solvent such as benzene. (Similar solvent effects are seen in lipase catalysed ester synthesis [108].) The lactonization reaction was

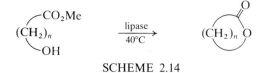

SCHEME 2.14

also dependent on the chain length of the substrate; for example, with $n = 18$ a yield of 80% was obtained and this decreased almost linearly with chain length. Intermolecular lactonization, resulting in the production of diolides, is also observed to occur in this process; however, diolide formation was observed to decrease with increasing chain length and was almost negligible when methyl-16-hydroxyhexadecanoate was employed as the substrate. Preliminary investigations into the asymmetric nature of the reaction have shown that Lipase P can catalyse stereospecific intramolecular transesterification leading to the formation of only the (R) isomer (154) from the racemic hydroxy ester (153).

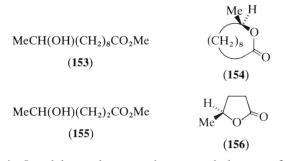

Workers in Israel have also recently reported the use of PPL for the transformation of a number of γ-hydroxy esters into γ-lactones [109]. The reactions were carried out in anhydrous organic solvents and the process was shown to be highly stereoselective as exemplified by the synthesis of (S)-($-$)-γ-methylbutyrolactone (156) from the racemic hydroxy ester (155). In order to optimize the formation of the lactone (156), the reaction was stopped at 36% conversion by removal of the enzyme. By continuing the conversion to 60%, the optically active unreacted starting ester (155) could be obtained, and by simple acid-catalysed lactonization this compound could be converted into the enantiomer of (156), i.e. (R)-($+$)-γ-methylbutyrolactone. Both these lactones were shown to be of high optical purity (e.e. $>94\%$). Under similar reaction conditions β-, δ- and ε-hydroxyacid methyl esters generally undergo oligomerization rather than lactonization [110].

L-Carnitine is essential in regulating the level of blood lipids, while D-carnitine is inactive. In response to the growing interest in the possible therapeutic applications of L-carnitine, Dropsy and Klibanov [111] have described a method for the preparation of this material by enantioselective cleavage of the acetate group from racemic acetyl-carnitine (**157**). Incubation of this acetate with immobilized electric eel acetylcholinesterase at

$$(Me)_3\overset{\oplus}{N}CH_2\underset{\underset{OCOMe}{|}}{C}HCH_2CO_2H$$

(**157**)

pH 7.4 allowed the isolation of D-carnitine in good yield along with the recovery of unreacted acetyl-L-carnitine in 52% yield. Simple acid hydrolysis of this material afforded L-carnitine with an optical purity of 88%. In contrast, horse serum butyrylcholinesterase readily hydrolysed only the L isomer of racemic butyrylcarnitine, producing the biologically active enantiomer in good chemical and optical yields. Acetyl carnitine was not a substrate for this enzyme.

Incubation of racemic methyl N-acetylphenylalanine with the serine proteinase, Subtilisin Carlsberg (sold under the trade name Alcalase®) afforded (S)-N-acetylphenylalanine in 96% yield (e.e. = 98%) along with the recovered (R) ester in 98% yield [112]. (S)-N-Acetylphenylalanine was readily converted into (S)-phenylalanine on treatment with acid.

Alcalase® has also been used for the selective hydrolysis of the dibenzyl esters of both aspartic and glutamic acids. In each case only the benzyl ester in the α-position was hydrolysed affording the corresponding monoesters in high yield (>82%) [113].

Cantacuzène et al. [114] have shown that papain is an excellent catalyst for obtaining a N-Boc-amino acid ester from the corresponding acid using a biphasic system. A wide range of N-Boc-amino acids was studied and the synthesis was shown to be highly dependent on the ratio of the volumes of the organic and aqueous phases; for each amino acid of interest the optimum ratio was investigated. In general, high yields were obtained but the sterically hindered amino acids valine, isoleucine and threonine gave only low yields, whilst the basic amino acids lysine, arginine and histidine could not be esterified at all. Interestingly, for dicarboxylic acids such as aspartic acid and glutamic acid, only the α-carboxyl group was esterified. The reactions were all performed in pH 4.2 citrate–phosphate buffer during 8 h.

2.1.2. Micro-organisms

The use of bakers' yeast for the enantioselective reduction of ketones is well documented and is discussed elsewhere in this book (see Chapter 3). The advantages gained by using bakers' yeast as a catalyst for hydrolysis reactions are often less clear-cut. However, a study of this property has recently been described by Glänzer *et al.* [115] whereby D-*N*-acetylamino acid esters were obtained via enantioselective hydrolysis of their racemates using fermenting yeast. A variety of amino acids was investigated and in all cases only the carboxylic ester was hydrolysed, the amide group remaining unchanged. Only the natural L derivatives were cleaved, the D enantiomers being recovered in varying degrees of optical purity. Enantiomeric excesses for the acids were of the order of 90% in almost all cases, the only disappointing results being obtained when the amino acid esters contained an additional polar substituent. Similarly, low enantioselectivity was observed on the hydrolysis of cyclic derivatives (e.g. proline). In general, anaerobic conditions were used for the fermentations, and reaction rates varied depending on the substrate employed.

Ziffer and co-workers [116] have studied the hydrolysis of acetates derived from a range of racemic aryl alkanols and have shown that the enantiomer (**158**) is hydrolysed more rapidly by *Rhizopus nigricans* than is its mirror

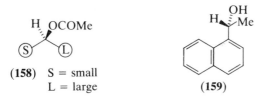

(**158**)   S = small
            L = large

(**159**)

image. The enantiomeric excesses obtained varied from 8 to 100% and the larger enantiomeric excesses were found in reactions involving those alcohols in which the relative sizes of the two groups (small and large) were very different. Put another way, the smallest enantiomeric excesses were obtained when the two groups were almost identical in size. Thus, 1-(1-naphthalenyl)ethanoyl acetate was hydrolysed to give the alcohol (**159**) in 33% yield (e.e. = 100%) along with recovered optically active acetate. This method has been carried out on a multigram scale.

The same workers have also investigated the hydrolysis of the esters of a range of cyclic carbinols [117] and, in all but one of these cases, the alcohol obtained had the configuration predicted by the guidelines postulated in the earlier work. No explanation could be offered for the one anomaly, but obviously where an alcohol is obtained in low enantiomeric excess, as was the case, revision of the guidelines would seem to be unwarranted. This work has since been extended to the study of substituted tetrahydronaphthalenes [118] and similar results were obtained. It is claimed that the stereochemistry of the alcohol formed is primarily dependent on the relative size of substituents on the carbinol carbon atom and that the presence of remote functionality has little effect on the course of the hydrolysis.

Rhizopus nigricans has also been used to produce the prostaglandin intermediate (160) in optically active form by hydrolysis of the corresponding benzoate [119]. The alcohol was shown to be 95% optically pure by conversion into the corresponding Mosher ester.

Bacillus subtilis var. *niger* has been employed for the enantioselective hydrolysis of racemic acetates. For example, incubation of this microorganism with the appropriate racemic acetate gave the hydroxy-decalin (161) in 43% yield (e.e = 54%) [120].

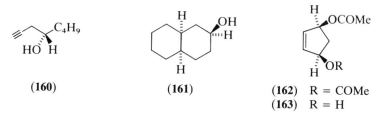

(160)   (161)   (162) R = COMe
                                            (163) R = H

The same workers have described the use of this organism for the asymmetric hydrolysis of esters derived from various terpene alcohols [121] as well as the hydrolysis of acetoxy-alkynes and -alkenes [122]. The same organism has been employed in the partially selective hydrolysis of the *meso*-diacetate (162) [123]. The acetate (163), an early-stage intermediate in prostaglandin synthesis, was obtained in 56% yield (e.e. = 35%).

Mori and Akao [124] have also shown that *B. subtilis* is a useful organism for the hydrolysis of acetates of both racemic alkynyl alcohols and α-hydroxy esters. For example, both (+) and (−) forms of highly optically pure (e.e. = 97%) mandelic acid could be obtained by hydrolysis of the racemic acetate of ethyl mandelate followed by recrystallization.

Iriuchijima and Keiyu [125] have shown that *B. subtilis* preferentially hydrolyses the (*S*) isomer of the ester (164) (e.e. = 83%), whereas *Aspergillus sojae* shows a preference for hydrolysis of the (*R*) isomer (the acid in this case being obtained in 71% enantiomeric excess).

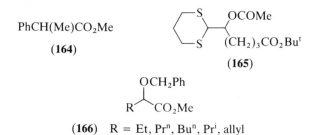

$PhCH(Me)CO_2Me$

(164)

(165)

(166) R = Et, Pr^n, Bu^n, Pr^i, allyl

Takaishi *et al.* [126] have also employed *Bacillus* sp. for the hydrolysis of the racemic acetate (165). The (*R*) alcohol was obtained in 32% yield (e.e. = 98%) along with 46% of the recovered (*S*) acetate (e.e. = 94%). Both these compounds are proposed intermediates in syntheses of the corresponding enantiomers of 5-HETE and 5-HPETE.

Rhizopus oryzae has been shown to hydrolyse both (−)-PGE$_1$ methyl ester and (−)-PGE$_2$ methyl ester in good yield [127]. More recently, Ohta and co-workers [128] have described the enantioselective hydrolyses of protected α-hydroxy alkanoates using *Corynebacterium equi* (IFO 3730). A range of substituted esters (166) was investigated and in each case the unreacted optically active esters were recovered in high optical purity (e.e. > 90%). The hydrolysed acid could not be recovered and this was assumed to be because of further degradation in the whole-cell system. In each case only the (*R*) ester was hydrolysed, allowing recovery of the (*S*) ester in high optical yield. In complete contrast, when the corresponding derivative, where R = benzyl, was subjected to hydrolysis using the above organism, a complete reversal of enantioselectivity was observed, with the (*S*) ester being preferentially hydrolysed. Presumably the C-benzyl moiety displaces the O-benzyl group from its position within the active site of the enzyme.

2.2. CLEAVAGE AND FORMATION OF THE AMIDE BOND

2.2.1. Preparation and Hydrolysis of Simple Amides

Whilst there is a wealth of literature governing the use of esterases and lipases in organic synthesis (see Section 2.1), the use of amidases is less well documented. The term amidase encompasses all enzymes which catalyse either cleavage or formation of amides, i.e. it covers carboxypeptidases, *N*-acylases, acid proteases and acyl transferases.

Carboxypeptidases hydrolyse only those amide functions which are adjacent to a free carboxyl group and the rate of hydrolysis is usually increased if the adjacent group, R, is an aromatic or large aliphatic moiety (see Scheme 2.15) [129]. The amino acid being attacked by the enzyme must have the

SCHEME 2.15

L configuration; the D enantiomer is not a substrate. The potential of carboxypeptidase has been realized in its use as a catalyst for the resolution of many amino acids [3]. For example, *threo*-β-phenylserine (168) has been success-

PhCHCHCO₂H   PhCHCHCO₂H
  | | | |
 HO NHCOCF₃   HO   NH₂

(167)   **(168)**

fully resolved by incubation of the racemic N-trifluoroacetate **(167)** with carboxypeptidase-A [130]. The optically pure L enantiomer was obtained in 61% yield after crystallization. Some acylases are like carboxypeptidases in that they require the presence of a free α-carboxylic acid in the substrate; the best known of the acylases are those which have been isolated from hog kidney. For example, hog renal acylase I has been used to determine the configuration of isomeric γ-hydroxy glutamic acids [131]. In addition, a wide range of amino acids has successfully been resolved via acylase catalysed hydrolysis [3].

Hog acylase I has also been used by Baldwin as part of the process involving the conversion of a peptide into a penicillin [132]. It was shown that hog acylase I, on incubation with the chloroacetyl amide **(169)** and separation of unreacted amide followed by acid hydrolysis, gave D-iso-dehydrovaline **(170)** in 60% yield. The material obtained was identical in all respects to a sample obtained by degradation of a penicillin.

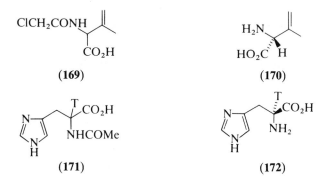

(169)   **(170)**

(171)   **(172)**

It has also been shown that hog kidney acylase can effect the resolution of the tritiated histidine derivative **(171)** affording the amino acid **(172)** [133]. The incubation was performed at 37°C during ca. 20 h and the pH was maintained at 7.2 by the addition of lithium hydroxide. The deuterium analogue of compound **(172)** was also successfully resolved using the above method.

An alternative strategy for obtaining deuterated amino acids has been described by Fujihara and Schowen [134] who also employed hog kidney acylase I as the enzymic catalyst. For example, N-acetyl-DL-phenylalanine **(173)** was deuterated using deuterium oxide and the resulting N-acetate was

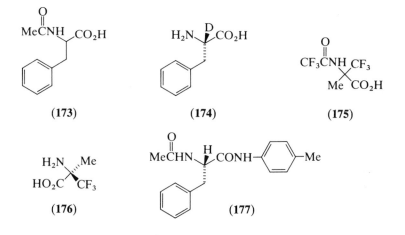

(173)   (174)   (175)

(176)   (177)

hydrolysed using the porcine acylase. The hydrolysis was performed in the presence of lithium hydroxide at pH 7.5 for ca. 24 h whereupon the required, optically pure, deuterated L amino acid (174) was obtained in 51% yield. The generality of the process was demonstrated by the successful deuteration and hydrolysis of leucine, glutamic acid, methionine and tyrosine-*N*-acetates. In all cases there was only a minimal loss of deuterium; the deuterium content of the optically active amino acid was shown to be >97% by NMR spectroscopy.

More recently, hog kidney acylase has been used for the catalytic hydrolysis of racemic *N*-trifluoroacetyl-2-trifluoromethylalanine [135]. Typically, the racemic trifluoroacetate (175) was incubated with the acylase at pH 7.5 for 67 h at 25°C. After work up, chromatography over Dowex resin, and crystallization of the optically pure (e.e. = 99%) amino acid (176) was obtained in 53% yield. Earlier, it had been shown that hog kidney acylase displayed a stereochemical preference for the hydrolysis of amino acid amides having the larger C-2 substituent in the pro-(*S*) position [136]; the hydrolysis of compound (175) conforms to this pattern.

A mould aminoacylase (produced from wheat bran and rice hulls after inoculation with *Aspergillus oryzae*, and immobilized by ionic binding to DEAE-Sephadex) has been used for the production of L-amino acids from racemic *N*-acetyl precursors [137]. The process was developed to the extent whereby simply the passage of a solution of the racemic acetyl amino acids through a column containing the immobilized enzyme at a pre-determined rate effected the hydrolysis. The column effluent, after evaporation and processing, gave the required L-amino acids in high yield (e.g. alanine, 35%; methionine, 40%; phenylalanine, 43%; tryptophan, 36%; valine, 41%). In addition, the unreacted acetyl-D-amino acids could be recovered readily. The process is now operated on an industrial scale such that up to 17 000 kg of the requisite amino acid may be processed within a period of 20 days.

Resolution of phenylalanine has also been achieved via the papain catalysed synthesis of acetyl-L-phenylalanine-p-toluidide (177) from acetyl-DL-phenylalanine (173) and p-toluidine [138]. Incubation of (±)-(173) and p-toluidine with papain in pH 4.6 buffer at 37°C for 7 days afforded, after work up and recrystallization, the p-toluidide (177) in ca. 43% yield. Acid hydrolysis of (177) afforded the resolved L-phenylalanine in 82% yield. Papain was also the catalyst of choice for the resolution of N-carbobenz-oxy-γ-methyl glutamic acid (178), a starting material in the synthesis of (+)-marmelolactones (Scheme 2.16) [139].

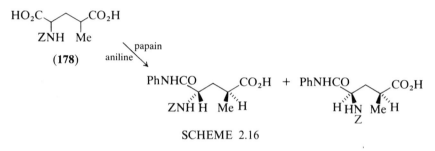

SCHEME 2.16

An acylase obtained from *Aspergillus* sp. has been used for the catalytic hydrolysis of the racemic N-chloroacetyl amino acid (179) [140]. The hydrolysis was carried out at pH 7.1 in the presence of a cobalt(II) salt for 2.5

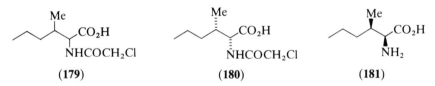

days. Concentration of the reaction mixture gave the amino acid (181) in 51% yield, whilst acidification of the mother liquor gave partially resolved starting material (180). The amino acid so obtained was shown to be of high optical purity and the recovered starting amide, by being subjected to further enzyme catalysed hydrolysis, could also be obtained in a highly optically pure state. These amino acid enantiomers were converted into optically active forms of *threo*-4-methylheptan-3-ol, a pheromone component of the elm-bark beetle.

Mori and Otsuka [141] have used the same *Aspergillus* amino acylase (Tokyo Kasei Co.) for the resolution of an unnatural long-chain α-amino acid, (±)-(182; R = C$_{14}$H$_{29}$). Thus, treatment with the acylase at pH 7.3 in the presence of a small amount of cobalt(II) chloride for 2 days at 37°C gave

Preparation 2.8.   Enzymic resolution of ($\pm$)-(**179**)

($\pm$)-(**179**) (95 g) was dissolved in deionized water (4000 ml) by adding 2 M sodium hydroxide up to pH 7.1. *Aspergillus* amino acylase (Amano Pharmaceutical Co., 20 g) and cobalt(II) chloride hexahydrate (40 mg) were added to the solution and the mixture was left to stand at 37°C for 2.5 days. This was decolourized with Norit, filtered and concentrated *in vacuo* to a volume of ca. 150 ml. The separated crystals were collected on a filter, and dried to give (2S,3R)-(**181**) (31.8 g, 51.2%). An analytical sample was obtained by recrystallization from water, m.p. 237–241°C, $[\alpha]_D^{20}$ + 42.1° (c = 0.275, 5 M HCl); ν_{max} (Nujol) 3050 (mb), 2700 (m), 2602 (m), 1580 (s), 1510 (s) cm^{-1}. The filtrate, after removing (**181**), was then acidified with 5 M HCl to pH 2. The precipitates were collected on a filter and dried to give (2R,3S)-(**180**) (40.6 g, 42.3%). An analytical sample was obtained by recrystallization from acetone, m.p. 129–130°C; $[\alpha]_D^{20}$ − 30.7 (c = 1.020, EtOH); ν_{max} (Nujol) 3420 (m), 2670 (m), 2630 (m), 1715 (m), 1630 (s), 1540 (m) cm^{-1}, ^{13}C-NMR δ (CDCl$_3$) 14.1, 15.1, 19.8, 34.5, 35.2, 42.4, 55.6, 166.4, 172.9.

the hydrolysed amino acid (**183**) along with the unhydrolysed (*R*) amide. Both species were obtained in satisfactory yield and were shown by a combination of processes to be about 92% optically pure. No attempt was made to optimize these values; the amino acid was required as a starting material for the synthesis of a naturally occurring cerebroside.

RCHCO$_2$H	R CO$_2$H
NHCOCH$_2$Cl	H NH$_2$
(**182**)	(**183**)

Amidases have also been shown to be extremely useful catalysts in the synthesis of semi-synthetic penicillins and cephalosporins. For example, the cleavage of the 6-acyl group from penicillin G (**184**) to give 6-aminopenicillanic acid (**185**) has been carried out using an acylase from *Escherichia coli*. The process originated in industrial laboratories in 1959 and has since been

Preparation 2.9.   Enzymic resolution of (**182**, R = C$_{14}$H$_{29}$)

Aspergillus amino acylase (Tokyo Kasei Co., 5 g) and cobalt(II) chloride hexahydrate (10 mg) were added to a solution of ($\pm$)-(**182**, R = C$_{14}$H$_{29}$) (36.0 g) in water (4000 ml) adjusted to pH 7.3 by the addition of sodium hydroxide. The solution was left to stand for 44 h at 37°C. The precipitated crystalline (*S*)-(**183**, R = C$_{14}$H$_{29}$) was collected on a filter, washed with methanol and ether, and dried over P$_2$O$_5$ to give (*S*)-(**183**, R = C$_{14}$H$_{29}$) (14.0 g, 50%), m.p. 234–236°C, $[\alpha]_D^{26}$ + 21.8° (c = 0.1, AcOH); ν_{max} (Nujol) 1575 (s), 1510 (s) cm^{-1}. The filtrate obtained after removal of (*S*)-(**183**, R = C$_{14}$H$_{29}$) was acidified with 3 M HCl. The precipitated solid was collected on a filter, and dissolved in ethyl acetate (1000 ml). The insoluble material was filtered off and the filtrate was concentrated *in vacuo*. The residue was recrystallized from n-hexane to give (*R*)-(**182**, R = C$_{14}$H$_{29}$) (15.5 g, 43%), m.p. 87.0–88.0°C, $[\alpha]_D^{21}$ − 28.0° (c = 0.5, CHCl$_3$).

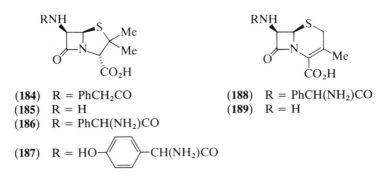

(184) R = PhCH₂CO   (188) R = PhCH(NH₂)CO
(185) R = H   (189) R = H
(186) R = PhCH(NH₂)CO

(187) R = HO—⟨◯⟩—CH(NH₂)CO

investigated by a number of research groups. For example, Lagerlöf et al. [142] have shown that, by immobilization of the enzyme on Sephadex G-200, potassium benzyl penicillin can be converted into 6-aminopenicillanic acid in approximately 90% yield on a 3-kg scale.

Many other techniques for immobilizing the acylase have been described [143]. Alternatively the use of the free acylase in an aqueous two-phase system for the attachment of side-chains to 6-APA has been described by workers in Sweden [144]. The yields of both ampicillin (186) and cephalexin (188) as formed under catalysis by the amidase from *E. coli* have been shown to be increased by recycling the substrate [145]. A variety of other penicillins and cephalosporins have been prepared using enzymic methods [146].

Whole cells of *Xanthomonas citri* effectively catalyse the formation of amoxicillin (187) from D-α-(p-hydroxyphenyl)glycine methyl ester and 6-aminopenicillanic acid (185) [147]. The yield of amoxicillin could be increased by reducing the ionic strength of the reaction mixture and by the addition of 5% (v/v) 2-butanol. The same system has been shown to be an excellent catalyst for the synthesis of cephalexin (188) from 7-aminodeacetoxycephalosporanic acid (189) [148].

Recently, the use of acyltransferases for the conversion of cephalosporin-C (190) into 7-phenoxyacetamidocephalosporanic acid (191) has been investigated [149]. Unfortunately, the yields were too low to be of immediate use.

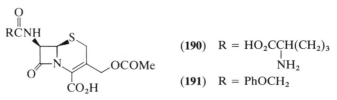

(190) R = HO₂CCH(CH₂)₃
                    |
                    NH₂
(191) R = PhOCH₂

2.2.2. Peptide Synthesis

The formation of the peptide bond should proceed quickly and without racemization of an adjacent chiral centre. Ideally, equimolar amounts of the carboxylic acid and amine components should combine together in quantitative yield; undesired side reactions must be avoided. Since most enzymes catalyse just one reaction, the potential use of such specific catalysts for the formation of a peptide bond was recognized long ago. As early as 1937, Bergmann and Fraenkel-Conrat [150] presented data to suggest that the reversibility of protease catalysed reactions could be utilized in the formation of peptide bonds. However, it was not until the 1970s that proteases were successfully employed as catalysts in the preparative-scale synthesis of peptides. Since then, there has been a wealth of reports on the enzymic synthesis of the peptide bond. This report will only deal with some selected examples; a more comprehensive survey can be obtained from the numerous reviews and publications on the subject [151]. For example, Kullman [152] has demonstrated the use of enzyme catalysis for the synthesis of the opioid peptides, leu- and met-enkephalin (192) and (193).

H·Tyr·Gly·Gly·Phe·LeuOH   H·Tyr·Gly·Gly·Phe·MetOH

(192)   (193)

In each case, all the peptide bonds were formed by enzymic catalysis using either papain or α-chymotrypsin; papain in sodium acetate buffer in the presence of 2-mercaptoethanol was used for the construction of the Gly–Phe bond whilst the Gly–Gly bond was formed with papain in the presence of ethanol/McIlvain buffer with added 2-mercaptoethanol. All the other peptide bonds were obtained by catalysis with α-chymotrypsin in sodium carbonate buffer/dimethylformamide. The yields were better than 60%, apart from the Phe–Leu and the Phe–Met couplings which were reported to be 50% and 45%, respectively. The results of this work complement those of earlier studies [153] in that, in addition to the preference of papain for substrates having phenylalanine in position "P_2", substrates with tyrosine at the same site can be equally well processed.

Kullmann [154] has also employed proteolytic enzymes for the synthesis of the COOH terminal octapeptide amide of cholecystokinin (194). Each tetrapeptide portion was constructed enzymically as shown in Scheme 2.17 with the final coupling being performed chemically. All attempts to couple the two tetrapeptides via papain, thermolysin, ficin, or bromelain catalysis failed: in each case one or more of the reagents was degraded by the enzyme. (For simplicity, all protecting groups, etc., have been removed from Scheme 2.17.) It was shown that tyrosine derivatives were inappropriate acceptor nucleophiles for thermolysin catalysed reactions and such couplings had to be performed using papain.

Another advantage of enzyme catalysis over chemical methods for the synthesis of peptide bonds has been elegantly demonstrated by Kullmann [155] who studied the synthesis of the sub-sequence (1–8) of dynorphin (195)

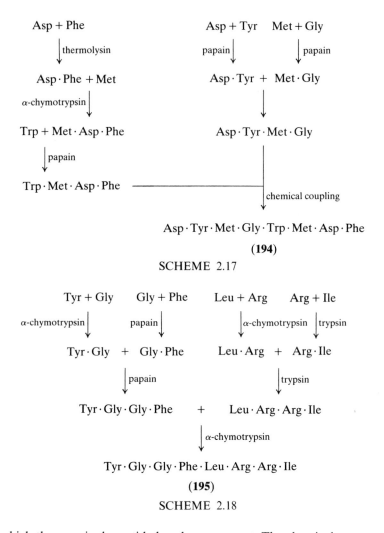

Asp + Phe

↓ thermolysin

Asp · Phe + Met

α-chymotrypsin ↓

Trp + Met · Asp · Phe

↓ papain

Trp · Met · Asp · Phe

Asp + Tyr   Met + Gly

papain ↓   ↓ papain

Asp · Tyr + Met · Gly

↓

Asp · Tyr · Met · Gly

chemical coupling

Asp · Tyr · Met · Gly · Trp · Met · Asp · Phe

(194)

SCHEME 2.17

Tyr + Gly   Gly + Phe   Leu + Arg   Arg + Ile

α-chymotrypsin ↓   papain ↓   α-chymotrypsin ↓   trypsin ↓

Tyr · Gly + Gly · Phe   Leu · Arg + Arg · Ile

↓ papain   ↓ trypsin

Tyr · Gly · Gly · Phe   +   Leu · Arg · Arg · Ile

↓ α-chymotrypsin

Tyr · Gly · Gly · Phe · Leu · Arg · Arg · Ile

(195)

SCHEME 2.18

in which three arginyl–peptide bonds are present. The chemical preparation of such bonds is often accompanied by undesirable side reactions. Trypsin can be used to overcome this problem because this enzyme promotes regio- and stereo-controlled synthesis of peptide bonds in which the carbonyl moiety is associated with a basic amino acid residue. As in the earlier work, each peptide bond was constructed enzymically (Scheme 2.18); the yields were ca. 60%. (All protecting groups, etc., have been removed from Scheme 2.18 for simplicity.) With α-chymotrypsin as the catalyst, the reaction is usually best carried out in sodium carbonate buffer containing dimethylformamide, whilst trypsin catalysis is best performed in the absence of co-solvent. Papain catalysed reactions were performed in either sodium acetate buffer or McIlvain buffer in the presence of 2-mercaptoethanol.

α-Chymotrypsin, immobilized on tresyl chloride activated Sepharose CL-4B, has also been used successfully for peptide synthesis in water/water-miscible organic solvent mixtures [156]. These investigations used Ac–Phe·OMe as the ester portion and Gly–NH$_2$ or Ala–NH$_2$ as the amine moiety. It was shown that the yield of peptide increased with increasing concentration of organic solvent and that 97% yield of Ac–Phe·Ala could be obtained by employing a six-fold excess of alanine. Similarly, immobilized thermolysin has been used as the catalyst for coupling L-phenylalanine methyl ester with N-carbobenzoxy-L-aspartic acid to afford the Aspartame precursor (**196**) [157]. The use of an immobilized enzyme helps the reaction to be driven in the direction required and the yields are

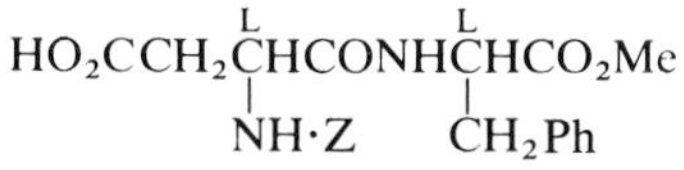

$$\overset{L}{\underset{NH·Z}{HO_2CCH_2\overset{|}{C}HCONH}}\overset{L}{\underset{CH_2Ph}{\overset{|}{C}HCO_2Me}}$$

(196)

often quite high. It is unnecessary to protect the side-chain carboxyl group of the aspartic acid derivative because thermolysin is highly specific in its catalytic activity.

A study of the synthesis of peptides using nagarse, papain, pepsin and thermolysin showed differing specificities for each enzyme [158]. For example, for the couplings shown in Scheme 2.19, thermolysin was found to

Z—X—OH + H—Phe · Val · OBu^t $\longrightarrow$ Z—X · Phe · Val · OBu^t

X = Gly; Val; Tyr; Ser; Arg (NO$_2$)

SCHEME 2.19

be a highly efficient catalyst, whereas papain was less effective and nagarse and pepsin were inactive. On the other hand, for the coupling shown in Scheme 2.20, pepsin was ineffective while the other enzymes all exhibited synthetic activity to different degrees.

Z · Phe · X—OH + H—Phe · Val · OBu^t $\longrightarrow$ Z · Phe · X · Phe · Val · OBu^t

X = Gly; Val; Tyr; Ser; Arg (NO$_2$)

SCHEME 2.20

Papain proved to be a highly efficient catalyst for the coupling of carbobenzoxy dipeptide acids with phenylalanine diphenylmethyl (DPM) ester (Scheme 2.21). Pepsin was also shown to be a useful catalyst for this process, whereas thermolysin and nagarse were ineffective.

Z—X·Y—OH + H—Phe·ODPM $\longrightarrow$ Z—X·Y·Phe·ODPM

X·Y = Leu·Phe; Phe·Tyr; Val·Tyr

SCHEME 2.21

Recently, peptide synthesis has been investigated in a biphasic system of water and water-immiscible organic solvent [159]. The coupling of Ac–Trp and Leu–NH_2 was carried out in a mixture of ethyl acetate and water using α-chymotrypsin in both its free and immobilized forms. The conditions were optimized by a study of volume ratios of the phases, pH and concentration of the reactants. Under the most favourable conditions the dipeptide was synthesized on the preparative scale in almost 100% yield.

Although proteolytic enzymes have many advantages as catalysts for peptide synthesis, they do suffer some shortcomings such as narrow substrate specificity, unfavourable thermodynamic equilibrium and the possibility of proteolysis of the peptide chain. Problems with the unfavourable thermodynamic equilibrium have been overcome to some extent by working in biphasic mixtures [159] or nonaqueous media [157]. Similarly, reverse micelles have been employed because of their capability to solubilize enzymes without causing any loss in catalytic activity [160]. By using this method, Swiss chemists were able to show that the reaction shown in Scheme 2.22 could be carried out by using α-chymotrypsin as the catalyst. A good yield of the water-insoluble tripeptide (197) was obtained [161].

Z—Ala·Phe—OMe + H—Leu $\longrightarrow$ Z—Ala·Phe·Leu

(197)

SCHEME 2.22

Very recently, Klibanov and co-workers [162] have shown that esterases possessing no visible amidase activity can be used successfully as catalysts for peptide synthesis. The synthesis of a wide range of dipeptides was investigated using porcine pancreatic lipase in organic solvents and this protein was shown to be a very effective catalyst for the synthesis of a range of peptides. The reactions were generally performed in either toluene or tetrahydrofuran, although other dry organic solvents were investigated: reaction yields were often better than 60%. The use of such a catalyst in an organic solvent helps to surmount the problems normally encountered with solubility and thermodynamic equilibrium. Most importantly, proteolysis of the peptide bond does not occur or takes place extremely slowly when using a lipase as the catalyst. Wong and co-workers [163] have also recently reported work on the synthesis of peptides using PPL, PLE and *Candida cylindracea* lipase as catalysts.

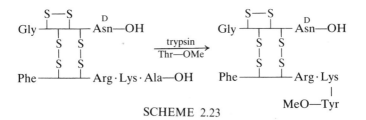

SCHEME 2.23

A commercially viable preparation of an ester of human insulin has been developed using a trypsin-mediated exchange of threonine for the terminal alanyl residue of porcine insulin [164]. This transformation (Scheme 2.23) marks the first example of an enzymic semi-synthesis of a protein for use in medicine [165]. An alternative synthesis of human insulin, involving the coupling of a derivative of deoctapeptide-(B23–B30)-insulin (derived from porcine hormone) and human octapeptide by trypsin-mediated catalysis has also been described [166].

2.3. CLEAVAGE AND FORMATION OF PHOSPHATE ESTERS

2.3.1. Phosphate Hydrolysis

Hydrolysis of phosphates is frequently a trivial synthetic step which can be performed by chemical methods. However, the use of phosphatases is to be advocated for phosphate hydrolysis where the substrate is sensitive; it should be noted that many phosphatases are substrate specific. As illustrated below, on some occasions advantage may also be taken of the enantiospecificity of some phosphate hydrolases.

Acid-catalysed hydrolysis of polyprenyl pyrophosphates is unsuccessful due to the acid lability of the resulting alcohols. However, commercially available potato acid phosphatase hydrolyses these phosphates readily to afford the corresponding alcohols, but only in the presence of large amounts of a simple alcohol [167]. Studies have revealed that the optimum concentration of the alcohol for phosphate hydrolysis is in inverse proportion to the size of the alkyl group on the alcohol. It was thought that the resistance of polyprenyl phosphates to many phosphatases was due to the propensity of these compounds to form micelles; it has also been postulated that the presence of large amounts of alcohol inhibit the formation of micelles. Hence it would be interesting to see if a normally inactive phosphatase would hydrolyse a polyprenyl phosphate in the presence of large amounts of alcohol.

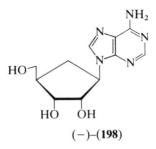

(−)–(198)

One of the enantiomers of the carbocyclic analogue of adenosine has been isolated from a *Streptomyces* strain and named (−)-aristeromycin [(−)-(198)]. The corresponding racemic mixture [(±)-(198)] has been synthesized chemically by several groups, and its resolution by chromatography on cellulose has been reported. However, neither of the melting points of the isolated enantiomers matched that of the natural product: this result, in combination with other evidence, strongly suggested that the resolved material was of the wrong isomeric constitution. This confusing situation was clarified by De Clercq and co-workers [168] who used 5′-ribonucleotide phosphohydrolase to hydrolyse a racemic mixture of the derived 5′-phosphates. This gave a high yield of optically active carbocyclic adenosine which was identical to the natural (−)-aristeromycin in all respects. The recovered optically active phosphate was hydrolysed to unnatural (+)-aristeromycin by calf intestinal alkaline phosphatase. Under identical reaction conditions, this alkaline phosphatase hydrolysed racemic aristeromycin 5′-phosphate to give [(±)-(198)].

2.3.2. Preparation of Phosphorylated Species other than Nucleoside Triphosphates

The introduction of the phosphate group into a polyfunctional molecule by normal chemical methods frequently involves many protection and deprotection steps. The use of enzymes to introduce a phosphate group can eliminate many of these additional steps and can make a synthesis much more efficient. An additional bonus is that in certain cases the introduction of a phosphate moiety is enantiospecific. In several of the cases cited below, alternative chemical procedures are available, but they are invariably poorer than the enzyme-catalysed procedure.

5-Phospho-D-ribosyl-l-pyrophosphate (199) (PRPP) is a key intermediate in the biosynthesis of nucleosides and nucleotides. Although it is commercially available, the use of this material on a large scale is impractical because of its high cost; the inherent instability of PRPP both thermally and at pH values much removed from 7 is another problem.

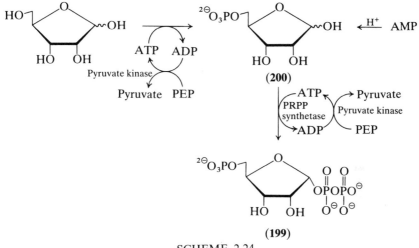

(199)

SCHEME 2.24

The biosynthesis of PRPP by the enzyme PRPP-synthetase has been utilized in the laboratory to prepare sub-mmol quantities of PRPP from ribose-5-phosphate. Whitesides and co-workers [169] have improved and scaled up these procedures by the use of immobilized PRPP synthetase and *in situ* ATP regeneration. This has not only provided a means of preparing useful quantities of PRPP, but also points a way to *in situ* preparation of PRPP by a multiple enzyme system. The ribose-5-phosphate (**200**) that was needed could be prepared by ribokinase catalysed phosphorylation of ribose or by various chemical means, the best being the acid catalysed hydrolysis of AMP (Scheme 2.24). The subsequent preparation with PAN-immobilized PRPP-synthetase could be readily performed on such a scale as to give aqueous solutions containing approximately 75 mmol of PRPP. The enzymic preparation of ribose-5-phosphate could be coupled with the PRPP-synthetase step to provide a one-pot synthesis of PRPP directly from D-ribose, but this method requires further refinement in order to be made viable for less-experienced hands. The power of this method, however, is evident from the direct synthesis of uridine-5'-monophosphate from D-ribose in 73% overall yield [169]. PRPP-synthetase has also been used to prepare the pyrophosphate (**201**) [170].

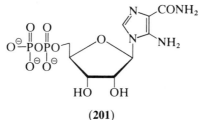

(201)

Another key biosynthetic intermediate is glucose-6-phosphate (202). The best preparation of this compound is the hexokinase catalysed phosphorylation of glucose with ATP regeneration (see later), but other routes have also been examined (Scheme 2.25) [171]. The conversion of a suitable polysaccharide (soluble starch or dextrin are recommended on the basis of cost) to glucose-1-phosphate (203) with phosphorylase (E.C.2.4.1.1) followed by isomerization to glucose-6-phosphate (202) with phosphoglucomutase (E.C.2.7.5.1) is a viable method but is less convenient than the hexokinase procedure, as the product is less easily isolated. An alternative procedure uses the action of phosphoglucose isomerase on fructose-6-phosphate (204),

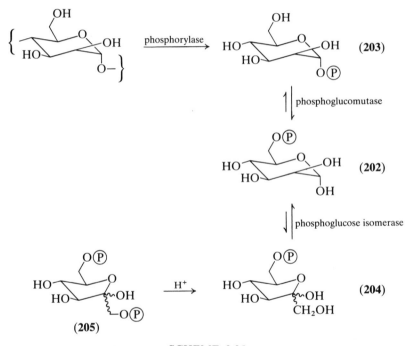

SCHEME 2.25

which is obtained by mild acid hydrolysis of the commercially available fructose-1,6-diphosphate (205). This method yields an equilibrium mixture containing approximately 70% glucose-6-phosphate, but it is not applicable to large-scale preparations, as fructose-1,6-diphosphate is relatively expensive. In all these methods, the enzymes are immobilized and can be recycled. Other routes to glucose-6-phosphate by normal chemical procedures are unsatisfactory because of low yields and difficulties with product isolation.

A PEP/pyruvate kinase system was also used in conjunction with yeast hexokinase to investigate the phosphorylation of a range of fluorinated hexapyranoses, as well as glucose analogues with sulphur and nitrogen in the ring [172]. This work illustrated that such fluorinated sugars can be accepted as substrates for the hexokinase, thereby providing a good method for the selective phosphorylation of the primary hydroxyl group without the necessity of resorting to functional group protection.

The preparation of substantial quantities of dihydroxyacetone phosphate by glycerol kinase catalysed phosphorylation of dihydroxyacetone is discussed in the section dealing with aldolase catalysed reactions (see Section 5.1). The former enzyme is cheap and stable in immobilized form and is highly enantioselective towards some substrates. For these reasons additional studies have been made of this enzyme to define its substrate specificity [173]. These studies demonstrated that glycerol kinase from a variety of sources will accept a range of unnatural substrates, including DL-3-aminopropane-1,2-diol which is phosphorylated on nitrogen to give a phosphoramidate (Scheme 2.26). All these substrates are potentially capable of being used in syntheses on at least a 10-g scale. A caveat is necessary, however, in some cases when a racemic mixture is used as a starting material. For example, D-propane-1,2-diol is phosphorylated by glycerol kinase but the racemic mixture is not transformed. This is because the D enantiomer is only a poor substrate and the reaction is prevented by the strong inhibitory activity of the L enantiomer.

R_1 = OH, OMe, OEt, OAc, SH, SMe, SEt, Me, Et, CH_2OH, CH_2CN, NH_2, Cl, Br
R_2 = H, OH, F
R_3 = H, OH (as hydrated ketone), Me (depending on enzyme source)
X = NH, O

SCHEME 2.26

An earlier example of the preparative use of glycerol kinase to make *sn*-glycerol-3-phosphate [174] was followed by further preparative examples which also exemplify the enantiospecificity of the enzyme [175]. For these kinetic resolutions, the substrates listed in Scheme 2.27 were reacted with ca. 0.05 eq ATP and 0.5 eq of PEP as the phosphate donor. These conditions are designed to eliminate over-phosphorylation and to ease the work-up procedure by ensuring that no PEP remained at the end of the reaction.

In each case the phosphorylated D enantiomers were isolated in yields ⩾75% and in high enantiomeric excess. The reaction is easily performed on a large scale, as exemplified by the ready preparation of 203 g of D-3-chloropropane-1,2-diol-l-phosphate, as its barium salt, in a total of

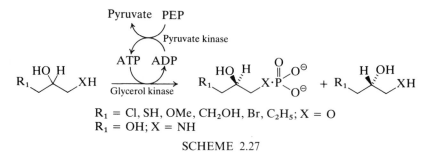

$$R_1 = Cl, SH, OMe, CH_2OH, Br, C_2H_5; X = O$$
$$R_1 = OH; X = NH$$

SCHEME 2.27

2.2 l of water. Enantiomerically pure non-phosphorylated material could be obtained from the phosphorylated products by hydrolysis with alkaline phosphatase as catalyst.

Preparation 2.10.   Preparation of D-3-chloropropane-1,2-diol-1-phosphate, L- and D-3-chloropropane-1,2-diol

(D)-*3-Chloropropane-1,2-diol-1-phosphate (large scale)*. PEP⁻K⁺ (127 g, 0.600 mol), ATP (4.0 g, 7.0 mmol), magnesium chloride (15.3 g, 75.0 mmol), 2-mercaptoethanol (0.2 ml, 3 mmol) and (DL)-3-chloropropane-1,2-diol (133 g, 1.20 mol) were dissolved in 2 l of deoxygenated water. The pH of the solution was adjusted to 7.2 with sodium hydroxide (ca. 30 g of solid NaOH and 4 M NaOH). The reaction was started by the addition of immobilized glycerol kinase (GK) (10 000 U determined with glycerol as substrate) and pyruvate kinase (PK) (1000 U). The reaction was monitored by ³¹P-NMR: aliquots (3 ml) were removed from the reactor and added to 1.0–2.0 ml of deuterium oxide in a 12 mm NMR tube. After 7 days the reaction was complete, but the reaction was continued for two additional days. The supernatant was separated from the immobilized enzymes by centrifugation and decantation, and the enzyme-containing gels were washed twice with 200 ml portions of deoxygenated water. The wash solutions and the supernatant were combined (total 2.6 l) and passed through charcoal (ca. 100 g) to remove ADP and ATP. Barium chloride dihydrate (159 g, 0.65 mol) was added and a precipitate formed. Ethanol (95%, 8 l) was added to this suspension and the precipitate was allowed to settle overnight. The solid was isolated by filtration and dried at 1 torr overnight over CaSO₄. A total of 203 g of white solid was isolated. The solid contained 92% of the title compound as the barium salt as determined by enzymic assay (90% of the salt as determined by quantitative ³¹P-NMR) and 3% inorganic phosphate. The yield was 93% based on D-3-chloropropane-1,2-diol: ¹H-NMR (D₂O, pH 7) δ (DSS) 4.0–4.1 (1H, m), 3.6–3.9 (4H, m); ¹³C-NMR (D₂O, pH 7) δ (DSS) 75.1 (d, ³J_{CCOP} = 7.3 Hz), 69.6 (d, ²J_{COP} = 3.6 Hz), 50.7 (s); ³¹P-NMR (D₂O, pH 7.2) δ 3.9 (t, ³J_{POCH} = 6.6 Hz); [α]$_D^{25}$ + 3.3° (c = 2.5, H₂O, pH ≃0). The turnover number for ATP during the synthesis was 85. The recovered enzyme activities were as follows: GK, 76%; PK, 81%.

L-3-Chloropropane-1,2-diol was recovered from the supernatant. The supernatant (10 l) was concentrated by evaporation to ca. 1 l. To the solution was added ca. 2 l of 95% ethanol and the resulting solid was removed by filtration and washed with absolute ethanol. The solution was concentrated to ca. 200 ml and added to ca. 1 l of ethanol. The precipitate was again removed by filtration and the solution was concentrated to dryness. The resulting oil was dissolved in 200 ml of absolute ethanol and the precipitate was removed by filtration. After removing most of the ethanol by evaporation, the residual oil was distilled [b.p. 98°C, 1 torr (lit. b.p. 213°C)]. This procedure yielded 21 g of colourless liquid (190 mmol, 31% yield): ^{1}H-NMR (D$_2$O) δ (DSS) 3.85–4.0 (1H, m), 3.5–3.7 (4H, m); ^{13}C-NMR (D$_2$O) δ (DSS) 75.7 (s), 67.1 (s), 50.4 (s); [α]$_D^{22}$ −6.8° (c = 5, H$_2$O). The enantiomeric excess was also determined to be 94% using Eu(hfc)$_3$ in CD$_3$CN.

D-*3-Chloropropane-1,2-diol.* Barium D-3-chloropropane-1,2-diol-1-phosphate (0.80 g, 2.1 mmol) was reconstituted in 6 ml of water by addition of Dowex-50 until the pH fell below 4. After removal of the resin, the pH of the solution was raised to ca. 9 by addition of sodium carbonate. Alkaline phosphatase (20 U) was added in soluble form, and the solution was left at ambient temperature. After 30 min the solution was cloudy (due to inorganic phosphate). Hydrolysis was complete by the following day as observed by ^{31}P-NMR. The inorganic phosphate was removed by first adding barium chloride dihydrate (0.5 g, 2 mmol) to the solution and then adding 20 ml of absolute ethanol. The precipitate was removed by centrifugation and the supernatant was concentrated to ca. 5 ml. Another 20 ml of absolute ethanol was added and the precipitated solid was removed by centrifugation. After concentrating the supernatant to ca. 0.5 ml, 1 ml of absolute ethanol was added, the solid was removed by centrifugation, and the supernatant concentrated. D-3-Chloropropane-1,2-diol was not further purified. The 132 mg of colourless liquid corresponded to a yield of 57%: ^{1}H-NMR (acetonitrile-d_3) 3.8–4.0 (1H, m), 3.5–3.7 (4H, m); [α]$_D^{22}$ + 7.1° (c = 5, H$_2$O) (corresponding to e.e. ca. 95%). The enantiomeric purity was determined using Eu(hfc)$_3$ in CD$_3$CN and found to be e.e. >97% (no L enantiomer was observed).

In three cases the residual non-phosphorylated L enantiomer was isolated from the glycerol kinase reaction, but the yield was low because of the difficulty in isolating low molecular weight diols and triols from dilute aqueous solutions.

Glycerol kinase has also been used to transfer chiral [^{16}O, ^{17}O, ^{18}O]phosphate groups from the corresponding ADP-phosphate of known absolute configuration [176]. Other chiral phosphates have been prepared using adenylate kinase [177], nuclease P$_1$ and nucleotide phosphoesterase [178].

3-Deoxy-D-*arabino*-heptulosonic acid (DAHP) (**205**) (Scheme 2.28) has been prepared by a multiple-step chemical synthesis but, in order to provide a faster and more efficient synthesis, Frost and co-workers [179] turned to a coupled-enzyme system. Hexokinase catalysed phosphorylation of D-fructose, with ATP regeneration by PEP/pyruvate kinase, gave D-fructose-6-phosphate (**204**) which was converted into D-erythrose-4-phosphate (**207**) by transketolase. The latter was then converted, in 85% overall yield, into the required product by DAHP-synthetase catalysed coupling with more PEP. Because all the enzymes were immobilized, they could be recovered and recycled with only a small loss in activity, although prolonged storage resulted in loss of DAHP-synthetase activity. A complementary whole-cell process was also used with success. Although this particular coupled-

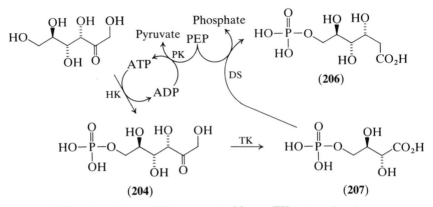

HK = hexokinase; PK = pyruvate kinase; TK = transketolase;
DS = DAHP synthetase

SCHEME 2.28

enzyme synthesis is not, as it stands, of general utility, it is an interesting example of how enzyme catalysed synthesis can successfully compete with other synthetic routes.

2.3.3. Regeneration of Nucleoside Triphosphates

Many enzyme catalysed reactions require nucleoside triphosphates (especially ATP) as coenzymes for their catalytic activity; these must be constantly regenerated in order for catalysis to continue. In a living organism the regenerative process occurs by normal metabolic processes, but for reactions which are performed *ex vivo* with purified or immobilized enzymes this does not occur. Addition of stoichiometric amounts of coenzyme to a reaction mixture would be undesirable; many coenzymes are prohibitively expensive, some are unstable and the accumulation of inactivated coenzyme may induce the displacement of the equilibrium in the undesirable direction. Therefore, it is necessary to use catalytic quantities of the coenzyme and continuously regenerate it during the course of the reaction. For some coenzymes, such as pyridoxal phosphate, flavin mononucleotide and flavin adenine dinucleotide, this is no problem because under normal aqueous reaction conditions or in the presence of oxygen they are automatically regenerated. However, other coenzymes, such as the very important adenosine triphosphate, require continuous regeneration by an auxiliary system. In this way, the reaction can be carried to completion using catalytic quantities of coenzyme, recycling enzymes and a stoichiometric quantity of an ultimate phosphate donor.

In order to make enzyme catalysed synthesis viable on a practical scale, the development of coenzyme recycling systems has been a topic of major importance and the focus of much work over the last 15 years. For ATP, at least, the problem has largely been solved by the use of inexpensive phosphate donors and kinase enzymes.

The enzyme carbamoyl phosphokinase catalyses the transfer of a phosphate group from carbamoyl phosphate (prepared *in situ* from potassium cyanate and potassium phosphate) to ADP (Scheme 2.29) [180]. Although this enzyme is commercially available, this regeneration method has not found favour because carbamoyl phosphate hydrolyses very rapidly and generates ammonium ions which remove the magnesium ions from solution. Unfortunately, these magnesium ions are essential for kinase activity.

$$KH_2PO_4 + KOCN + H_2O \longrightarrow H_2N \cdot CO_2 \cdot PO_3^{\ominus}H + KOH$$

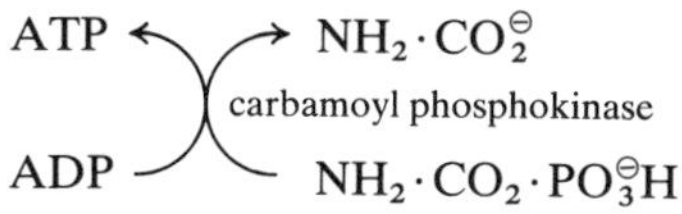

$$ATP \longleftarrow \qquad \longrightarrow NH_2 \cdot CO_2^{\ominus}$$
carbamoyl phosphokinase
$$ADP \longrightarrow \qquad \longleftarrow NH_2 \cdot CO_2 \cdot PO_3^{\ominus}H$$

SCHEME 2.29

A common ATP regeneration system is based upon the acetyl phosphate/acetate kinase system (Scheme 2.30). This system has been popularized by the work of Whitesides and co-workers [181,182], with the facile, large-scale (> 1 M) preparation of aqueous solutions of acetyl phosphate as its disodium or dipotassium salt. This procedure is much more practically convenient than previous procedures for preparing diammonium acetyl phosphate [183], and the disodium salt has the advantage of being appreciably more stable to hydrolysis than the diammonium salt. The necessary acetate kinase is available from two sources, *E. coli* and *Bacillus stearothermophilus*; the latter kinase is preferred for synthetic use because it is stable to auto-oxidation. Both these preparations are commercially available. By the use of this acetyl phosphate (AcP)/acetate kinase system, large-scale preparations of compounds such as glucose-6-phosphate (**202**) (using hexokinase; Scheme 2.31) [184], arginine phosphate (**208**) (Scheme 2.32) [182] and creatine phosphate (**209**) (Scheme 2.33) [185] are possible with no requirement for special apparatus. This latter example is particularly interesting; ATP is not a strong phosphorylating agent and the phosphorylation of creatine does not have a favourable equilibrium constant. The use of acetyl phosphate as the ultimate phosphorylating agent, however, displaces the equilibrium of the overall phosphorylation reaction so as to produce useful quantities of product because it has a greater phosphate-donor ability than ATP. The overall equilibrium for the coupled-enzyme system was also maximized in favour of the product by the use of a mixed solvent system (10% v/v aqueous ethylene glycol).

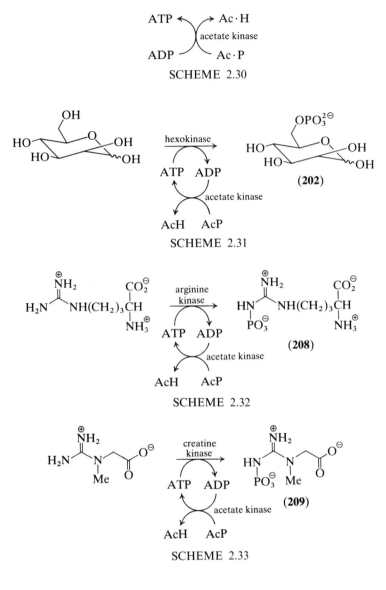

SCHEME 2.30

SCHEME 2.31

SCHEME 2.32

SCHEME 2.33

Many enzyme catalysed reactions convert ATP to AMP rather than ADP; in some cases adenosine is produced. These reactions can still be performed with *in situ* regeneration of ATP by the use of a three-enzyme-coupled system with adenosine kinase, adenylate kinase and acetate kinase, using acetyl phosphate as the ultimate phosphate donor (Scheme 2.34) [186].

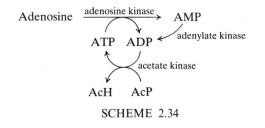

SCHEME 2.34

The slow, spontaneous hydrolysis of acetyl phosphate to acetate and phosphate can present problems in some preparations, because acetate kinases are subject to inhibition by acetate ions [187]. This can be an important factor in any preparation where the overall rate of reaction is low, as in the synthesis of arginine phosphate, described above, where product inhibition considerably slows down the later stages of the reaction. In such cases, the use of a phosphoenol pyruvate (PEP)/pyruvate kinase system to regenerate ATP is superior because PEP has greater hydrolytic stability (the half life for hydrolysis at 25°C and pH 7 is: PEP 1000 h, AcP 21 h) [188]. Further advantages of PEP are that it is a stronger phosphorylating agent than acetyl phosphate (as measured by the free energy of hydrolysis: acetyl phosphate $= -10.1 \, \text{kcal} \, \text{mol}^{-1}$; PEP $= -12.8 \, \text{kcal} \, \text{mol}^{-1}$). In addition, due to its greater stability to hydrolysis, PEP can be added in one portion at the start of the reaction and it is possible to use lower concentrations of ATP (or ADP) as the K_m of ADP for pyruvate kinase is lower than that for acetate kinase. Unfortunately, pyruvate kinase is subject to inhibition by pyruvate, but this can be circumvented by carrying out the reaction in dilute solution or, more conveniently, by utilizing a high concentration of PEP to minimize the effects of the inhibition.

Although potassium PEP is commercially available (other salt forms are available but the potassium salt is preferred) it is too expensive for large-scale syntheses. Fortunately, Whitesides and co-workers [189] have yet again come to the rescue with a facile, multi-molar synthesis of potassium PEP in two steps from pyruvic acid. This provides a cheap source of potassium PEP as a stable, crystalline salt.

The ready availability of PEP has led to its use as part of an ATP-regenerating system in a range of large-scale preparative procedures. The preparation of unnatural saccharides and saccharide analogues from dihydroxyacetone phosphate is discussed later (see Section 5.1 on carbon–carbon bond formation and the aldolases). Other notable preparations using a PEP-recycling system include those of glucose-6-phosphate (on a $> 350 \, \text{g}$ scale!) [189, 190], ribose-5-phosphate [169], ribulose-1,5-diphosphate [169], (S_p)-ATP-α-S [191], uridine-5′-diphosphate-glucose [192], and sn-glycerol-3-phosphate and analogues [175]. The previously described synthesis of arginine phosphate is much superior with PEP/pyruvate kinase rather than AcP/acetate kinase as the ATP-regenerating system [182].

Kazlauskas and Whitesides [188] have described the preparation of methoxycarbonyl phosphate (MCP) as an unnatural substrate for both acetate and carbamate kinase, but not pyruvate kinase. Methoxycarbonyl phosphate is readily prepared by the reaction of aqueous phosphate with methyl chloroformate; the resulting solution can be used immediately or the methoxycarbonyl phosphate can be isolated as its solid dilithium salt. This is a strong phosphorylating agent (free energy of hydrolysis = 12.4 kcal mol^{-1}) and has the advantage that the byproduct remaining after phosphate transfer (methyl carbonate) decomposes spontaneously to the relatively innocuous materials, methanol and carbon dioxide (which can be flushed out of the solution by purging with nitrogen). The price to be paid for these advantages is that MCP hydrolyses even more rapidly than acetyl phosphate (the half-life for hydrolysis at 25°C and pH 7 is 0.3 h); this could present problems in preparations where the product is difficult to separate from inorganic phosphate.

Despite its hydrolytic lability, methoxycarbonyl phosphate can advantageously replace acetyl phosphate in some preparations because of its higher phosphate donor potential. An example of this is the previously cited preparation of creatine phosphate (Scheme 2.33). With MCP as the ultimate phosphate donor, the use of a mixed solvent system to displace the equilibrium in favour of the products is no longer necessary and a 55% yield of creatine phosphate is readily obtained in an aqueous reaction medium. Purification of the product is also easier as the by-products from MCP are easily removed.

Regeneration of the other nucleoside triphosphates (GTP, UTP and CTP) and the corresponding 2'-deoxynucleoside triphosphates from their diphosphates with acetate or pyruvate kinase is as facile as the process for ATP regeneration [181,193].

2.4. HYDROLYSIS OF EPOXIDES

2.4.1. Liver Microsomal Epoxide Hydrolase

Enzymes catalysing the regiospecific and enantiospecific hydrolyses of epoxides have been carefully studied in recent years. Particular attention has been paid to liver microsomes from various species, as the epoxide hydrolases they contain tend to be catholic in their choice of substrate and yet demonstrate high regioselectivity and, frequently, high enantioselectivity. The former property is commensurate with the metabolic role of the liver, which has to detoxify a wide range of xenobiotic materials.

The microsomal preparations used in the preparative experiments described below are versatile and potentially synthetically useful, but it must be borne in mind that access to the appropriate facilities and expertise is necessary to prepare the microsomal fractions. Once prepared they are stable at low temperatures for some months [194].

Both (R) and (S) enantiomers of monosubstituted and 1,1-disubstituted epoxides with at least one large lipophilic substituent are among the best substrates, while highly substituted epoxides are not hydrolysed [195]. It has been suggested that attack by water on epoxides which are symmetrical, or nearly symmetrical, occurs at the carbon atom which has (S) chirality (Scheme 2.35) [196].

SCHEME 2.35

The first epoxide hydrolase reaction which was studied from a stereochemical point of view was the hydrolysis of cyclohexene oxide [197]. This compound gave $(-)$-$(1R,2R)$-dihydroxy cyclohexane with an enantiomeric excess of about 70%. Since then a considerable amount of data have been accumulated to determine the effect on the reaction of the stereochemical environment of the epoxide group. Hydrolyses of cis- and trans-stilbene oxide with rabbit liver microsomal epoxide hydrolase indicated that, even for the cis isomer, there was a highly selective, stereospecific trans addition of water across the epoxide group [198]. Results from racemic trans-stilbene oxide indicated that the (S,S) diastereoisomer was hydrolysed 2.3 times faster than the (R,R) diastereoisomer and that this stilbene oxide isomer was much more slowly hydrolysed than the corresponding cis isomer; this effect was also noted by Oesch in the enzymic hydrolysis of cis- and trans-4-methyl-2-pentene-oxide [195]. Similar results to those described above were obtained for cis-9,10-epoxystearic acid. Wistuba and Schurig [199] have followed the hydrolysis of cis- and trans-2,3-epoxypentane by complexation gas chromatography. They observed, for the racemic cis isomer, a completely regio- and enantio-specific reaction whereby only the (2S,3R) isomer (210) was biotransformed to give threo-(2R,3R)-2,3-pentanediol (211). For the trans epoxide, little enantioselectivity was observed, although, yet again, there was a marked difference in the rates of hydrolysis between the diastereoisomers, with the (S,S) diastereoisomer being metabolized faster; a similar effect was noted for trans-2,3-epoxybutane. For simple monosubstituted epoxides the (S) enantiomer was preferentially metabolized.

(210)　　　　　　　　　　　(211)

The topology of the active site of rabbit liver microsomal epoxide hydrolase has been carefully studied by a group at the University of Pisa. Hydrolysis of the *cis* and *trans* isomers of 4- and 3-t-butyl-1,2-epoxycyclohexanes demonstrated that the enantiomers in which the cyclohexane ring has 3,4-M helicity [200] are preferentially hydrolysed unless steric factors intervene [201]. This was illustrated in the case of the *trans*-4-t-butyl epoxide (**212**) by the isolation, in high enantiomeric excess, of the (1*R*,2*R*,4*S*)-diol (**214**) after 50% hydrolysis. The (1*R*,2*S*,4*R*)-epoxide was hydrolysed at a much slower rate. A similar result was noted for the *cis*-3-t-butyl epoxide (**213**) which was hydrolysed to the diol (**215**). Further confirmation of the "helicity theory" came from a study of the hydrolysis of *trans*-4,5-dimethyl-1,2-epoxycyclohexane [202].

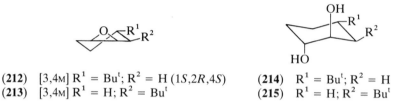

(**212**) [3,4M] R^1 = Bu^t; R^2 = H (1*S*,2*R*,4*S*)   (**214**)   R^1 = Bu^t; R^2 = H
(**213**) [3,4M] R^1 = H; R^2 = Bu^t (**215**)   R^1 = H; R^2 = Bu^t

While the helicity theory works well for cyclohexane epoxides it does not seem to be the whole story. When 3,4-epoxytetrahydropyran was hydrolysed, the product was exclusively the (−)-(3*R*,4*S*)-diol (**216**), with an enantiomeric excess of at least 96%; each enantiomer of the tetrahydropyran epoxide reacted at a similar rate (**203**). If this is rationalized in terms of the helicity theory (see Scheme 2.36), with a diaxial opening of the epoxide, it

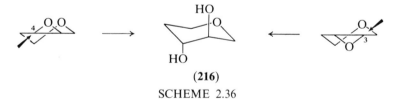

(**216**)

SCHEME 2.36

can only be by assuming that each enantiomer is exclusively attacked with a different regiospecificity, i.e. one enantiomer is attacked at C-3 and the other at C-4. In contrast, results obtained on the hydrolysis of *cis*- and *trans*-2-methyl-3,4-epoxytetrahydropyran demonstrated a reversion to regiospecific opening at C-4 [204]. Thus, hydrolysis of the racemic *trans* epoxide (**217**) (Scheme 2.37) gave exclusively the (2*R*,3*R*,4*R*) diol (**218**) (e.e. >98%) at up to 50% hydrolysis. The conversion stopped entirely at about 70% reaction at which stage the enantiomeric excess of the product was only 66%. The (2*S*,3*S*,4*R*) epoxide (**219**) was thus a much poorer substrate for the hydrolase. This is entirely in accord with the hypothesis that the most favoured substrate is one where there is a lipophilic substituent to the right of the epoxide group and the helicity is correct (the substrate is viewed with the epoxide group in front and on the topside of the molecule). If the usual assumption is made that diaxial opening of the epoxide takes place, then the enzyme catalysed reaction proceeds through the conformation of compound (217) having the methyl group in an axial situation.

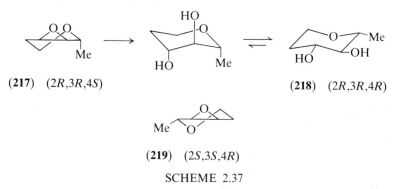

(217) (2R,3R,4S)   (218) (2R,3R,4R)

(219) (2S,3S,4R)

SCHEME 2.37

These stereochemical studies have been put to use in the synthesis of a deoxy-sugar [205]. Reaction of the racemic *lyxo*-epoxide (220) with epoxide hydrolase gave, after 50% conversion, almost diastereomerically pure (ca. 96%) samples of the L diol (221) and the D epoxide (220). This is, again, consistent with the above-mentioned working hypothesis concerning the required conformation for the substrate in the active site of the enzyme.

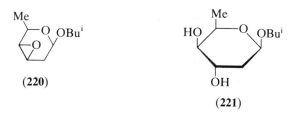

(220)   (221)

By use of the [^{18}O]-labelled epoxides, Hanzlik *et al.* [206], using epoxide hydrolase from rat liver, demonstrated that the hydrolysis of monosubstituted or 1,1-disubstituted epoxides always gave the product corresponding to hydroxyl attack at the least hindered epoxide carbon atom. For the less stereochemically biased 1,2-disubstituted epoxides, e.g. compounds (222) and (223), it was shown that the epoxide oxygen was retained as the C-3 or benzylic oxygen to the extent of at least 85% (it should be noted that the epoxide hydrolase treatment of racemate (223) displayed insignificant enantioselectivity) [297]. The regioselectivity of the epoxide opening is thus very clear, as is the fact that these are not simple acid catalysed hydrations. Addition of metal binding agents or nucleophiles to the epoxide hydrolase reactions did not affect the rate of reaction, nor did it give rise to any product other than the diol [206]. This argues for a mechanism whereby

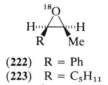

(222)   R = Ph
(223)   R = C₅H₁₁

general base catalysis at the active site gives rise to hydroxide ion which immediately attacks the bound epoxide. Evidence for concomitant acid catalysed activation of the epoxide by a study of deuterium isotope effects proved to be inconclusive [208].

Liver microsomes also contain an NADPH-dependent oxygenase which can epoxidize alkenes (see Section 4.5). Thus, when alkenes are incubated with microsomes in the presence of NADPH and oxygen, there occurs simultaneous epoxidation and hydrolysis to produce a *trans*-1,2-diol [209]. Mono-oxygenase enzymes are difficult to handle and this must be added to the initial difficulty in obtaining the requisite microsomes: these drawbacks render the direct enzyme catalysed synthesis of diols from alkenes of limited value at present.

Throughout this discussion, microsomal epoxide hydrolases from various species have been treated as functionally identical for synthetic purposes. While this is generally true for pigs, guinea-pigs and rabbits, one study has shown that the microsomal epoxide hydrolases from other species, such as rats and houseflies, are relatively inactive [210]. Although this paper only dealt with the hydrolysis of the epoxides of several cyclodiene-type insecticides, the general conclusion is probably a fair one.

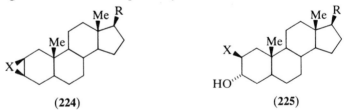

(224)   (225)

Hydrolysis of aziridines and episulphides by rabbit liver microsomes has been studied by Watabe and co-workers [211]. The unsubstituted aziridine (224, X = NH) and its *N*-methyl homologue (224, X = NMe) were hydrolysed exclusively to the aminols (225, X = NH₂ and NHMe, respectively) in the same stereochemical sense as for the corresponding epoxide, while the acyl-aziridine (224, X = NCOMe) was first hydrolysed to (224, X = NH) which then gave the amine (225, X = NH₂). The episulphide (224, X = S) was not hydrolysed even with extended reaction times.

2.4.2. Other Methods

The epoxide hydrolase from *Alcaligenes levotartaricus* converts *cis*-epoxy-succinic acid into D-tartaric acid while *Acremonium tartarogenes* produces L-tartaric acid from the same substrate [212]. D-Tartaric acid was also prepared from the latter epoxide in yields of 81% or more by several strains of *Achromobacter* and *Alcaligenes* [213]. This is an eminently practical process, being performed at room temperature at relatively high concentrations.

Note, in this connection, that fungi such as *Corynespora cassiicola*, *Diplodia gossypina* and *Gibberella cyanea*, have proved useful in dihydroxylation of terpenoids [214]. In these cases the toxicity and volatility of terpenoids requires the use of special reactors to remove products as soon as they are formed and to keep the concentration of substrate down.

2.5. HYDROLYSIS OF NITRILES

This is an area of increasing interest. Enzymes are available that convert a wide range of aliphatic and aromatic nitriles into the corresponding amides. Some of the processes are of commercial importance. For example, the Nitto Chemical Industry Co. Ltd have developed a process for the conversion of acrylonitrile into acrylamide using immobilized *Corynebacterium* cells [215]. Other types of micro-organism that are suitable for this bioconversion are Gram-positive bacteria of the genera *Nocardia*, *Bacillus*, *Bacteridium*, *Micrococcus*, and *Brevibacterium*. Thus Harper found that benzonitrile is hydrolysed to benzoic acid by *Nocardia* sp. [216], while more recently a bacterium of the genus *Rhodococcus* has been shown to convert 2,6-difluorobenzonitrile into the corresponding benzamide [217]. 2,6-Difluoro-benzamide is a known intermediate for the preparation of insecticidal compounds such as diflubenzuron.

The micro-organism *Brevibacterium* sp. (R312) appears to possess a nitrile hydratase and an amidase with wide substrate selectivities. Thus highly functionalized nitriles such as γ-alkoxy-α-amino acids are obtained from the corresponding nitrile in good yield [218].

A nitrilase possessing a broad substrate specificity is commercially available, and this biocatalyst has been shown to convert methyl cyanoacetate into methyl malonate and adipodinitrile into corresponding half-amide in good yield [219].

REFERENCES

1. C. Niemann, *Science* **143**, 1287 (1964).

2. L. Cunningham, in *Comprehensive Biochemistry*, Vol. 16 (eds. M. Florkin and E. H. Stotz), p. 85. Elsevier, Amsterdam, 1965.

3. J. B. Jones, C. J. Sih and D. Perlmann, *Tech. Chem.* **10**, 107 (1976).

4. N. C. Dirlam, B. S. Moore and F. J. Urban, *J. Org. Chem.* **52**, 3287 (1987).

5. S. G. Cohen, *Trans. N.Y. Acad. Sci.* **31**, 705 (1969).

6. M. L. Bender and J. V. Killheffer, *Crit. Rev. Biochem.* **1**, 149 (1973).

7. G. E. Clement and R. Potter, *J. Chem. Educ.* **48**, 695 (1971).

8. C. Niemann, D. W. Thomas and R. V. MacAllister, *J. Am. Chem. Soc.* **73**, 1548 (1951).

9. C. Niemann and H. T. Huang, *J. Am. Chem. Soc.* **73**, 1541 (1951).

10. A. Berger, M. Smolarsky, N. Kurn and H. R. Bosshard, *J. Org. Chem.* **38**, 457 (1973).

11. S. G. Cohen and E. Khedouri, *J. Am. Chem. Soc.* **83**, 1093 (1961).

12. S. G. Cohen, Y. Sprinzak and E. Khedouri, *J. Am. Chem. Soc.* **83**, 4225 (1961).

13. S. G. Cohen and E. Khedouri, *J. Am. Chem. Soc.* **83**, 4228 (1961).

14. R. Roy and A. W. Rey, *Tetrahedron Lett.* **28**, 4935 (1987).

15. A. S. Gopalan and C. J. Sih, *Tetrahedron Lett.* **25**, 5235 (1984).

16. S. G. Cohen, J. Crossley and E. Khedouri, *Biochemistry* **2**, 820 (1963).

17. S. G. Cohen and A. Milovanović, *J. Am. Chem. Soc.* **90**, 3495 (1968).

18. S. G. Cohen, J. Crossley, E. Khedouri and R. Zand, *J. Am. Chem. Soc.* **84**, 4163 (1962); **85**, 1685 (1963).

19. M. L. Bender, F. J. Kézdy and C. R. Gunter, *J. Am. Chem. Soc.* **86**, 3714 (1964).

20. J. B. Jones and Y. Y. Lin, *Can. J. Chem.* **50**, 2053 (1972).

21. Y. Y. Lin and J. B. Jones, *J. Org. Chem.* **38**, 3575 (1973).

22. E. L. Eliel, *Conformational Analysis*, p. 237. Interscience, New York, 1965.

23. G. E. Clement and M. L. Bender, *Biochemistry* **2**, 836 (1963).

24. V. Daffe and J. Fastrez, *J. Am. Chem. Soc.* **102**, 3601 (1980).

25. C. Schregenberger and D. Seebach, *Liebigs Ann. Chem.*, 2081 (1986).

26. T. T. Herskovitz, G. Gadegbeku and H. Jaillet, *J. Biol. Chem.* **245**, 2588 (1970).

27. H. Kise and H. Shirato, *Tetrahedron Lett.* **26**, 6081 (1985).

28. D. Tarquis, P. Monsan and G. Durand, *Bull. Soc. Chim. France II*, 76 (1980).

29. F. C. Huang, L. F. Hsu Lee, R. S. D. Mittal, P. R. Ravikumar, J. A. Chan, C. J. Sih, E. Caspi and C. R. Eck, *J. Am. Chem. Soc.* **97**, 4144 (1975).

30. W. K. Wilson, S. B. Baca, Y. J. Barber, T. J. Scallen and C. J. Morrow, *J. Org. Chem.* **48**, 3960 (1983).

31. P. Mohr, L. Rösslein and C. J. Tamm, *Helv. Chim. Acta* **70**, 142 (1987).

32. D. W. Brooks and J. T. Palmer, *Tetrahedron Lett.* **24**, 3059 (1983).

33. T. Rosen, M. Watanabe and C. H. Heathcock, *J. Org. Chem.* **49**, 3657 (1984).

34. P. Mohr, N. Waespe-Sarčević, C. Tamm, K. Gawronska and J. K. Gawronski, *Helv. Chim. Acta* **66**, 2501 (1983).

35. C.-S. Chen, Y. Fujimoto, G. Girdaukas and C. J. Sih, *J. Am. Chem. Soc.* **104**, 7294 (1982).

36. P. Herold, P. Mohr and C. Tamm, *Helv. Chim. Acta* **66**, 744 (1983).

37. P. Mohr, M. Tori, P. Grossen, P. Herold and C. Tamm, *Helv. Chim. Acta* **65**, 1412 (1982).

38. M. Ohno, S. Kobayashi, T. Limori, Yi-Fong Wang and T. Izawa, *J. Am. Chem. Soc.* **103**, 2405 (1981); M. Ohno, S. Kobayashi and K. Adachi, in *Enzymes as Catalysts in Organic Synthesis* (ed. M. P. Schneider), p. 123, and references therein. D. Reidel, Dordrecht, Holland, 1986.

39. K. Adachi, S. Kobayashi and M. Ohno, *Chimia* **40**, 311 (1986).

40. K. Okano, Y. Kyotani, H. Ishihama, S. Kobayashi and M. Ohno, *J. Am. Chem. Soc.* **105**, 7186 (1983); K. Okano, T. Izawa and M. Ohno, *Tetrahedron Lett.* **24**, 217 (1983).

41. T. Limori, Y. Takahashi, T. Izawa, S. Kobayashi and M. Ohno, *J. Am. Chem. Soc.* **105**, 1659 (1983).
42. Yi-Fong Wang, T. Izawa, S. Kobayashi and M. Ohno, *J. Am. Chem. Soc.* **104**, 6465 (1982).
43. H. Kotani, Y. Kuze, S. Uchida, T. Miyabe, T. Limori, K. Okano, S. Kobayashi and M. Ohno, *Agric. Biol. Chem.* **47**, 1363 (1983).
44. C. J. Francis and J. B. Jones, *J. Chem. Soc., Chem. Commun.*, 579 (1984).
45. L. K. P. Lam, R. A. H. F. Hui and J. B. Jones, *J. Org. Chem.* **51**, 2047 (1986).
46. Ching-Shih Chen, Y. Fujimoto and C. J. Sih, *J. Am. Chem. Soc.* **103**, 3580 (1981).
47. G. M. Ramos Tombo, H. P. Schar, W. Zimmerman and O. Ghisalba, *Chimia* **39**, 313 (1985).
48. G. Sabbioni, M. L. Shea and J. B. Jones, *J. Chem. Soc., Chem. Commun.*, 236 (1984).
49. J. B. Jones, M. A. W. Finch and I. J. Jakovac, *Can. J. Chem.* **60**, 2007 (1982).
50. M. Schneider, N. Engel, P. Hönicke, G. Heinemann and H. Görisch, *Angew. Chem. Int. Ed. Engl.* **23**, 67 (1984).
51. H. J. Gais and K. L. Lukas, *Angew. Chem. Int. Ed. Engl.* **23**, 142 (1984).
52. S. Kobayashi, K. Kamiyama, T. Limori and M. Ohno, *Tetrahedron Lett.* **25**, 2557 (1984).
53. G. Sabbioni and J. B. Jones, *J. Org. Chem.* **52**, 4565 (1987).
54. F. Björkling, J. Boutelje, S. Gatenbeck, K. Hult and T. Norin, *Appl. Microbiol. Biotechnol.* **21**, 16 (1985).
55. J. B. Jones, R. S. Hinks and P. G. Hultin, *Can. J. Chem.* **63**, 452 (1985).
56. M. Kurihara, K. Kamiyama, S. Kobayashi and M. Ohno, *Tetrahedron Lett.* **26**, 5831 (1985).
57. Y. Ito, T. Shibata, M. Arita, H. Sawai and M. Ohno, *J. Am. Chem. Soc.* **103**, 6739 (1981). M. Ohno, Y. Ito, M. Arita, T. Shibata, K. Adachi and H. Sawai, *Tetrahedron* **40**, 145 (1984).
58. M. Arita, K. Adachi, Y. Ito, H. Sawai and M. Ohno, *J. Am. Chem. Soc.* **105**, 4049 (1983); *Nucleic Acids Symp. Ser.* **11**, 13 (1982).
59. R. Bloch, E. Guibe-Jampel and C. Girard, *Tetrahedron Lett.* **26**, 4087 (1985).
60. S. Iriuchijima, K. Hasegawa and G. Tsuchihashi, *Agric. Biol. Chem.* **46**, 1907 (1982).
61. F. Björkling, J. Boutelje, S. Gatenbeck, K. Hult and T. Norin, *Tetrahedron Lett.* **26**, 4957 (1985).
62. F. Björkling, J. Boutelje, S. Gatenbeck, K. Hult, T. Norin and P. Szmulik, *Tetrahedron* **41**, 1347 (1985).
63. M. Schneider, N. Engel and H. Boensmann, *Angew. Chem. Int. Ed. Engl.* **23**, 66 (1984).
64. M. Luyten, S. Müller, B. Herzog and R. Keese, *Helv. Chim. Acta* **70**, 1250 (1987).
65. S. Ramaswamy, R. A. H. F. Hui and J. B. Jones, *J. Chem. Soc., Chem. Commun.*, 1545 (1986). For an alternative strategy to produce chiral allenes see G. Gil, E. Ferre, A. Meou, J. Le Petit and C. Triantaphylides, *Tetrahedron Lett.* **28**, 1647 (1987).
66. M. Levy and P. Ocken, *Arch. Biochem. Biophys.* **135**, 259 (1969).
67. Y.-F. Wang and C. J. Sih, *Tetrahedron Lett.* **25**, 4999 (1984).
68. K. Laumen and M. Schneider, *Tetrahedron Lett.* **26**, 2073 (1985). See also H. Hemmerle and H.-J. Gais, *Tetrahedron Lett.* **28**, 3471 (1987).
69. G. Guanti, L. Banfi, E. Narisano, R. Riva and S. Thea, *Tetrahedron Lett.* **27**, 4639 (1986).
70. M. Schneider and K. Laumen, *Tetrahedron Lett.* **25**, 5875 (1984).
71. K. Laumen, E. H. Reimerdes and M. Schneider, *Tetrahedron Lett.* **26**, 407 (1985).
72. D. R. Deardorff, A. J. Matthews, D. S. McMeekin and C. L. Craney, *Tetrahedron Lett.* **27**, 1255 (1986).
73. C. R. Johnson and T. D. Penning, *J. Am. Chem. Soc.* **108**, 5655 (1986).

74. A. J. Pearson, H. S. Bansai and Yen-Shi-Lai, *J. Chem. Soc., Chem. Commun.*, 519 (1987).
75. U. Burger, D. Erne-Zellwegger and C. Mayeri, *Helv. Chim. Acta* **70**, 587 (1987).
76. J. A. Jongejan and J. A. Duine, *Tetrahedron Lett.* **28**, 2767 (1987).
77. S. Sicsic, M. Ikbal and F. LeGoffic, *Tetrahedron Lett.* **28**, 1887 (1987).
78. B. Camboie and A. M. Klibanov, *J. Am. Chem. Soc.* **106**, 2687 (1984).
79. A. Zaks and A. M. Klibanov, *Proc. Natl. Acad. Sci., USA* **82**, 3192 (1985).
80. P. Desnuelle, *Adv. Enzymol.* **23**, 129 (1961).
81. G. Kirchner, M. P. Scollar and A. M. Klibanov, *J. Am. Chem. Soc.* **107**, 7072 (1985).
    A. M. Klibanov, *CHEMTECH*, 354 (1986). M. Therisod and A. M. Klibanov, *J. Am. Chem. Soc.* **109**, 3977 (1987).
82. A. Zaks and A. M. Klibanov, *Science* **224**, 1249 (1984).
83. A. Belan, J. Bolte, A. Fauve, J. G. Gourcy and H. Veschambre, *J. Org. Chem.* **52**, 256 (1987).
84. T. M. Stokes and A. C. Oehlschlager, *Tetrahedron Lett.* **28**, 2091 (1987).
85. M. Degueil-Castaing, B. de Jeso, S. Drouillard and B. Maillard, *Tetrahedron Lett.* **28**, 953 (1987).
86. P. Sonnet, *J. Org. Chem.* **52**, 3477 (1987).
87. B. Cambou and A. M. Klibanov, *Appl. Biochem. Biotechnol.* **9**, 255 (1984).
88. J. Lavayre, J. Verrier and J. Baratti, *Biotechnol. Bioeng.* **24**, 2175 (1982).
89. E. Guibé-Jampel, G. Rousseau and J. Salaün, *J. Chem. Soc., Chem. Commun.*, 1080 (1987).
90. N. A. Porter, J. D. Byers, K. M. Holden and D. B. Menzel, *J. Am. Chem. Soc.* **101**, 4319 (1979).
91. C.-H. Lin, D. L. Alexander, C. G. Chidester, R. R. Gorman and R. A. Johnson, *J. Am. Chem. Soc.* **104**, 1621 (1982).
92. W. Kasel, P. G. Hultin and J. B. Jones, *J. Chem. Soc., Chem. Commun.*, 1563 (1985).
93. Zhuo-Feng Xie, H. Suemune and K. Sakai, *J. Chem. Soc., Chem. Commun.*, 838 (1987).
94. G. Eichberger, G. Penn, K. Faber and H Griengl, *Tetrahedron Lett.* **27**, 2843 (1986).
95. Y. F. Wang, C.-S. Chen, G. Girdaukas and C. J. Sih, *J. Am. Chem. Soc.* **106**, 3695 (1984).
96. W. E. Ladner and G. M. Whitesides, *J. Am. Chem. Soc.* **106**, 7250 (1984).
97. S. Iriuchijima, A. Keiyu and N. Kojima, *Agric. Biol. Chem.* **46**, 1593 (1982).
98. G. Langrand, M. Secchi, G. Buono, J. Baratti and C. Triantaphylides, *Tetrahedron Lett.* **26**, 1857 (1985).
99. G. Langrand, J. Baratti, G. Buono and C. Triantaphylides, *Tetrahedron Lett.* **27**, 29 (1986).
100. C. Marlot, G. Langrand, C. Triantaphylides and J. Baratti, *Biotech. Lett.* **9**, 647 (1985).
101. S. Iriuchijima and N. Kojima, *J. Chem. Soc., Chem. Commun.*, 185 (1981).
102. T. Kitazume, T. Sato and N. Ishikawa, *Chem. Lett.*, 1811 (1984).
103. T. Inukai, *Jpn. Kokai Tokyo Koho* 78, 104, 792; *Chem. Abs.* **90**, 101958g (1979).
104. S. Iriuchijima and N. Kojima, *Agric. Biol. Chem.* **46**, 1153 (1982).
105. S. Hamaguchi, J. Hasegawa, H. Kawaharada and K. Watanabe, *Agric. Biol. Chem.* **48**, 2055 (1984). S. Hamaguchi, M. Asada, J. Hasegawa and K. Watanabe, *Agric. Biol. Chem.* **48**, 2331 (1984).
106. H. M. Sweers and C.-H. Wong, *J. Am. Chem. Soc.* **108**, 6421 (1986).
107. A. Makita, T. Nihara and Y. Yamada, *Tetrahedron Lett.* **28**, 805 (1987).
108. Y. Inada, H. Nishimura, K. Takahashi, T. Yoshimoto, A. R. Saha and Y. Saito, *Biochem. Biophys. Res. Commun.* **122**, 845 (1984).
109. A. L. Gutman, K. Zuobi and A. Boltansky, *Tetrahedron Lett.* **28**, 3861 (1987).
110. A. L. Gutman, D. Oren, A. Boltansky and T. Bravdo, *Tetrahedron Lett.* **28**, 5367 (1987).
111. E. P. Dropsy and A. M. Klibanov, *Biotechnol. Bioeng.* **26**, 911 (1984).

112. J. M. Roper and D. P. Bauer, *Synthesis* **12**, 1041 (1983).
113. S.-T. Chen and K.-T. Wang, *Synthesis* **6**, 581 (1987).
114. D. Cantacuzène, F. Pascal and C. Guerreiro, *Tetrahedron* **43**, 1823 (1987).
115. B. I. Glänzer, K. Faber and H. Griengl, *Tetrahedron* **43**, 771 (1986); *Tetrahedron Lett.* **27**, 4293 (1986).
116. K. Kawai, M. Imuta and H. Ziffer, *Tetrahedron Lett.* **22**, 2527 (1981). H. Ziffer, K. Kasai, M. Kasai, M. Imuta and C. Froussios, *J. Org. Chem.* **48**, 3017 (1983). M. Charton and H. Ziffer, *J. Org. Chem.* **52**, 2400 (1987).
117. M. Kasai, K. Kawai, M. Imuta and H. Ziffer, *J. Org. Chem.* **49**, 675 (1984).
118. M. Kasai, H. Ziffer and J. V. Silverton, *Can. J. Chem.* **63**, 1287 (1985).
119. W. J. McGahren, K. J. Sax, M. P. Kuntsmann and G. A. Ellestad, *J. Org. Chem.* **42**, 1659 (1977).
120. T. Oritani, K. Yamashita and C. Kabuto, *J. Org. Chem.* **49**, 3689 (1984).
121. T. Oritani and K. Yamashita, *Agric. Biol. Chem.* **37**, 1687, 1691, 1695 (1973); **38**, 1961, 1965 (1974); **39**, 89 (1975). T. Oritani, H. Kondo and K. Yamashita, *Agric. Biol. Chem.* **51**, 263 (1987) and references therein.
122. T. Oritani and K. Yamashita, *Agric. Biol. Chem.* **44**, 2407 (1980).
123. S. Takano, K. Tanigawa and K. Ogasawara, *J. Chem. Soc., Chem. Commun.*, 189 (1976).
124. K. Mori and H. Akao, *Tetrahedron* **36**, 91 (1980).
125. S. Iriuchijima and A. Keiyu, *Agric. Biol. Chem.* **45**, 1389 (1981).
126. Y. Takaishi, Y.-L. Yang, D. DiTullio and C. J. Sih, *Tetrahedron Lett.* **23**, 5489 (1982).
127. C. J. Sih, J. B. Heather, R. Sood, P. Price, G. Peruzzotti, L. F. Hsu Lee and S. S. Lee, *J. Am. Chem. Soc.* **97**, 865 (1975).
128. Y. Kato, H. Ohta and Gen-ichi Tsuchihashi, *Tetrahedron Lett.* **28**, 1303 (1987).
129. J. A. Hartsuck and W. N. Lipscomb, *The Enzymes*, Vol. 3 (ed. P. D. Boyer), p. 1. Academic Press, New York, 1971.
130. W. S. Fones, *J. Biol. Chem.* **204**, 323 (1953).
131. L. Benoiton, M. Winitz, S. M. Birnbaum and J. P. Greenstein, *J. Am. Chem. Soc.* **79**, 6192 (1957).
132. J. E. Baldwin, M. A. Christie, S. B. Haber and L. I. Kruse, *J. Am. Chem. Soc.* **98**, 3045 (1976).
133. A. R. Battersby, M. Nicoletti, J. Staunton and R. Vleggaar, *J. Chem. Soc., Perkin Trans. 1*, 43 (1980).
134. H. Fujihara and R. C. Schowen, *J. Org. Chem.* **49**, 2819 (1984).
135. J. W. Keller and B. J. Hamilton, *Tetrahedron Lett.* **27**, 1249 (1986).
136. J. P. Greenstein and M. Winitz, *Chemistry of the Amino Acids*, Vol. 2, p. 1753. Wiley, New York, 1961.
137. I. Chibata, T. Tosa, T. Sato and T. Mori, *Methods Enzymol.* **44**, 746 (1976). I. Chibata, in *Asymmetric Reactions and Processes in Chemistry* (eds. E. L. Eliel and S. Otsuka). American Chemical Society, Washington, 1982. For information on large-scale resolutions see C. Wandrey, in *Enzymes as Catalysts in Organic Synthesis* (ed. M. Schneider), p. 263. Reidel, Dordrecht, 1986.
138. H. T. Huang and C. Niemann, *J. Am. Chem. Soc.* **73**, 475 (1951).
139. Y. Nishida, H. Ohrui and H. Meguro, *Agric. Biol. Chem.* **47**, 2123 (1983).
140. K. Mori and H. Iwasawa, *Tetrahedron* **36**, 2209 (1980).
141. K. Mori and T. Otsuka, *Tetrahedron* **41**, 547 (1985).
142. E. Lagerlöf, L. Nathorst-Westfelt, B. Ekström and B. Sjöberg, *Methods Enzymol.* **44**, 759 (1976).
143. See, for example, J. M. Park, C. Y. Choi, B. L. Seong and M. H. Han, *Biotechnol. Bioeng.* **24**, 1623 (1982).

96   2. HYDROLYSIS AND CONDENSATION REACTIONS

144. E. Andersson, B. Mattiasson and B. Hahn-Hägerdal, *Enzyme Microbiol. Technol.* **6**, 301 (1984).
145. V. Kasche, *Biotechnol. Lett.* **7**, 877 (1985).
146. See, for example, P. B. Mahajan, *Appl. Biochem. Biotechnol.* **9**, 537 (1984). T. A. Savage, in *Biotechnology of Industrial Antibiotics* (ed. E. J. Vandamme), p. 171. M. Dekker, New York, 1984.
147. K. Kato, K. Kawahara, T. Takahashi and S. Igarasi, *Agric. Biol. Chem.* **44**, 821 (1980).
148. W. G. Choi, S. B. Lee and D. D. Y. Ryu, *Biotechnol. Bioeng.* **23**, 361 (1981).
149. R. B. Frederiksen and C. Emborg, *Biotechnol. Lett.* **6**, 549 (1984).
150. M. Bergmann and H. Fraenkel-Conrat, *J. Biol. Chem.* **119**, 707 (1937); **124**, 1 (1938).
151. J. D. Glass, *Enzyme Microbiol. Technol.* **3**, 2 (1981). H.-D. Jakubke, P. Kuhl and A. Könnecke, *Angew. Chem. Int. Ed. Engl.* **24**, 85 (1985). I. M. Chaiken, A. Komoriya, M. Ohno and F. Widmer, *Appl. Biochem. Biotechnol.* **7**, 385 (1982). K. Oyama and K. Kihara, *Chem. Technol.* **14**, 100 (1984). W. Kullman, *J. Protein. Chem.* **4**, 1 (1985); *Enzymatic Peptide Synthesis.* CRC Press, Boca Raton, Fl., 1987.
152. W. Kullmann, *J. Biol. Chem.* **255**, 8234 (1980).
153. I. Schechter and A. Berger, *Biochem. Biophys. Res. Commun.* **32**, 898 (1968); M. Smolarsky, *Biochemistry* **19**, 478 (1980).
154. W. Kullmann, *Proc. Natl. Acad. Sci., USA* **79**, 2840 (1982).
155. W. Kullman, *J. Org. Chem.* **47**, 5300 (1982).
156. K. Nilsson and K. Mosbach, *Biotechnol. Bioeng.* **26**, 1146 (1984).
157. K. Oyama, S. Nishimura, Y. Nonaka, K. Kihara and T. Hashimoto, *J. Org. Chem.* **46**, 5241 (1981). See also: C. Fuganti, P. Grasselli and P. Casati, *Tetrahedron Lett.* **27**, 3191 (1986).
158. Y. Isowa, M. Ohmori, T. Ichikawa, H. Kurita, M. Sato and K. Mori, *Bull. Chem. Soc., Jpn.* **50**, 2762 (1977).
159. Y. L. Khemel'nitski, F. K. Dien, A. N. Semenov, K. Martinek, B. Veruovič and V. Kubánek, *Tetrahedron* **40**, 4425 (1984).
160. J. H. Fendler, *Membrane Mimetic Chemistry*, p. 55. Wiley, New York, 1982. F. M. Menger, J. A. Donohue and R. F. Williams, *J. Am. Chem. Soc.* **95**, 286 (1973).
161. P. Lüthi and P. L. Luisi, *J. Am. Chem. Soc.* **106**, 7285 (1984).
162. A. L. Margolin and A. M. Klibanov, *J. Am. Chem. Soc.* **109**, 3802 (1987). A. M. Klibanov, *Chem. Technol.* **6**, 354 (1986).
163. J. B. West and C.-H. Wong, *Tetrahedron Lett.* **28**, 1629 (1987). J. R. Matos, J. B. West and C.-H. Wong, *Biotechnol. Lett.* **9**, 233 (1987).
164. K. Morihara, T. Oka, H. Tsuzuki, Y. Toshino and T. Kanaya, *Biochem. Biophys. Res. Commun.* **92**, 396 (1980). R. Obermeier and G. Seiphe, *Proc. Biochem.* **19**, 29 (1984).
165. J. Markussen, Semisynthesis of Human Insulin, in *Methods in Diabetes Research*, Vol. 1 (eds. J. Larner and S. Pohl), p. 403. Wiley, New York, 1984.
166. K. Inouye, K. Watanabe, K. Morihara, Y. Tochino, Y. Kanaya, J. Emura and S. Sakakibara, *J. Am. Chem. Soc.* **101**, 751 (1979).
167. H. Fujii, T. Koyama and K. Ogura, *Biochim. Biophys. Acta* **712**, 716 (1982) and references therein.
168. P. Herdewijn, J. Balzarini, E. De Clercq and H. Vanderhaeghe, *J. Med. Chem.* **28**, 1385 (1985).
169. A. Gross, O. Abril, J. M. Lewis, S. Geresh and G. M. Whitesides, *J. Am. Chem. Soc.* **105**, 7428 (1983).
170. R. L. Sabina, E. W. Holmes and M. A. Becker, *Science* **223**, 1193 (1984).
171. C.-H. Wong and G. M. Whitesides, *J. Am. Chem. Soc.* **103**, 4890 (1981).
172. D. G. Drueckhammer and C.-H. Wong, *J. Org. Chem.* **50**, 5912 (1985).

173. D. C. Crans and G. M. Whitesides, *J. Am. Chem. Soc.* **107**, 7008 (1985).
174. V. M. Rios-Mercadillo and G. M. Whitesides, *J. Am. Chem. Soc.* **101**, 5828 (1979); for an improved preparation see [181].
175. D. C. Crans and G. M. Whitesides, *J. Am. Chem. Soc.* **107**, 7019 (1985).
176. W. A. Blattler and J. R. Knowles, *Biochemistry* **18**, 3927 (1979).
177. J. P. Richard and P. A. Frey, *J. Am. Chem. Soc.* **100**, 7757 (1978).
178. M. R. Webb, *Methods Enzymol.* **87**, 301 (1982).
179. L. M. Reimer, D. L. Conley, D. L. Pompliano and J. W. Frost, *J. Am. Chem. Soc.* **108**, 8010 (1986).
180. D. L. Marshall, *Biotechnol. Bioeng.* **15**, 447 (1973).
181. D. C. Crans and G. M. Whitesides, *J. Org. Chem.* **48**, 3130 (1983).
182. J. Bolte and G. M. Whitesides, *Bioorg. Chem.* **12**, 170 (1984).
183. See, for example, C.-H. Wong, A. Pollak, S. D. McCurry, J. M. Sue, J. R. Knowles and G. M. Whitesides, *Methods Enzymol.* **89**, 108 (1982) and previous procedures.
184. A. Pollak, R. L. Baughn and G. M. Whitesides, *J. Am. Chem. Soc.* **99**, 2366 (1977).
185. Y.-S. Shih and G. M. Whitesides, *J. Org. Chem.* **42**, 4165 (1977).
186. R. L. Baughn, O. Adalsteinsson and G. M. Whitesides, *J. Am. Chem. Soc.* **100**, 304 (1978).
187. F. Cramer, E. Scheiffele and A. Vollmar, *Chem. Ber.* **95**, 1670 (1962).
188. R. J. Kazlauskas and G. M. Whitesides, *J. Org. Chem.* **50**, 1069 (1985).
189. B. L. Hirschbein, F. P. Mazenod and G. M. Whitesides, *J. Org. Chem.* **47**, 3765 (1982).
190. C.-H. Wong and G. M. Whitesides, *J. Am. Chem. Soc.* **105**, 5012 (1983).
191. J. R. Moran and G. M. Whitesides, *J. Org. Chem.* **49**, 704 (1984).
192. C.-H. Wong, S. L. Haynie and G. M. Whitesides, *J. Org. Chem.* **47**, 5416 (1982). C. Auge, C. Mathieu and C. Merienne, *Carbohydr. Res.* **151**, 147 (1986).
193. C.-H. Wong, S. L. Haynie and G. M. Whitesides, *J. Am. Chem. Soc.* **105**, 115 (1983); K. M. Plowman and A. R. Krall, *Biochemistry* **4**, 2809 (1965); in *Methods of Enzymatic Analysis* (ed. H. U. Bergmeyer), 2nd Edn. p. 2081. Academic Press, New York, 1974.
194. For a review see, G. Berti, in *Enzymes as Catalysts in Organic Synthesis* (ed. M. P. Schneider), pp. 349–354. Reidel, Dordrecht, 1986.
195. F. Oesch, *Xenobiotica* **3**, 305 (1972); *Biochem. J.* **139**, 77 (1974). F. Oesch, N. Kaubisch, D. M. Jerina and J. W. Daly, *Biochemistry* **10**, 4858 (1971).
196. P. M. Dansette, V. B. Makedonska and D. M. Jerina, *Arch. Biochem. Biophys.* **187**, 290 (1978).
197. Th. Posternak, D. Reymond and H. Friedli, *Helv. Chim. Acta* **38**, 205 (1955).
198. T. Watabe, K. Akamatsu and K. Kiyonaga, *Biochem. Biophys. Res. Commun.* **44**, 199 (1971). T. Watabe and K. Akamatsu, *Biochim. Biophys. Acta* **279**, 297 (1972).
199. D. Wistuba and V. Schurig, *Angew. Chem. Int. Ed. Engl.* **25**, 1032 (1986).
200. R. S. Cahn, C. Ingold and V. Prelog, *Angew. Chem. Int. Ed. Engl.* **5**, 385 (1966).
201. G. Bellucci, G. Berti, G. Ingrosso and E. Mastrorilli, *J. Org. Chem.* **45**, 299 (1980). G. Bellucci, G. Berti, R. Bianchini, P. Cetera and E. Mastrorilli, *J. Org. Chem.* **47**, 3105 (1982).
202. G. Bellucci, G. Berti, M. Ferretti, E. Mastrorilli and L. Silvestri, *J. Org. Chem.* **50**, 1471 (1985).
203. G. Bellucci, G. Berti, G. Catelani and E. Mastrorilli, *J. Org. Chem.* **46**, 5148 (1981).
204. G. Catelani and E. Mastrorilli, *J. Chem. Soc., Perkin Trans. 1*, 2717 (1983).
205. P. Barili, G. Berti, G. Catelani, F. Colonna and E. Mastrorilli, *J. Chem. Soc. Chem. Commun.* **7**, (1986); *J. Org. Chem.* **52**, 2886 (1987).
206. R. P. Hanzlik, M. Edelman, W. J. Michaely and G. Scott, *J. Am. Chem. Soc.* **98**, 1952 (1976).

207. R. P. Hanzlik, S. Heideman and D. Smith, *Biochem. Biophys. Res. Commun.* **82**, 310 (1978).
208. R. B. Westkaemper and R. P. Hanzlik, *Arch. Biochem. Biophys.* **208**, 195 (1981).
209. H. Breuer and R. Knuppen, *Biochem. Biophys. Acta* **49**, 620 (1961). E. W. Maynert, R. L. Foreman and T. Watabe, *J. Biol. Chem.* **245**, 5234 (1970). T. Watabe, Y. Ueno and J. Imazumi, *Biochem. Pharmacol.* **20**, 912 (1971).
210. G. T. Brooks, A. Harrison and S. E. Lewis, *Biochem. Pharmacol.* **19**, 255 (1970). See also, G. T. Brooks, S. E. Lewis and A. Harrison, *Nature* **220**, 1034 (1968).
211. T. Watabe, K. Kiyonaga and S. Hara, *Biochem. Pharmacol.* **20**, 1700 (1971). T. Watabe and S. Suzuki, *Biochem. Biophys. Res. Commun.* **46**, 1120 (1972).
212. M. Ohno, *Ferment. Ind.* **37**, 836 (1979).
213. Toray Industries Inc., GB Patent 1,445,326 (1975).
214. For a review see, W. R. Abraham, H. M. R. Hoffman, K. Kieslich, G. Reng and B. Stumpf, in *Enzymes in Organic Synthesis (Ciba Foundation Symposium* 111), pp. 146–160. Pitman, London, 1985.
215. I. Watanabe, K. Sakashita and Y. Ogawa, U.K. Patent GB 2086376B (1984). I. Watanabe, U.S. Patent 4,343,900 (1982).
216. D. B. Harper, *Biochem. J.* **165**, 309 (1977). See also, E. A. Linton and C. J. Knowles, *J. Gen. Microbiol.* **132**, 1493 (1986).
217. European Patent Application No. 87201275.2 (1986).
218. Y. Vo-Quang, D. Marais, L. Vo-Quang, F. Le Goffic, A. Thiéry, M. Maestracci, A. Arnaud and P. Galzy, *Tetrahedron Lett.* **28**, 4057 (1987).
219. S. E. Godtfredsen, *World Biotech Report 1988*. Blenheim Online Ltd, London, 1988.

—3—

ENZYME CATALYSED REDUCTION REACTIONS

3.1. REDUCTION OF ALDEHYDES AND KETONES

Alcohol dehydrogenases (E.C. 1.1.X.Y) are enzymes that catalyse the oxidation and reduction of hydroxyl and carbonyl compounds [1,2]. They may be employed as purified or crude enzymes or as components of whole cells [3,4]. The two most frequently used systems are bakers' yeast and horse liver alcohol dehydrogenase. In total there are about 200 known alcohol dehydrogenases which use nicotinamide adenine dinucleotide (NAD) or the corresponding 2′-phosphate (NADP) as a cofactor to accept or donate hydrogen (as NADH or NADPH). About another 20 enzymes use other cofactors, including an increasing number that use pyrroloquinoline quinone (PQQ, methoxatin) [5].

As discussed in Chapter 1, the fundamental difference between the use of purified enzymes and whole cells is that in the former case provision must be made for the regeneration of the NAD(P)/NAD(P)H cofactors, whilst in the latter case the cells contain the necessary enzymes to achieve cofactor recycling. It is not practical (except in enzyme assays or the most unusual cases) to use the cofactors stoichiometrically.

The disadvantage of using whole cells is that they frequently contain more than one oxidoreductase and the activity expressed depends on the structure of the substrate, the metabolic state of the cell and a host of other factors. Despite these problems, oxidoreductases rival the hydrolases as the most widely used enzyme systems, and impressive results are frequently reported [6]. In general, reductions are more easily achieved than oxidations and they usually show high stereoselectivity, i.e. prochiral substrates are often reduced from one enantiotropic face with high selectivity (Scheme 3.1), whilst racemates either undergo conversion of one enantiomer alone or both enantiomers may be converted to different diastereoisomers [7]. It is a good general "rule of thumb" that the faster the reaction, the higher the enantio-selectivity observed.

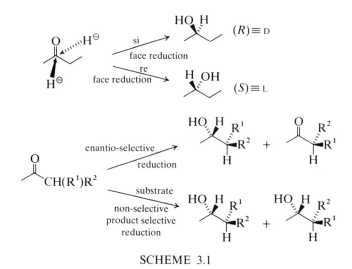

SCHEME 3.1

3.1.1. Yeast Reductions

Yeasts are a heterogeneous group of fungi which live either as saprophytes or as parasites and are mostly unicellular [8]. The word yeast is derived from an Anglo-Saxon verb "to froth or foam". The genus *Saccharomyces* is composed of species which ferment a variety of sugars to ethanol and carbon dioxide. The most important species is *Saccharomyces cerevisiae* [9], strains of which are used as bakers', brewers' or wine yeast. Bakers' yeast was developed from brewers' yeast in the 19th century [10].

In publications prior to the 1960s, biotransformations involving yeast or other micro-organisms were referred to as "phytochemical reactions" [11,12].

3.1.1.1. Acyclic β-Ketoesters

Perhaps the easiest of all biotransformations is the yeast reduction [13] of β-ketoesters, first described in 1918 [11]. In an *Organic Syntheses* preparation, a handful of bakers' yeast and table sugar in tapwater suffices to reduce 40 g of ethyl acetoacetate (ethyl-3-oxobutanoate) to ethyl (*S*)-(+)-3-hydroxy-butanoate (**1**) with high enantioselectivity (e.e. = 85–95%) and in good yield

(**1**)

TABLE 3.1

Reduction of some β-ketoesters with yeast

$$R^2 \overset{O\quad O}{\diagdown\diagup\diagup} OR^1 \longrightarrow \overset{HO\ H\ O}{R^2 \diagdown\diagup\diagup} OR^1$$

R^1	R^2	Configuration of product	e.e. (%)	Yield (%)	Ref.
Me	Me	(S)	87	23	27
Me	(CH$_2$)$_2$CH=CH$_2$	(R)	92	30	27
Me	(CH$_2$)$_2$C(Me)=CH$_2$	(R)	67	18	27
Me	(CH$_2$)$_2$CH=CMe$_2$	(R)	92	73	27
Me	CH$_2$Cl	(S)	36,63	—	34
Me	Bu^{t}OCH$_2$	(R)	82	70	31
Me	PhCH$_2$O(CH$_2$)$_2$	(S)?	30	65	30
Me	Bu^{t}OCOCH$_2$	?	24	36	38
Me	MeOCO(CH$_2$)$_2$	(S)	48	19	38
Me	PhCH$_2$OCO(CH$_2$)$_2$	(S)	2	24	38
Et	Me	(S)	⩾96	32,56	14,15,27
Et	CF$_3$	(R)	45,96	70,92	28,29
Et	PhCH$_2$O(CH$_2$)$_2$	(S)	56	60	30
Et	Bu^{t}OCH$_2$	(R)	97	72	31
Et	PhCH$_2$OCH$_2$	(R)	56	58	31,32
Et	CH$_2$N$_3$	(R)	80	70–80	33
Et	Et	(R)	40	67	23
Et	CH$_2$Cl	(S)	55	—	34
Et	Ph	(S)	—	70	35,36
Et	CCl$_3$	(S)	85	70–80	28
Et	(CH$_2$)$_2$CH=CH$_2$	(R)	80	54	27
Et	(CH$_2$)$_2$C(Me)=CH$_2$	(R)	18	15	27
Et	(CH$_2$)$_2$CH=CMe$_2$	(R)	50	12	27
Et	(CH$_2$)$_3$OCH$_2$Ph	—	—	0	27
Et	Me	(S)	96	32	27
Et	CH$_2$Br	(S)	70,100	40–50	33
Bu	(CH$_2$)$_2$CH=CH$_2$	(R)	81	66	27
Bu	(CH$_2$)$_2$C(Me)=CH$_2$	—	0	22	27
Bu	(CH$_2$)$_2$CH=CMe$_2$	—	0	40	27
Bu	CH$_2$Cl	(S)	28	—	34
Bu	PhCH$_2$O(CH$_2$)$_2$	(S)	70	55	30
Bu^i	PhCH$_2$O(CH$_2$)$_2$	(S)	90	70	30
C$_8$H$_{17}$	CH$_2$Br	(R)	100	—	33
C$_8$H$_{17}$	CH$_2$N$_3$	(R)	100	70–80	33
C$_8$H$_{17}$	CH$_2$Cl	(R)	96	low	34
C$_8$H$_{17}$	Et	(S)	95	75	37
C$_8$H$_{17}$	PhCH$_2$O(CH$_2$)$_2$	—	—	0	30

(59–76%) [14,15]. The process is akin to brewing beer [16]. The results obtained on yeast reduction of a variety of methyl-, ethyl-, butyl- and octyl-3-oxoalkanoates are presented in Table 3.1. 3-Hydroxybutyric esters produced in this way have been used in the synthesis of β-lactams [17], griseoviridin [18], several insect pheromones [19,20] and carotenoids [21] and have been converted into a variety of useful chiral building blocks [22–25]. (*R*)-3-Hydroxybutanoates can also be obtained by other means [15,17,20,22,26]. Perusal of Table 3.1 will indicate that the enantioselectivity of yeast reductions of β-ketoesters depends on the relative sizes of the 1,3-dicarbonyl unit and the alkoxy moiety that, together, make up the ester. The preferred direction of hydride attack for β-ketoesters of differing size is diagrammatically represented in Scheme 3.2. This situation is due to competition between the several oxidoreductase enzymes that are present in yeast.

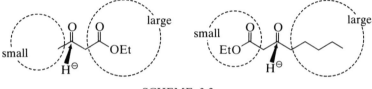

SCHEME 3.2

It is therefore important to note that yeasts *do not* simply employ yeast alcohol dehydrogenase (YAD) for these reactions, as is implied in many of the early papers. YAD is fairly specific for short-chain alcohols/aldehydes and, although it may play a role in some reductions, it clearly does not have a major involvement in most cases.

Preparation 3.1   Preparation of octyl (*S*)-3-hydroxypentanoate

Dry bakers' yeast (Oriental Yeast Co., 800 g) was added to a solution of sucrose (2 kg) in tap water (13 l) at 30°C. The mixture was stirred for 10 min at 30°C. Octyl 3-oxopentanoate (81 g) was added to the stirred yeast suspension. After 2 h, sucrose (2 kg) and yeast (100 g) was added and the fermentation was continued for 5 h. The culture broth was mixed with ether and benzene, filtered through Celite and the filter cake was extracted with acetone. The extract was concentrated *in vacuo*. The filtrate was saturated with NaCl and extracted with EtOAc. This was concentrated *in vacuo* and the residue was combined with the extract from cells of yeast to give crude product (67.4 g). This was chromatographed over SiO$_2$ followed by distillation to give octyl (*S*)-3-hydroxypentanoate (49.2 g, 47.4%): b.p. 112–122°C/0.66 torr, n_D^{23} 1.4371 $[\alpha]_D^{21}$ + 22.1° (*c* = 1.08, CHCl$_3$), ν_{max} (film) 3470 (bs), 1740 (s), 1175 (s) cm^{-1}.

In fact, three enzymes capable of reducing β-ketoesters have been purified to homogeneity from the cystolic fraction of bakers' yeast. Two have D selectivity and one has L selectivity. [The designation of D and L is used for clarity in some places in this chapter so that, for example, 3-hydroxybutyr-ates and 4-chloro-3-hydroxybutyrates are placed in the same series. The Cahn–Ingold–Prelog nomenclature places them in opposite series. L is equivalent to (S) for (+)-ethyl-3-hydroxybutyrate.] The two most "active" enzymes* respond differently to variation in size of the alkoxy group. For example, the activity of the L-selective enzyme increases, whereas that of the D-selective enzyme decreases, for a series of γ-chloro-β-ketoesters of increas-ing molecular weight. Thus, in the reduction of ethyl 4-chlorobutanoate it is the D-selective enzyme which is most active. The D-selective enzyme, how-ever, cannot accommodate the long chain alkoxy group of octyl 4-chlorobu-tanoate and so the activity of the L-selective enzyme becomes predominant (Scheme 3.3). The strategy of using longer chain alkoxy groups to increase the preference for L products seems to be general [12,30,37,38], although the rate of reduction may be reduced [34], and an exception has been observed [33].

Yeast reduction of β-ketoacids or salts of these acids results in products having the D configuration in many cases with high optical purity (Table 3.2) [27,39].

TABLE 3.2

Reduction of β-ketoacids (potassium salts) with yeast

$$\underset{R^1}{\overset{O\quad O}{\diagdown}}\!\!\!\diagup\!\!\!\diagup OR^2 \longrightarrow \underset{R^1}{\overset{HO\quad O}{\diagdown}}\!\!\!\diagup\!\!\!\diagup OR^2$$

R^1	R^2	Configuration of product	e.e. (%)	Yield (%)	Ref.
Me	K	(S)	>96	34	12
$CH_2CH_2CH{=}CH_2$	K	(R)	>98	35	39
$CH_2CH_2CH{=}CH_2$	K	(R)	99	38	27
$CH_2CH_2C(Me){=}CH_2$	K	(R)	99	55	27
$CH_2CH_2CH{=}CMe_2$	K	(R)	99	59,67	27,40
$CH_2CH_2CH_2OCH_2Ph$	K	(R)	99	51	27
Pr^n	K	(R)	—	16	12
Bu^n	K	(R)	—	26	12
$C_{15}H_{31}{}^n$	K	(R)	98	40	43

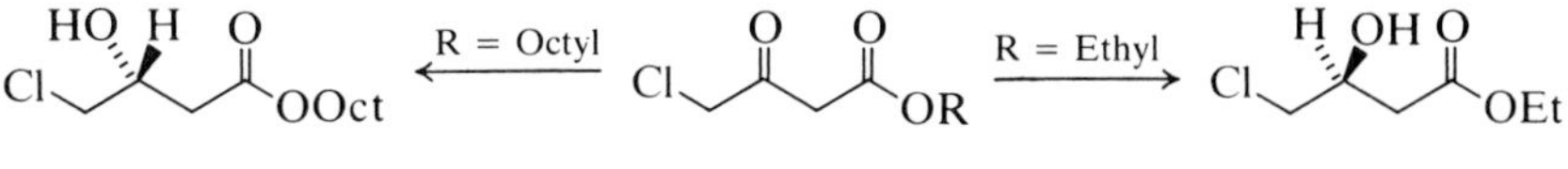

SCHEME 3.3

* "Activity" is defined as the turnover number divided by the Michaelis constant (K_{cat}/K_m).

Because the variation in stereoselectivity of the reduction of β-keto-esters, -acids, and -salts depends on a competition between two enzymes, it is possible to alter the selectivity by reducing the activity of one or other of the enzymes. This can be accomplished by varying the substrate concentration [3,30,34,42,43] (if the K_m values are different), or by using mutants lacking one of the enzymes [42], or by employing other strains of yeast [44–47]. Immobilization of yeast cells [48] makes them more convenient to use but may affect the enantioselectivity [45]. The addition of allyl alcohol or one of several α,β-unsaturated ketones to biotransformations involving bakers' yeast gives rise to the D-configuration alcohols of high optical purity, presumably by inhibition of the L-selective enzyme [49] through "suicide" or competitive inhibition [50]. The enantioselectivity may even be reversed simply by changing the weight ratio of yeast to substrate [36]. The use of anaerobic or aerobic conditions alters the balance between oxidative and reductive (fermentative) metabolism which can modify the stereoselection expressed in the bioreduction [51,52]. Fermentation may also be stimulated by the addition of ammonium ions [53], mineral supplements, in particular zinc salts [54] and thiamine (vitamin B_1) [55], and depressed by the addition of ethanol [36] or sodium chloride [56]. However, it is important to note that fermentation is not a prerequisite for reduction to take place [52,57].

A further complicating factor is that yeasts also show enantioselective esterase activity, which may augment or reduce the enantioselectivity of the reduction and/or result in reduced yields [58].

3.1.1.2.   Yeast Reduction of Acyclic 2-Substituted β-Ketoesters and β-Diketones

Yeast reduction of the 2-methyl β-ketoester (2) [59] is completely diastereoselective for formation of the 3(S) alcohol, giving the (2R,3S) (3) and (2S,3S) products (4) in the ratio 3:1 in 65% overall yield and >98% enantiomeric excess for both compounds [23–25,60,61]. Similar results were obtained for other 2-alkyl and 2-thioalkyl β-ketoesters (Table 3.3) [62]. It be seen that the selectivity is reversed when the 2-substituent is made larger, so that the (2S,3S) diastereoisomer predominates [25]. As expected, the preference for products possessing the (2R,3S) configuration (L configuration) is enhanced as the size of the ester alkoxy group is increased [60]. In some cases, other micro-organisms were employed to increase the stereoselectivity [63]. The products obtained were used in a synthesis of stegobinone, and other natural products [61,64].

TABLE 3.3

Yeast reduction of some 2-substituted β-ketoesters

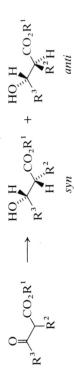

R¹	R²	R³	Syn configuration	e.e. (%)	Anti configuration	e.e. (%)	Ratio Syn : Anti	Yield (%)	Ref.
Et	Me	Me	(2R,3S)	100	(2S,3S)	100	80 : 20	59–75	25,60,61
n-Octyl	Me	Me	(2R,3S)	100	(2S,3S)	high	95 : 5	82	60
CH₂Ph	Me	Me	(2R,3S)	100	(2S,3S)	80	67 : 33	44	72
Me	Me	CH₂OCH₂Ph	(2R,3R)	68	(2S,3R)	low	10 : 90	50	65
Et	Me	CH₂OCH₂Ph	(2R,3R)	36	(2S,3R)	30	54 : 46	46	65
Prⁱ	Me	CH₂OCH₂Ph	(2R,3R)	96	(2S,3R)	50	24 : 76	64	65
Me	Me	CO₂Me	(2R,3R)	70	(2S,3R)	low	53 : 47	57	72
Et	Me	CO₂Et	(2R,3R)	79	(2S,3R)	31	43 : 57	85	41
Et	CH₂CH=CH₂	Me	(2R,3S)	100	(2S,3S)	100	25 : 75	84	25
Et	CH₂Ph	Me	(2R,3S)	100	(2S,3S)	100	35 : 65	21	25
Me	SMe	Me	(2R,3S)	>96	(2S,3S)	>96	72 : 28	72	62
Et	SMe	Me	(2R,3S)	>96	(2S,3S)	>96	59 : 41	62	62
Buᵗ	SMe	Me	(2R,3S)	>96	(2S,3S)	>96	41 : 59	75	62
Me	SPh	Me	(2R,3S)	>96	(2S,3S)	>96	83 : 17	40	62
Et	SMe	(CH₂)₂OCH₂Ph	(2R,3S)	>96	(2S,3S)	>96	32 : 68	30	62

Preparation 3.2   Reduction of octyl 2-methyl-3-oxobutanoate

To a suspension of bakers' yeast (100 g) in 125 ml of water, 1.14 g (5 mmol) of the title compound was added and the suspension was stirred at 30°C. Glucose (2.5 g × 6 portions) was added at 6 h intervals. After two days, ethyl acetate and then Celite (Hyflo Super-Cel) were added to the suspension and the mixture was filtered. The Celite layer was washed with ethyl acetate. The combined ethyl acetate layers were washed with water, dried over anhydrous sodium sulphate and evaporated *in vacuo.* The *syn/anti* ratio was measured by NMR spectroscopy. The residue was subjected to a column chromatography on silica gel using hexane/ethyl acetate (9 : 1) as eluent, giving 0.93 g (82%) of *syn*-octyl (2*R*,3*S*)-3-hydroxy-2-methylbutanoate: $[\alpha]_D^{24}$ + 3.33 (*c* = 3.00, CHCl₃).

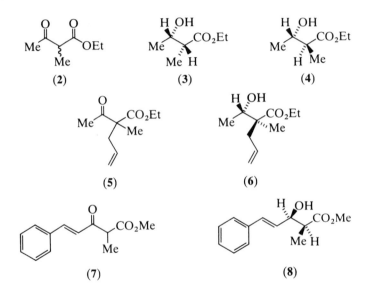

The β-ketoester **(5)** gave the (2*S*,3*S*) hydroxyester **(6)** (20% yield, e.e. = 100%) on incubation with yeast [25]. There have been several attempts to reduce 2-methyl β-keto esters having benzyloxymethyl [65], furanyl [66] or thienyl groups [67] attached to the ketone unit. In general, the diastereo-selectivities are similar to the shorter chain analogues, but the enantioselec-tivities and yields are poorer. The styryl derivative **(7)** unexpectedly gave the diastereoisomer **(8)** (e.e = 95%) but with only 7% yield [68]. However, several other organisms are capable of effecting this transformation with higher yields and improved enantioselectivities [65,66,68,69]. Similarly, only the yeast-like fungus *Geotrichum candidum* is capable of reducing the β-diketone **(9)** [51]. The enantioselectivity depends on the culture conditions, with aerobic metabolism giving the hydroxyketone **(10)** while the epimer **(11)** is produced under anaerobic fermentative conditions [51,70].

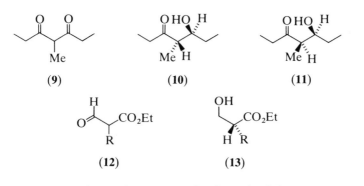

(9)   (10)   (11)

(12)   (13)

Under the conditions of a yeast reduction, the β-ketoester group can enolize, which for 2-substituted compounds results in racemization. So, in principle, a racemic 2-substituted 3-oxobutanoate can be converted into a single enantiomer of the corresponding 2-substituted 3-hydroxybutanoate [71,78]. The reduction of ester (12, R = Me) has cryptic diastereoselectivity but observable enantioselectivity. As might be expected the predominant enantiomer (13, R = Me) has the (R) configuration (50–80% yield), but the enantiomeric excess is somewhat disappointing (e.e. = 60–83%) [31,73]. In contrast, ethyl 2-formylbutanoate (12, R = Et) is reduced by yeast to give the (S)hydroxyester in 88% yield (e.e. = 91%). Note that (S)-3-hydroxy-isobutyric acid is available through hydroxylation of isobutyric acid using *Pseudomonas putida* [26,74] (see Section 4.3.3).

Preparation 3.3   Preparation of ethyl (S)-2-hydroxymethylbutanoate

In a 2-l Erlenmeyer flask with indentation, 125 g bakers' yeast were suspended at 30°C in 1000 ml H₂O/EtOH 95:5. After four days shaking at 120 rpm, 3.7 g (26 mmol) ethyl 2-formylbutanoate were added and the reduction was followed by GC. After completion (2 or 3 days) the mixture was centrifuged (20 min, 7000 rpm) and the supernatant was extracted continuously with ether (4 days). The organic layer was dried over MgSO₄, filtered, evaporated and purified by bulb-to-bulb distillation (air-bath temperature 90–105°C/15 torr) to give ethyl (S)-2-hydroxymethylbutanoate, 3.3 g (88%) as a colourless liquid with an optical purity of e.e. = 91%: [α]$_D^{RT}$ = +2.1° (c = 4, MeOH).

3.1.1.3.   Yeast Reduction of Cyclic β-Ketoesters β-Diketones, and β-Ketolactones

The yeast reduction of cyclic ketones bearing an α-carboalkoxy group reliably yields the (2R,3S) products (Scheme 3.4) [70]. This is analogous to the yeast reduction of 2-substituted acyclic ketoesters (*vide supra*). However, the reaction involving the cyclic substrate is often more diastereo- and enantioselective than the reduction of the corresponding acyclic compound

(Table 3.4) and it may be worthwhile temporarily to create a ring in the substrate in order to enjoy this benefit [75].

TABLE 3.4

Yeast reduction of some β-ketoesters

R¹	R²	e.e. of (2R,3S) isomer (%)	Yield (%)	Ref.
(CH₂)₂	Et	ca. 99	44,80	35,52,78
(CH₂)₃	Et	86–99	64–85	24,35,52,77,78
SCH₂	Me	85	—	75
CH₂S	Me	>95	—	75
CH₂SCH₂	Me	70	71	75
CH₂C(OCH₂CH₂O)CH₂	Et	98	67,86	80,52
N(Et)CH₂	Et	73	65	52

SCHEME 3.4

Some of the β-hydroxyesters so formed have been elaborated to cyclo-hexanes bearing three contiguous stereogenic centres [76] and the southern region of avermectin A_{2b} [77]. Compounds with a second ring appended to the cyclic β-ketoester [for example compounds (14)–(17)] are also reduced in the same manner with high selectivity [52,58,75]. The reaction of ketoester (14) is particularly interesting; Brooks *et al.* [58] showed that the two major products obtained were the hydroxyester (18) and the ketone (19). The latter compound is obviously formed by hydrolysis and decarboxylation. The conversion of the ketoester (20) into the hydroxyester (21) demonstrates that 2-substitution does not seem to affect the stereochemistry generated in the product [24].

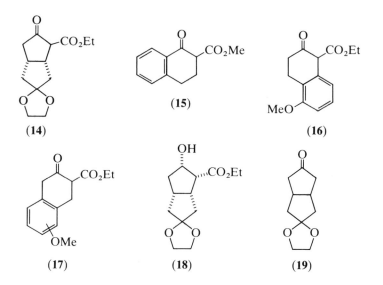

(14)   (15)   (16)

(17)   (18)   (19)

The stereochemical fidelity of the yeast catalysed reaction motivated a search for organisms capable of producing the other stereoisomers [78]. Four were able to produce the "normal" (2R,3S) stereoisomer stereospecifically, whilst three others gave the (2S,3S)-(anti) stereoisomer with good diastereoselectivity and total enantioselectivity. No organism capable of producing an enantiomeric excess of a (3R) stereoisomer was discovered.

Preparation 3.4   Preparation of ethyl (1R,2S)-5,5-ethylenedioxy-2-hydroxycyclohexane-1-carboxylate

Reduction of ethyl 5,5-ethylenedioxy-2-oxocyclohexane-1-carboxylate (15 g) with dry bakers' yeast (Oriental Yeast Co., 200 g) in sucrose solution (300 g in 2000 ml of tap water) for 2 days at 30°C gave 10.1 g (67% or 74% considering the recovery of 1.55 g of starting material) of the title compound after extraction with ethyl acetate, chromatography over silica gel and distillation: b.p. 117–118°C/0.35 torr, n_D^{25} 1.4695; $[\alpha]_D^{23}$ +51.1° (c = 1.02, CHCl$_3$); ν_{max} (film) 3500 (m), 1725 (s) cm^{-1}.

α-Oxalylcyclic ketones are reduced by yeast with fair to moderate enantio-selectivity at the *exo*-cyclic ketone group (Table 3.5) [79]. Early interest in the reduction of cyclic β-diketones centred around the conversion of diones such as (22) into chiral intermediates, such as the hydroxyketone (23) [81] (using micro-organisms such as *Bacillus* spp. and *Saccharomyces* spp.) for the total synthesis of steroids, for example estradiol 3-methyl ether (24) [82–85].

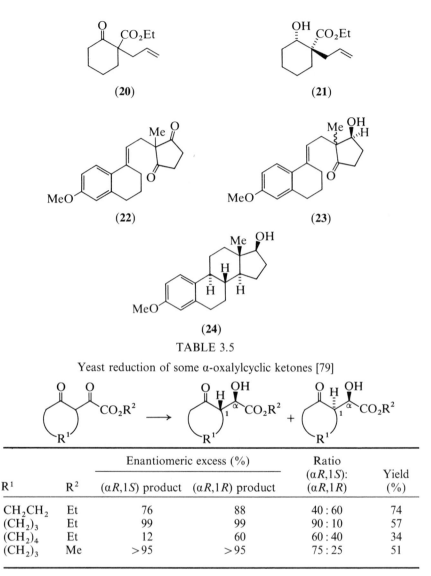

(20)

(21)

(22)

(23)

(24)

TABLE 3.5

Yeast reduction of some α-oxalylcyclic ketones [79]

		Enantiomeric excess (%)		Ratio ($\alpha R,1S$):	Yield
R^1	R^2	($\alpha R,1S$) product	($\alpha R,1R$) product	($\alpha R,1R$)	(%)
CH_2CH_2	Et	76	88	40 : 60	74
$(CH_2)_3$	Et	99	99	90 : 10	57
$(CH_2)_4$	Et	12	60	60 : 40	34
$(CH_2)_3$	Me	>95	>95	75 : 25	51

More recently Brooks *et al.* have investigated the yeast reduction of a wide range of simple cyclic β-diketones, and have employed them in syntheses of anguidine [86,87] and zoapatanol [88]. In all cases, the hydroxy group of the product has the (*S*) configuration, whereas the configuration of the 2-centre (if it bears two different substituents) is variable. In virtually all cases studied, one of the substituents at the 2-centre is methyl, which enables Cahn–Ingold–Prelog notation to be consistently used for correlating the stereochemistry. In general, cyclopentanediones give predominantly the (2S,3S) stereoisomer [86,89] (Table 3.6); the sole exception to this observation is provided by a 2-ethyl substituted cyclopentanone (25) which was transformed into the hydroxyketone (26) (53% yield, e.e. = 100%) by an unusual strain of yeast [83].

Preparation 3.5 Preparation of (2S,3S)-3-hydroxy-2-methyl-2-propylcyclopentanone

To a solution of D-glucose (15 g) and yeast extract (0.5 g) in distilled water (100 ml) at 35–40°C was added with stirring dry active bakers' yeast (10 g, Fleischmann's, Standard Brands Inc.). The mixture was stirred open to the air for 30 min maintaining the temperature between 30 and 35°C, after which 2-methyl-2-propylcyclopenta-1,3-dione (1 g, neat or dissolved in 1 ml DMSO) was added dropwise over 5 min. The mixture was vigorously stirred open to the air at 23–25°C for 24 h and then diluted with 150 ml of water and extracted with dichloromethane (250 ml) in a continuous extractor for 48 h. The organic extract was evaporated to provide a crude product consisting of ketol, unreacted dione, and a trace of diol. The components of the mixture were purified by chromatography over silica gel, using 30% ethyl acetate in hexane as eluent to give the title compound (60%): $[\alpha]_D$ +68.7° (c = 1.94, CHCl$_3$); IR cm^{-1} (neat) 3400 (bs), 1720 (s).

TABLE 3.6

Yeast reduction of some 2,3-disubstituted cyclopenta-1,3-diones

R	Ratio of (2S,3S):(2R,3S)	Enantiomeric excess of (2S,3S) isomer (%)	Yield (%)	Ref.
H$_2$C=<⟨Me	100:0	98	69	89
H$_2$C—//	94:6	>90	65	89
	90:10	>98	75	86,89,90
CH$_2$C≡CH	100:0	>90	65	89
	67:33	>98	60	89,90
CH$_2$COOH	100:0	—	64	89
CH$_2$COOMe	100:0	>98	9	90
Pr	100:0	>98	60	89,90
CH$_2$CH$_2$CN	96:4	>98	71	90
CH$_2$CH$_2$COOMe	100:0	>98	52	90

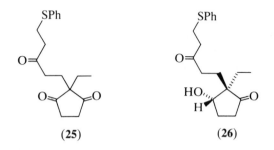

(25)   **(26)**

In contrast, cyclohexanediones preferentially yield the (2R,3S) stereo-isomer [88,90] (Table 3.7). Finally, as the ring size of the cyclic 1,3-dione is increased to seven, eight and nine membered, the yields drop and the stereochemistry becomes less predictable; nevertheless, the products still exhibit high enantiomeric excesses (Table 3.8) [91].

TABLE 3.7

Yeast reduction of some 2,2-disubstituted cyclohexa-1,3-diones

R	Ratio of (2S,3S) : (2R,3S)*	Yield (%)	Ref.
(CH₂)₃—⟨Me	24 : 76	75	88
Pr	22 : 78	80	88,90
Me	—	58	90
CH₂—	45 : 55	80	88,90
CH₂—≡	27 : 73	75	88,90
CH₂—⟨Me	40 : 60	49	90
CH₂CH₂C≡N	30 : 70	49	90
CH₂CH₂CO₂Me	35 : 65	20	90

* All e.e. values reported to be >98%.

TABLE 3.8

Yeast reduction of some 2,2-disubstituted cyclohepta-, cycloocta-, and cyclonona-1,3-diones [91]

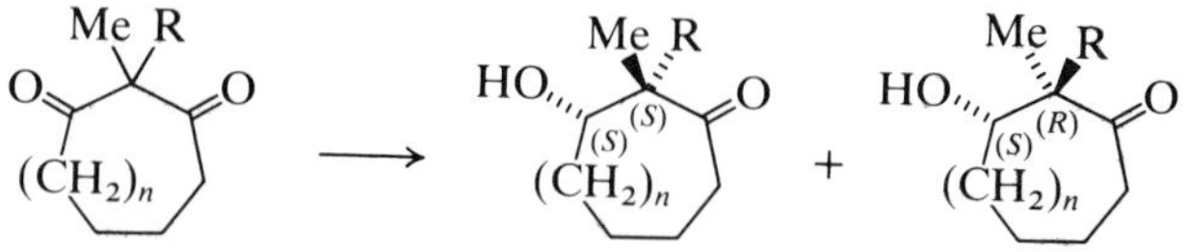

Me R   Me R   Me R

O ⎼ $(CH_2)_n$ ⎼ O $\longrightarrow$ $HO_{\cdots}$ $(S)(S)$ $(CH_2)_n$ O $+$ $HO_{\cdots}$ $(S)(R)$ $(CH_2)_n$ O

n	R	Ratio of (2S,3S) : (2R,3S)*	Yield (%) Product	Yield (%) Starting material
1	Pr	2 : 98	10	75
1	CH₂ ⌇	100 : 0	20	60
1	CH₂—≡	71 : 29	60	30
1	CH₂—⟨ Me	55 : 45	40	50
2	CH₂—	82 : 18	5	75
3	CH₂—	0 : 0	0	80

* All e.e. values reported to be >98%.

3.1.1.4. Yeast Reduction of Ketothioacetals and Related Compounds

A wide range of ketones and esters substituted with sulphur-containing functional groups has been reduced with yeast. Particular attention has been paid to α-ketothioacetals (27) [92,93], the synthetic equivalents of the labile α-ketoaldehydes [94] and β-ketothioacetals [46].

In virtually all cases, the newly-formed chiral centre possessed the (S) configuration and this preference is maintained for the corresponding sulphones (28) [95], sulphoxides (29) [96] and sulphides (30) and (31) [97]. A bewildering range of functional groups can be accommodated adjacent to the keto group [e.g. (27), R^3 = Me, CH_2F, CH_2OCH_2OMe, Et, CH_2Cl, CH_2OH, CF_3] without affecting the enantioselectivity (e.e. >95%), although the yield may be reduced [92]. The sulphoxides showed an interesting difference in behaviour: (S_S)-benzylsulphinylpropanone was reduced faster than its enantiomer, whereas (R_S)-phenylsulphinylpropanone was reduced more quickly than its enantiomer [96]. A *Corynebacterium* sp. gave a similar result with the phenylsulphoxide, except that some sulphone was also isolated [98]. An (R)-configuration alcohol has been obtained by reduction of a β-ketodithiane with *Aspergillus niger* IPV283 [46]. The α,β-diketodithiane (**32**) was reduced sequentially, first at the β-keto-group to give the mono-ol (**33**) and then at the α-keto-group to give the diol (**34**) [99]. Trost *et al.* [100] found that a δ-ketosulphone (or sulphide) was not reduced by yeast; however, it was reduced successfully using *C. guillermonde*.

Preparation 3.6 Preparation of t-butyl (5S)-hydroxy-5-(1,3-dithian-2-yl)pentanoate

To a suspension of 960 g of bakers' yeast (Red Star) in 1920 ml of water was added 16 g of t-butyl 5-oxo-5-(1,3-dithian-2-yl)pentanoate. After the mixture was incubated on a rotary shaker (250 rpm, 2-in. stroke) for 70 h at 25°C, the contents were extracted with ethyl acetate (4 × 1000 ml). The organic layer was separated from the aqueous layer by centrifugation, washed with brine, and dried over anhydrous Na_2SO_4. After removal of the solvent, under vacuum, the crude product was purified by silica gel column chromatography. Elution of the column with ethyl acetate/Skelly B (1:5) afforded the title compound 10.4 g (65%) as an oil: TLC (3:1 hexane/ethyl acetate) R_f 0.19; $[\alpha]_D^{25}$ −21.96° (c = 3.5, $CHCl_3$); IR ($CHCl_3$) 3450 (w), 1715 (s), 1150 (s); ^1H-NMR γ 1.43 (s, 9H), 1.60–2.20 (m, 6H), 2.30 (t, 2H), 2.58–3.1 (m, 4H), 3.7–4.0 (m, 2H).

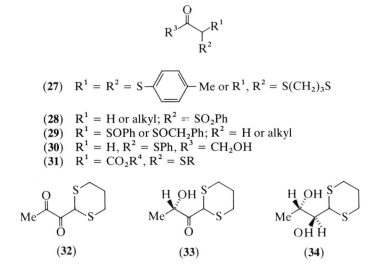

(**27**) $R^1 = R^2 = S-\langle\!\!\bigcirc\!\!\rangle- Me$ or $R^1, R^2 = S(CH_2)_3S$

(**28**) $R^1 = H$ or alkyl; $R^2 = SO_2Ph$
(**29**) $R^1 = SOPh$ or $SOCH_2Ph$; $R^2 = H$ or alkyl
(**30**) $R^1 = H, R^2 = SPh, R^3 = CH_2OH$
(**31**) $R^1 = CO_2R^4, R^2 = SR$

(**32**)   (**33**)   (**34**)

3.1.1.5. Aliphatic Ketones and Aldehydes

In the past few years there has been enormous interest in the enantioselective reduction of ketones [101] by chemical [102], biochemical [7] and other means [103]. In contrast, the reduction of ketones by yeast has been investigated from the beginning of the 20th century [11].

A definitive investigation of yeast reductions of simple ketones was made in the 1960s by Mosher and co-workers [55]. In virtually all cases, (S)-configuration alcohols were obtained with good optical purities (Table 3.9). Longer chain ketones such as n-propyl-n-butylketone and several phenyl ketones were not reduced. As might be expected the best chiral discrimination was achieved with groups of greatly different sizes, and at

TABLE 3.9

Yeast reaction of some aliphatic and aromatic ketones

R^1	R^2	Enantiomeric excess (%)	Major enantiomer	Yield (%)	Ref.
Me	Et	67	(S)	—	55
Me	Pr	64	(S)	20	55
Me	Bu	82	(S)	18	55
Et	Pr	23	(S)	12	55
Et	Bu	27	(S)	8	55
Pr	Bu	0		—	55
Me	Ph	89	(S)	45	55
Et	Ph	72	(S)	19	55
Pr	Ph	90	(S)	13	55
Bu	Ph	89	(S)	7	55
Me	(structure, (CH₂)₂ cyclic with Me, Me)	94	(S)	80	104
CF_3	β-Naphthyl	60	(R)	—	57
Me	$(CH_2)_3C\equiv$	>99	(S)	34	104
Ph	CH_2Cl	—	(R)	32	106
CF_3	CH_3	>80	(S)	68	57
CF_3	cyclohexyl	>95	(R)	88	57
CH_3	cyclohexyl	>95	(S)	35	57
CF_3	Ph	>80	(R)	87	57
CH_3	Ph	>80	(S)	36	57
CF_3	α-Naphthyl	>80	(R)	61	57
CH_3	α-Naphthyl	>80	(S)	12	57
CF_3	$BrCH_2$	>80	(S)	10–15	57
CH_3	$BrCH_2$	>80	(S)	20	57
Ph	CH_2COCH_3	>98	(S)	33	70,106
Me	$C(CH_3)_2NO_2$	>96	(S)	57	107
Me	CH_2 NPhthalyimido	>94	(S)	74	107
Me	CH_2NSucc	>96	(S)	48	107

Table 3.9 (continued)

R^1	R^2	Enantiomeric excess (%)	Major enantiomer	Yield (%)	Ref.
Me	CH$_2$NSacch	>96	(S)	81	107
Me	(CH$_2$)$_2$NPhthalylimido	>95	(S)	94	107
Me	(CH$_2$)$_3$NPhthalylimido	>45	(S)	41	107
Me	(CH$_2$)$_3$NSacch	>75	(S)	52	107
Et	CH$_2$NPhthalylimido	>15	(S)	9	107
Et	CH$_2$NSacch	>96	(S)	54	107
CH$_3$	CH$_2$OH	91	(R)	44	110
CH$_3$	CH$_2$OH	—	(R)	49–58	108
Et	CH$_2$OH	100	(R)	37	110
Et	CH$_2$OH	98	(R)	41	111
Pr^i	CH$_2$OH	100	(R)	65	110
Bu^t	CH$_2$OH	82	(R)	66	110
Pent	CH$_2$OH	>98	(R)	56	109
Ph	CH$_2$OH	100	(S)	—	110
(R)PhCHOH	Ph	100	(R)(R)	—	110
Me	CH$_2$OH	95	(R)	38	111
Me	C$_3$F$_7$	87	(−)	68	29
Me	C$_7$F$_{15}$	91	(−)	74	29
Me	(CH$_2$)$_2$COMe	>95	(S)(S)	57	105
Me	(CH$_2$)$_2$CO$_2$Et	No reaction			106
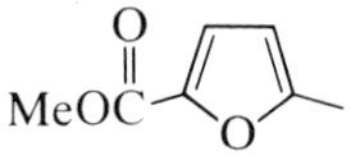 MeOC(=O)-furan	CH$_2$Ofm	—	(R)	45	106
Ph	CO$_2$H	PhCH$_2$OH		65	106
Ph	CO$_2$Et	—	(R)	68	106
Me	Pr^i	90	(S)	17	116
Me	Bu	—		11	116
Me	p-I-C$_6$H$_4$	96	(S)	30	47

least one long-chain group can be tolerated if the other group is methyl [104]. 1,3-Diketones can be reduced once [70], whilst 1,4-diketones may be doubly reduced [105] in both cases with high enantioselectivity for the production of (S) alcohols. As can be seen from Table 3.9, a wide range of functional groups can be tolerated adjacent to the ketone function [11], including chloro [106], bromo [57], perfluoroalkyl [29,57], nitro [107],

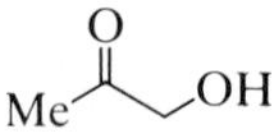

(35)

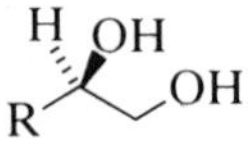

(36) R = Me
(37) R = Et
(38) R = C$_5$H$_{11}$

hydroxyl [108–111], dithianyl [46], phenyldimethylsilyl and *N*-imido groups [107]. The reduction of "acetol" (hydroxyacetone) (**35**) to (*R*)-1,2-propane-diol (**36**) [110] is one of the earliest biological procedures in *Organic Syntheses* [108]. The method is general and (*R*)-1,2-butanediol (**37**) [111,112] and (*R*)-1,2-heptanediol (**38**) [109] have been prepared in this way and used in synthesis [93]. Similar reductions with the same enantioselectivity have been achieved with glycerol dehydrogenase [111,113]. Note that (*R*)-1,2-propane-diol can also be obtained by a microbial degradation of glucose [114] and that α,ω-diols can be oxidized selectively [115].

The reduction of γ- and δ-ketoacids yields (*R*)-configuration alcohols (e.e. >98%) [117,118] (Table 3.10), a similar result to that obtained with β-ketoacids [27,39]. A particularly interesting example is the reduction of the γ-ketoester acids (**39**) (prepared by partial microbial hydrolysis [118]) to give the lactones (**40**) (e.e. >98%).

TABLE 3.10

Yeast reduction of some γ- and δ-ketoacids and the corresponding esters

$$R^1 \overset{O}{\underset{}{\big\|}} (CH_2)_n CO_2 R^2 \longrightarrow R^1 \overset{H \quad OH}{\underset{}{\diagdown\diagup}} (CH_2)_n CO_2 R^2 \text{ (or lactone)}$$

n	R^1	R^2	Configuration of product	e.e. (%)	Yield (%)	Ref.
2	$C_{11}H_{23}$	K	(*R*)	98	50	117
2	Et	H	(*R*)	>98	13	117
2	Bu	H	(*R*)	>98	42	117
2	C_8H_{17}	H	(*R*)	>98	71	117
2	Ph	Et	(*S*)	>95	31	36
3	Et	H	(*R*)	>98	6	117
3	Pr	H	(*R*)	>98	64	117
3	C_8H_{17}	H	(*R*)	>98	54	117
3	Et	Et	(*R*)	>98	11	117
3	Bu	Et	(*R*)	>98	71	117

Deuterated aldehydes are reduced with high enantioselectivity by yeast to furnish the corresponding (*S*) alcohols [119]. The reaction is tolerant of widely differing alkyl groups [120].

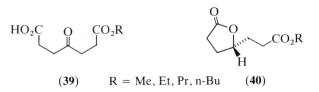

(**39**)   R = Me, Et, Pr, n-Bu   (**40**)

3.1.1.6. Aromatic Aldehydes and Ketones

In early work in this area, simple aromatic ketones were reduced with high enantioselectivities but low yields (Table 3.9) [55]. Aromatic ketones are less reactive to nucleophilic addition reactions than saturated ketones and, therefore, a saturated ketone can be reduced in the presence of an aromatic ketone [70]. However, electron withdrawing groups adjacent to a ketone moiety increase its reactivity and so good results were obtained with aromatic trifluoromethyl ketones [57], α-ketoacids, α-ketoesters [35,106] and imidazoyl ketones [121] and β-ketoesters [36]. Similarly, the improved results achieved with p-iodo [47] and p-nitro-acetophenone [116] over acetophenone and p-methoxyacetophenone [116] may be due to the electron-withdrawing effect of the groups rather than the steric arguments presented by the authors.

The restricted level of reactivity has resulted in the use of other organisms for the reduction of indanones [122–124], tetralones [122,123], acetophenone [123], aryloxyacetic acids [125,126], benzils [127], and pyridyl [128], imidazoyl [121], furanyl [66], thienyl [67] and phenothiazinyl ketones [129].

Some by-products in the reduction of an aromatic aldehyde, such as benzaldehyde, are the α-ketol (41) [130] and the (1R,2S)-erythro-diol (42) and a small amount of the (1S,2R)-threo-diol. The intermediate (41) (which can be formed in up to 76% yield) [130] is presumably formed in a benzoin type condensation; this carbon–carbon bond forming reaction is discussed further in Chapter 5.

(41)　　　　　　　　　　　　　(42)

3.1.1.7. Yeast Reduction of Cyclic and Polycyclic Ketones

The yeast reduction of 2-methylcyclohexanone was reported as early as 1923 [131], whilst cyclopentanone was bio-catalytically reduced for the first time in 1950 [11,132]. The reduction of single enantiomers of racemic 2-substituted cyclohexanones has been achieved in good yield (Table 3.11) and the recovered ketones were also obtained as single enantiomers [96]. The homochiral prostaglandin PGE_1 (43) and its methyl ester are both reduced to $PGF_{1\alpha}$ (44). The ester grouping is cleaved to the acid by the yeast prior to reduction. Racemic PGE_1 is reduced to $PGF_{1\alpha}$ and ent-$PGF_{1\beta}$ (45). The same results were achieved with racemic PGE_2 [133–135].

TABLE 3.11

Yeast reduction of some substituted cyclohexanones [96]

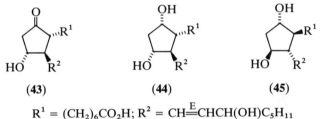

R¹	R²	Stereochemistry of product	e.e. (%)	Yield (%)
Cl	H	(1R,2S)	>95	32
OAc	H	(1R,2S)	54	40
OAc	Buᵗ	(1R,2S,4S)	>95	33

(43)　　　　**(44)**　　　　**(45)**

$R^1 = (CH_2)_6CO_2H; R^2 = CH\overset{E}{=}CHCH(OH)C_5H_{11}$

Similar results have been obtained with the bicyclo[3.2.0]heptenone (46) (Scheme 3.5) [136,137], an oxabicyclo[3.2.0]heptanone [138] and a bicyclo-[3.3.1]nonane dione [139]; in every case the alcohols formed have the (S) configuration at the centre carrying the new hydroxyl group [79,119]. Even molecules as large as steroids [82,84] may be reduced enantioselectively [85]. Fifty micro-organisms were screened for 17-keto steroid reductase activity. Of these, *Hansenula capsulata* (IFO 984) and *Saccharomyces chevalieri* reduced the natural enantiomer to the 17β-alcohol, leaving the unnatural enantiomer unchanged [140].

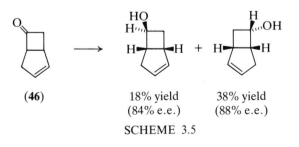

(46)　　　　18% yield　　　　38% yield
　　　　　　　(84% e.e.)　　　　(88% e.e.)

SCHEME 3.5

3.1.2. Reductions Using Enzymes

3.1.2.1. Horse Liver Alcohol Dehydrogenase

Horse liver alcohol dehydrogenase [1] (HLAD) (E.C. 1.1.1.1) is probably the most intensely studied oxidoreductase. It is a dimer consisting of near-identical sub-units each of which contains two zinc atoms (which may be replaced with other metal ions [141]) and 374 amino acids*, arranged around a narrow, barrel-like, hydrophobic active site [142,143]. The three-dimensional structure has been refined to 2.4 Å resolution using X-ray diffraction [144,145]. Although the primary sequence is quite different, the tertiary structure of HLAD is similar to that of yeast alcohol dehydrogenase [144,146,182].

HLAD accommodates a wide range of substrates from acetaldehyde to bicyclic ketones. Early workers used Prelog's diamond lattice model (which was developed for *Curvularia falcata*) [147] to explain the stereoselectivity of reductions catalysed by HLAD [148]. More recently, rationales based on a cubic-space section model [149,150], symmetry properties [151,152], an updated diamond lattice model [153] and a refined diamond lattice model based on molecular graphics [154] have been proposed.

The Prelog rule which predicts *re*-face attack is the most useful model for aldehydes [147], whereas Jones' cubic model is best applied to ketones [149] and, in particular, to those ketones in which the centre(s) of chirality is remote from the carbonyl group. Ketones which are chiral but not asymmetric are best dealt with using Nakazaki's quadrant rules [151]. This precise knowledge of the form of the active site has even enabled catalytically active non-protein analogues to be synthesized [155]. HLAD has been used in both the oxidation of alcohols and the reduction of carbonyl compounds. Oxidations are dealt with in Section 4.1.

The cofactor used with HLAD is NADH. The transfer of hydrogen from NADH to the substrate is specific for one face of the NADH molecule. This enables each face of the NADH molecule to be labelled specifically with deuterium or tritium and material of high activity can be isolated [156].

Unlike yeast, HLAD does not reduce alkenic double bonds conjugated to electron-withdrawing groups. It does, however, show high selectivity for the reduction of the carbonyl group of *trans*-α,β-unsaturated aldehydes over the *cis* isomers, because the former are accommodated more easily in the long barrel-like "active site" [157]. HLAD may be partially inactivated by pre-

---

* This is the form for which the X-ray diffraction data [144] were obtained. The subunits are designated as E (for ethanol active). At least one other sub-unit exists, the S (for steroid active) which differs at six amino acid residues [144]. So there are three possible forms of the enzyme (EE, ES and SS), each with different activity. The EE is the most common form.

incubation (i.e. before addition of NADH) with aldehydes, but this phenomenon does not occur when HLAD is immobilized on CNBr-activated Sepharose 4B [158]. Immobilization improves the long-term stability of HLAD and adenosine monophosphate protects the enzyme against heat inactivation [159]. The enzyme is also stable in low water systems when deposited on the surface of glass beads [160] and in aqueous solutions of alcohols and ethers [161]. The activity of HLAD towards the redox of short-chain alcohols/aldehydes/ketones is increased upon chemical modifications with ethyl acetimidate and 4-bromobutyramide; however, there is little improvement for more hydrophobic substrates [162]. The most useful activity of HLAD is in the redox of medium-ring and bicyclic alcohols and carbonyl compounds. Molecules larger than decalins are not readily accepted and acyclic ketones are not reduced enantioselectively [163].

An NADH regeneration system must be employed in tandem with the enzymic reduction process. The most frequently used ultimate reductant is ethanol, which is used in huge excess [164,165]. This use of ethanol has another benefit in that it also increases the solubility of hydrophobic substrates in aqueous media [161].

HLAD reduces all simple cyclic ketones from four to nine membered. The rate is fastest for four- and six-membered cyclic ketones [166,167] and parallels the rate of reduction by sodium borohydride in isopropanol [166,168]. The reductions of cyclohexanone [164] and 2- and 3-alkyl derivatives give either exclusively, or predominantly, the (1S), 2(S) or (3S) *trans* alcohols with high [169,170] or low enantioselectivity [171], plus the (2R) ketone [170,171] (Scheme 3.6).

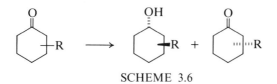

SCHEME 3.6

Formation of the *trans* (1S) alcohol is maximized by operating at pH 7, with a low enzyme to substrate ratio and short reaction times [171]. The order of the rates of reduction is cyclohexanone > 4-substituted ≫ 3-substituted > 2-substituted, except when linear chains longer than propyl are appended (Table 3.12) [172]. Alkoxycarbonyl groups are tolerated better than alkyl groups at the 3 position [173].

4-Keto-2-alkyltetrahydropyrans (**47**, X = O, R^1 = H, R^2 = alkyl) [174], alkylthiopyrans (**47**, X = S, R^1 = H, R^2 = alkyl) [163], 3-alkylthiopyrans (**47**, X = S, R^1 = alkyl, R^2 = H) [170,171,175] and piperidines (**47**, X = N-alkyl; R^1 = R^2 = H) [176] are reduced by HLAD. In general, amines are inhibitors because they complex to the zinc atoms [177]. The (S)-hydroxyl-*trans*-stereoisomer (**48**) is always the major product, together with a little (< 5%) of the (S)-hydroxyl *cis* stereoisomer (**49**) except for the 2-alkyltetra-hydrothiopyrans, which afford up to 30% of the *cis* isomer. Similarly, the 2-phenyl derivative (**47**, X = S, R^1 = H, R^2 = Ph), gives the *cis* isomer (**49**) as the major product [163]. 3-Cyanocyclohexanone is anomalous in that significant amounts of the (1R,3S) alcohol are formed in addition to the expected (1S,3S) alcohol [178]. Six-, seven- and eight-membered 2-thia-ketones are also reduced, but the enantioselectivity is low [167].

TABLE 3.12

Relative rate of reduction of cyclohexanone and derivatives using HLAD (total $K_{3\beta}$ katal^{-1} s^{-1} at 25°C)

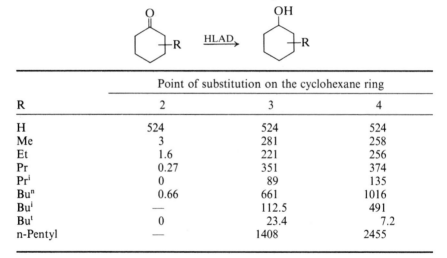

R	Point of substitution on the cyclohexane ring		
	2	3	4
H	524	524	524
Me	3	281	258
Et	1.6	221	256
Pr	0.27	351	374
Pr^i	0	89	135
Bu^n	0.66	661	1016
Bu^i	—	112.5	491
Bu^t	0	23.4	7.2
n-Pentyl	—	1408	2455

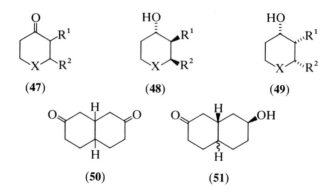

The reduction of prochiral diketones is particularly valuable because all of the substrate may be reduced, e.g. both the *cis* and *trans* isomers of the diketones (50) are reduced to the (*S*)-configuration ketoalcohols (51) (e.e. = 100%) [162]. Similarly, several cage-shaped polycyclic [151] and norbornyl [179,180] ketones were reduced with fair to good enantiomeric excesses. However, these molecules seem to represent the size limit for HLAD because 2-twistanone [151] and a bicyclo[3.2.0]ketone [181] are not reduced.

3.1.3. Miscellaneous Microbial and Enzymic Reductions

Yeast alcohol dehydrogenase (YAD, E.C. 1.1.1.1) [1] was the first pyridine nucleotide-dependent dehydrogenase to be crystallized but, to date, no X-ray structure determination has been achieved [182]. It is a tetramer containing varying amounts of zinc and probably has a similar tertiary structure to HLAD [143,144]. On an activity basis, it is easily the cheapest commercially available dehydrogenase, but it is rather unstable in solution and only accepts aldehydes and methyl ketones as substrates. Like HLAD it is normally *re*-face selective and so it generally yields (*S*)-configuration alcohols on catalysing the reduction of ketones. However, with pyruvate it gives (*R*)-lactic acid by *si*-face attack [183]. The most important use of YAD is for the preparation of alcohols that are chiral by virtue of an isotopic label [119]. The reduction of 1-D-acetaldehyde gives (1*S*)-1-D-ethanol; the other enantiomer can be prepared by reduction of acetaldehyde with NAD(^{2}H) [184]. Six grams of enantiospecifically labelled (*R*)- and (*S*)-[1-^{2}H]-propan-1-ols were prepared in one lot using YAD and diaphorase [185].

Three oxidoreductases have been investigated in detail by Prelog [147] and Dutler and co-workers [186,187]. *Curvularia falcata* alcohol dehydrogenase is *re*-face selective and has been used for the reduction of a wide range of decalones [147]. *Mucor javanicus* alcohol dehydrogenase [1] is a temperature sensitive *si*-face enzyme which reduces acyclic, monocyclic and bicyclic ketones and is also useful for oxidations. However, its stereoselectivity is not as reliable as many other alcohol dehydrogenases [186]. Pig liver alcohol dehydrogenase [1] is generally *re*-face selective but is temperature sensitive [187]. The stereoselectivities are, in general, well predicted by the diamond lattice model [2,147].

Sih has improved upon the selectivity achieved in the reduction of the diketone (52) by *Curvularia falcata* by using the fungus *Aureobasudium pullulans*, which gives the desired stereoisomer (53), a synthon for compactin, as a major product together with the isomers (54) and (55) [188]. The diketone (56), an intermediate in the synthesis of deltamethrine, was reduced by *Curvularia lunata* to the (*S*) enantiomer of the hydroxyketone (57) [189].

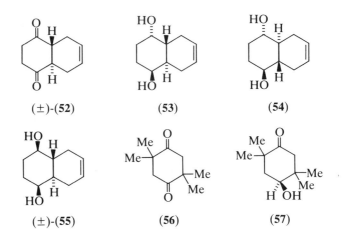

(±)-(52)   (53)   (54)

(±)-(55)   (56)   (57)

α-Hydroxycarboxylic acids [190] are valuable synthetic building blocks which are complementary to the β-hydroxyesters produced by yeast reduction of β-ketoesters. L-Lactic acid [(S) configuration] is readily available, but it cannot easily be homologated. Thirty-nine grades of L-lactate dehydrogenase (L-LDH) (E.C. 1.1.1.27) from 11 species are commercially available and several grades of D-lactate dehydrogenase (D-LDH) (E.C. 1.1.1.28) are also available. Both enzymes reduce pyruvate [179], chloropyruvate [191] and α-ketobutanoate [192] with high enantioselectivity.

Preparation 3.7.   Preparation of D-chlorolactic acid using D-lactate dehydrogenase

(i)   Immobilization of Enzymes

D-LDH and G-6-PDH were immobilized separately in cross-linked PAN gel with a loading of ca. 2.5 mg protein per gram of PAN-500. The reaction mixtures contained enzyme substrates to protect the active sites during the immobilization procedure. The procedure outlined below is representative. A commercially available suspension of ca. 20 mg of D-LDH in 4 ml of 3.2 M $(NH_4)_2SO_4$ was centrifuged at 4°C at 15 000 g for 10 min. The precipitate was dissolved in 3 ml of 0.3 M Hepes buffer (pH 7.5). This solution was dialysed against 1 l of a 50 mM Hepes buffer (pH 7.5, deoxygenated with a stream of argon) to decrease the concentration of $(NH_4)_2SO_4$. The resulting solution contained 1160 units of chlorolactic dehydrogenase activity (using 5 mM chloropyruvate, pH 7.6, 25°C). To buffer (pH 7.5, 0.05 M $MgCl_2$) containing 50 mg of sodium pyruvate and 50 mg of NADH. The mixture was stirred vigorously. After 1 min, 650 μl of a dithiothreitol solution (0.50 M) and 5.53 ml of a triethylenetetramine solution (0.50 M) were added. One minute later, the D-LDH-containing solution was added. The mixture gelled after an additional ca. 2 min of stirring. The gel was kept at room temperature for 1 h, ca. 200 ml of 0.005 M Hepes buffer (pH 7.5) containing 50 mM ammonium sulphate was added, and the gel was broken into small particles in a Waring blender at low speed for 3 min and then at high speed for 30 s. The gel particles were separated by centrifugation, washed with 200 ml of 50 mM Hepes buffer (pH 7.5), and again separated by centrifugation. The gel particles were suspended in H_2O to produce 20 ml of a suspension containing 405 units of chloroacetic dehydrogenase activity (35% immobilization yield).

(ii)   D-β-Chlorolactic Acid

A 2.5 l aqueous solution containing dithiothreitol (3.6 mM), EDTA (1.4 mM), $MgCl_2$ (7.0 mM), and glucose-6-phosphate (0.25 mol) was adjusted to pH 7.6 with KOH. The solution was transferred to a 5-l three-necked flask equipped with a magnetic stirrer and a pH electrode. The solution was degassed with argon and NAD (0.53 mmol) and an aqueous suspension of PAN-immobilized D-LDH (405 units) and G-6-PDH (386 units) was added. The reaction was carried out under an argon atmosphere at ambient temperature (ca. 20°C), and the pH was maintained at pH 7.4–8.0 by the pH-stat. controlled addition of 1.0 M KOH solution. An aqueous solution of chloropyruvic acid (1.0 M, maintained at 5°C) was added dropwise at an average rate of 3 ml h^{-1} by using a peristaltic pump. Several times during the course of the reaction, additional NAD was added to maintain a 0.05–0.20 mM concentration of NAD(H) in the reaction mixture. A total of 0.25 mol of chloropyruvic acid and 0.625 mmol of NAD were added over the course of the reaction. After three days the PAN was allowed to settle and the solution decanted. The PAN was washed with 0.05 M Hepes buffer (pH 7.5), separated by centrifugation and resuspended in fresh buffer. The resulting suspension (500 ml) contained 100% of the original G-6-PDH activity and 45% of the original D-LDH activity. To the decanted reaction solution was added 1 mol of $BaCl_2$, followed by 2 l of ethanol. The white precipitate which formed was filtered, washed with 50% aqueous ethanol and dried under vacuum, yielding barium 6-phosphogluconate-$x$$H_2O$ (100% yield by enzymic assay). The solution was made acidic with HCl, saturated with NaCl, and extracted in several fractions into a total of 5 l of ethyl acetate. The ethyl acetate solution was dried with Na_2SO_4 and filtered. An aliquot (40 ml) was saved for determination of optical purity. The remaining solution was concentrated by rotary evaporation and then dried under vacuum, yielding crude product (20.4, 66%). The chlorolactic acid was recrystallized from 1 : 9 toluene/benzene yielding white needles (14.0 g): m.p. 88–89°C, $[\alpha]_D^{25}$ 4.14 ± 0.08° (c = 9.05 g/100 ml, H_2O); ^{1}H-NMR (acetone-d_6), γ = 9.0–11.0 (b, variable and exchangeable with D_2O, 2, OH), 4.55 (m, 1, methine), 2.85 (d, 2, methylene); IR (Nujol) 3450 (OH), 2300–2600 (OH), 1718 cm^{-1} (carbonyl).

L-LDH reduces small and medium sized α-ketoacids but D-LDH is less versatile [192]. D-LDH [191–194] and L-LDH [191,192,194,195] have been used as catalysts in several NADH recycling systems [196]. A mixture of both enzymes was used to regenerate NADH with dihydrogen and a bis(phosphine) rhodium complex [179], whereas D-LDH has been used in conjunction with a bipyridylrhodium complex [197]. Malate dehydrogenase catalyses the related reduction of oxaloacetate to L-malate in a polyacrylamide matrix [198].

The massive effort devoted to prostaglandin (PG) synthesis [199] spawned efforts to resolve PGs [133] and their synthetic precursors [134,135,137,200]. Particular attention has been paid to bicyclo[3.2.0]heptanones which are particularly versatile intermediates for a wide range of natural products [201,202]. As discussed above, yeast reduced the bicycloheptenone (46) to a mixture of *endo*- and *exo*-alcohols (Scheme 3.5) of good optical purity [136]. Both stereoisomers were converted to natural PGs by different "enantio-complementary" routes [137,203]. However, separation of the alcohols was not easy on a large scale and so a search was made for organisms or enzymes that would reduce one enantiomer only. Nineteen yeasts and 13 strains of bacteria were screened; the best results were obtained with *Mortierella ramanniana* [204] which gave the *endo*-alcohol (Scheme 3.5) in up to 89% optical yield. In contrast, the dimethyl ketone (**58**) gave a mixture of *endo*- (**59**) and *exo*-alcohol (**60**) which were both used in syntheses of eldanolide and leukotriene-B_4 [205,206]. Reduction of the ketone (**58**) with $3\alpha,20\beta$-hydroxysteroid dehydrogenase (HSDH) (E.C. 1.1.1.53) gave the *endo* alcohol (**59**), plus recovered ketone [206]. This enzyme is capable of reducing a range of steroids that possess a 20-keto group and is also capable of oxidizing a few 3α-hydroxy steroids [207]. HSDH is a tetramer [208] with an essential lysyl residue at the active site [209] which is capable of highly enantioselective reduction of bicyclo[3.2.0]heptan-6-ones and bicyclo-[4.2.0]octan-7-ones with electron-withdrawing substituents adjacent to the carbonyl group [181,202,210]. If the electron-withdrawing substituents are absent, the rates decrease (increased K_m, decreased V_{max}) and the enantio-selectivity drops [211]. HSDH is active when immobilized or in organic solvent/water emulsions [210] and reversed phase micelles [212]. The closely-related enzyme 20β-hydroxysteroid dehydrogenase (E.C. 1.1.1.51) has been co-immobilized with NAD^+ so as to form a catalytically active membrane [196,213].

One of Sih's major contributions to the field of PG research involves the discovery that the micro-organism *Dipodascus uninucleatus* reduces the triones (**61**) and (**62**) to the corresponding PG precursors (**63**) and (**64**) (e.e. = 90–100%) in 75% and 48% yield, respectively. Reduction of the trione (**61**) with *Mucor rammanianus* gave the (*S*) hydroxydione (**65**) (e.e. >90%; 43% yield) [214].

Enzymes from thermophilic bacteria [215] are attractive for synthetic purposes because of their enhanced stability at high temperatures. The most intensively investigated thermostable oxidoreductase is that from *Thermoanaerobium brockii* (TBADH) [216], which is stable at 86°C for 70 min [217]. The requisite cofactor is NADP/NADPH. The enantioselectivity is particularly interesting because small ketones ($<C_5$) are reduced with *si*-facial selectivity to give (*R*) alcohols, whereas larger ketones ($>C_5$) are reduced from the *re*-face to give (*S*) alcohols [217,218]. The rates decrease as the ketones become more hydrophobic [150]. As might be expected, the rates of reduction increase with temperature, but the enantioselectivities decrease [218]. A broad range of substrates from aldehydes to secondary cyclic ketones is accepted by this enzyme [31,217,218,219]. Selected aliphatic chloroketones were reduced with **TBADH** to give (*S*) alcohols which were cyclized with base to afford tetrahydrofurans and pyrans [220]. A similar alcohol dehydrogenase has been isolated from *Thermoanaerobacter ethanolicus* [221]. Finally, an isopropylmalate dehydrogenase (E.C. 1.1.1.85) has been isolated from *Thermus thermophilus* [222].

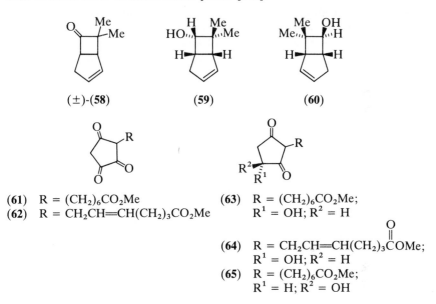

(±)-(**58**)   (**59**)   (**60**)

(**61**) R = $(CH_2)_6CO_2Me$
(**62**) R = $CH_2CH=CH(CH_2)_3CO_2Me$

(**63**) R = $(CH_2)_6CO_2Me$;
R^1 = OH; R^2 = H

(**64**) R = $CH_2CH=CH(CH_2)_3\overset{\text{O}}{\overset{\|}{C}}OMe$;
R^1 = OH; R^2 = H

(**65**) R = $(CH_2)_6CO_2Me$;
R^1 = H; R^2 = OH

3.2. REDUCTION OF CARBON–CARBON DOUBLE BONDS

Face-selective reduction of carbon–carbon double bonds by micro-organisms is now a well-established technique. It is applicable to a wide range of substrates and has a degree of specificity which is difficult to reproduce by the usual chemical techniques.

3.2.1. Reductions with Bakers' and Brewers' Yeasts

This type of reaction was first discovered by Fischer and Wiedemann [223] in a study of the action of fermenting brewers' yeast on unsaturated α-ketoacids. For example, they noted the formation of 3-phenylpropan-1-ol from 4-phenyl-2-oxobut-3-enoic acid in 75% yield. That decarboxylation was not a necessary prerequisite was demonstrated when, in this and later papers [224,225], the reaction was exemplified by the facile reduction of a range of α,β-unsaturated primary alcohols, aldehydes, and ketones (see

Table 3.13). Preliminary indications of the scope of the reaction were obtained. In all cases the aldehydes examined were further reduced to the corresponding primary alcohol, whereas the α,β-unsaturated ketones gave the saturated ketone as a primary product; extensive reduction of the ketone group was only observed after a prolonged reaction time [224]. Double bonds remote from the enone, enal or enol moiety were not reduced: even with an extended reaction time geraniol and citral were only reduced at the α,β-unsaturated bond and 3-buten-1-ol was not reduced [224]. This stricture applies even when the double bonds are conjugated; reduction of 2,4,6-octatrien-1-ol gave only 4,6-octadien-1-ol and 3,5-heptadien-2-one furnished hept-5-en-2-one [226].

TABLE 3.13

Examples of alkenals and alkenyl ketones reduced by brewers' yeast

Functional group adjacent to alkene unit	
Alcohol or aldehyde	Ketone
$CH_3(CH{=}CH)_nCH_2R$ (n = 1, 2 or 3)	$PhCH{=}CHCOCH_3$
	Furyl $CH{=}CHCOCH_3$
$CH_3CH{=}C(CH_3)R$	
$PhCH{=}CHR$	
(R = CH_2OH or CHO)	

Substitution on or near the double bond markedly affects the rate of reaction. 2-Methyl-2-buten-1-ol was only reduced at approximately half the rate of 2-buten-1-ol, and 4-methylpent-3-en-2-one was almost inert to the reaction conditions [224]. Where substituted double bonds were reduced it was noted that the resulting products were optically active, although in these early reports enantiomeric purity was not assessed [224].

Much of this work belies the idea that such reactions are only feasible on a small scale as many of the reductions were performed on a 25-g scale with yeast from the local brewery. The method employed in these reactions was to stir the yeast and the substrate vigorously in a sugar solution preferably buffered to a pH of about 8.5.

Δ-1-Androstene-3,17-dione and the corresponding 17-ol-3-one were reduced to iso-androstane-3,17-diol by fermenting yeast [227], and the β,γ-unsaturated ketone Δ-5-androstene-3,17-dione was reduced to iso-androstane-3,17-diol [228].

In the latter case, surprise was expressed that the non-conjugated bond was reduced, but it seems to be significant that a quantity of the conjugated isomer, Δ-4-testosterone, was also isolated from the reaction mixture. The implication is that the double bond moved into conjugation before reduction occurred in the usual way. It is noteworthy that Kergomard, commenting on some of Mamoli's other work, stated that "the original strain (of the micro-organism) has been lost", and that, "the bakers' yeast commonly used today (1981) in France is not necessarily the same as that in Germany in 1939" [228].

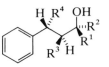

(68)	$R^1 = R^2 = H; R^3 = CDO$	(69)	$R^1 = D; R^2 = R^3 = R^4 = H$
(70)	$R^1 = H; R^2 = D; R^3 = CH_2OH$	(71)	$R^1 = R^2 = R^4 = H; R^3 = D$
(72)	$R^1 = D; R^2 = H; R^3 = CH_2OH$	(73)	$R^1 = R^2 = R^3 = H; R^4 = D$
(74)	$R^1 = R^2 = D; R^3 = CD_2OH$	(75)	$R^1 = R^3 = R^4 = D; R^2 = H$

Apart from reports by Protiva *et al.* [229,230] on the selective reductions of the dienone-ester (66) and the related dienone (67) [230] (Scheme 3.7), little further work on the use of brewers' yeast as a synthetic reagent was reported until 1975. In this year Fuganti *et al.* [231] described a valuable study of the stereochemical course of bakers' yeast reductions of cinnam-aldehyde and cinnamyl alcohol. Reduction of [formyl-^{2}H]-cinnamaldehyde (68) introduced a pro-(*R*) hydrogen atom at C-1 giving (1*S*)-1-deuterio-3-phenylpropanol (69) and reduction of (*E*)-[2-^{2}H]-cinnamyl alcohol (70) was shown to give the (2*S*) isomer of the saturated alcohol (71) indicating a stereospecific introduction of a pro-(*R*) hydrogen atom at C-2. Reduction of [2-^{3}H] cinnamaldehyde has also been reported [231]. A similar study of the reduction of (*E*)-[3-^{2}H]-cinnamyl alcohol (72) gave the compound (73) as the

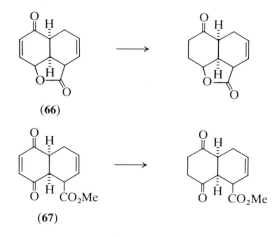

(66)

(67)

SCHEME 3.7

(3R) isomer. These results were confirmed by reducing $[1,1,2,3-^2H_4]$cinnamyl alcohol (74) to obtain $(1S,2S,3R)$-3-phenyl$[1,2,3-^2H_3]$propanol (75). The loss of a deuterium atom at the 1-position of the product (75) was studied by Gramatica et al. [232]. Reduction of (E)-3',4'-dimethoxycinnamyl alcohol doubly labelled at the 1-position gave a saturated alcohol which showed an isotopic composition corresponding to total loss of one deuterium atom; the small amount of unreduced alcohol that was recovered had a higher abundance of monodeuterated species than the starting material. This is consistent with a rapid equilibrium between unsaturated alcohol and unsaturated aldehyde induced by yeast alcohol dehydrogenase, and was taken to indicate that the reduction of cinnamyl alcohols must proceed through the corresponding aldehyde. However, this explanation is not certain because Fuganti et al. [231] have demonstrated that the action of yeast on 3-phenyl$[1-^2H_2]$propanol results in the exchange of a deuterium atom for a pro-(R) hydrogen atom. Thus the monodeuterated products obtained could be explained by the establishment of an alcohol–aldehyde equilibrium by substrate and/or product.

Fuganti's work demonstrates the stereochemical course of yeast reduction to be a formal trans-stereospecific addition of hydrogen across the double bond and introduction of a pro-(R) hydrogen atom at C-1 when aldehydes are reduced.

Substrate specificity of the yeast reduction of cinnamyl compounds has been investigated by Fuganti [231] and Gramatica [232] and their co-workers. Substitution at the 2-position of an (E)-cinnamyl alcohol with a methyl [231,232] or ethyl group [232] did not inhibit the reaction and, in the case of (E)-2-methyl-cinnamyl alcohol, gave (S) stereochemistry at C-2 [231]. (Yeast reduction of α-methyl-α,β-unsaturated systems will generally give (S) stereochemistry at C-2 of the resulting alcohol or ketone, vide infra.) In contrast, (E)-3-methylcinnamyl alcohol did not react and the corresponding aldehyde was reduced to the unsaturated alcohol [232]. This inhibition of reduction by 3-substituents may be unique to cinnamyl systems. In a closely related case, both 2- and 3-methyl-5-phenylpenta-2,4-dienals (76) and (77) were reduced by yeast to the pent-4-en-1-ols (78) and (79), albeit that compound (78) was contaminated by the alcohol formed by reduction of the carbonyl group alone. (Z)-Cinnamyl alcohols have been shown to be inert to reduction by yeast, even under forcing conditions [231–233]. Once again this restriction may be specific to the cinnamyl system.

The microbial reduction of geraniol (80), previously reported by Fischer and Weidemann [223], was studied in detail by Gramatica et al. [234]. By the use of a chiral lanthanide shift reagent, they were able to show by NMR studies that (R)-$(+)$-citronellol (81), which was produced in 25% yield, was "essentially enantiomerically homogeneous". The reduction of nerol (82) gave a 6 : 4 mixture of (*R*) and (*S*) enantiomers. The authors attributed this result to a partial *cis–trans* isomerization of the nerol induced by yeast alcohol dehydrogenase. Predictably, similar results were found with the corresponding aldehydes.

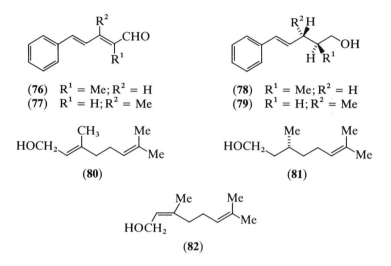

(**76**) $R^1 = Me; R^2 = H$
(**77**) $R^1 = H; R^2 = Me$

(**78**) $R^1 = Me; R^2 = H$
(**79**) $R^1 = H; R^2 = Me$

(**80**)

(**81**)

(**82**)

Selenium dioxide oxidation of (**81**) gave the (*E*) aldehyde (**83**) (Scheme 3.8) which was reduced by yeast to the diastereomerically pure (2*S*,6*R*) diol (**84**) [235]. Similar reduction of (*S*)-(−)-citronellol gave the (2*S*,6*S*) epimer of (**84**). In a one-pot double hydrogenation, Gramatica *et al.* succeeded in converting the geraniol-derived aldehyde (**85**) into (2*S*,6*R*)-(**84**). Gramatica *et al.* have utilized these yeast reductions of geraniol derivatives to prepare two natural products: a component of the secretion of the male African Monarch butterfly (**86**) and a pheromone of the azuki bean weevil (**89**) [236].

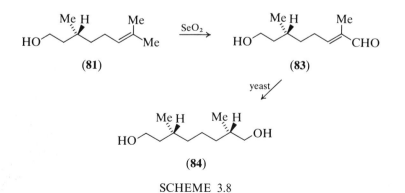

SCHEME 3.8

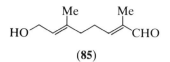

(85)

These two compounds are derived from monoreduced geraniol and thus protection of one of geraniol's double bonds is required. In one case this was achieved by acetylation of one of the two allylic oxygen functions which prevented the initial yeast alcohol dehydrogenase oxidation to the aldehyde and hence protected the adjacent double bond from reduction (Scheme 3.9). The diol (86) resulting from hydrolysis of the reduction product had an enantiomeric excess of >97%. Yeast reduction of the formyl ester (87) was similarly specific for the double bond adjacent to the aldehyde group to give the (7S)-hydroxy-ester (88) with a similar enantiomeric excess. Pyridinium dichromate oxidation of the compound (88) followed by hydrolysis gave the required pheromone (89).

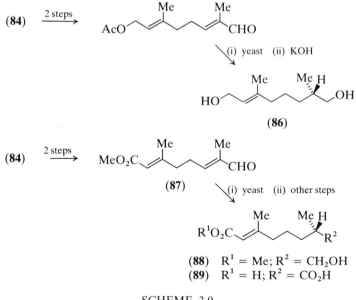

SCHEME 3.9

Leuenberger and his co-workers at Hoffmann-La Roche have used yeast reductions on a very large scale for the preparations of optically pure intermediates for the synthesis of tocopherol and hydroxylated carotenoids [237]. The acetal-ester (90) was reduced on a 133-g scale to give, after 56 h, 49% of the saturated hydroxy-ester (91) and 47% of the unsaturated ester

(**92**). A yeast-catalysed preparation of (*S*)-3-methyl-γ-butyrolactone (or the corresponding hydroxy-ester) as an intermediate in the preparation of (25*S*)-26-hydroxycholesterol has recently been described [238].

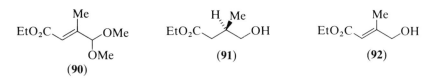

(**90**) (**91**) (**92**)

Halogen substitution on the α-carbon atom of the α,β-unsaturated ketones (**93**) altered the course of the biotransformation [239]. A time study of the conversion (Table 3.14) indicated that reduction of the double bond preceded reduction of the carbonyl function. Further studies demonstrated that increasing the chain length of the ketone did not affect the rate of double-bond reduction but did have a marked effect in slowing the rate of the reduction of the ketone group. The products were obtained in high optical purity; for example, reduction of the nonenone (**93**, R = n-C_5H_{11}) gave the saturated (3*S*) ketone (**94**, R = n-C_5H_{11}) (82% enantiomeric excess), while the two chlorohydrins (**95** *syn*, R = n-C_5H_{11}) and (**95** *anti*, R = n-C_5H_{11}), obtained in an 8:1 ratio, had an enantiomeric excess of 98%. These results imply that the reduction of the carbonyl group to the (2*S*) alcohol was stereospecific regardless of the configuration of the neighbouring asymmetric centre (see Section 3.1.1).

TABLE 3.14

Yeast reduction of 3-chloronon-3-en-2-one (**93**, R = n-C_5H_{11}) (300 mg)

Weight dry yeast (g)	Time (h)	Ratio of products (%)		
		(**93**) R = n-C_5H_{11}	(**94**) R = n-C_5H_{11}	(**95**) R = n-C_5H_{11} *syn + anti*
3	1	48	10	0
3	6	13	26	4
6	6	0	33	20
6	12	0	16	31

Utaka *et al.* [240] used a similar reduction of an α-halo-unsaturated ester in a synthesis of the unusual, naturally occurring amino acid L-armentomycin (**96**) [240]. Reduction of (*E*)-2,4,4-trichloro-2-butenoate [(*E*)-**97**] with immobilized bakers' yeast and re-esterification of the product with diazomethane gave (*R*)-(**98**) in 92% enantiomeric excess and 65% yield. Simple

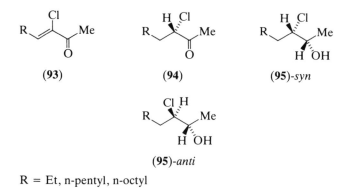

(93)   (94)   (95)-*syn*

(95)-*anti*

R = Et, n-pentyl, n-octyl

functional group transformations gave an enantiomerically pure sample of compound (96). A similar reduction of the (Z) isomer of (97) gave (S)-(98) (60%, e.e. = 97%). The product of each of these fermentations was obtained as the free carboxylic acid. As the corresponding ethyl ester of (Z)-(97) was not hydrolysed or reduced, the implication must be that it is the free acid that is the true substrate for the reducing enzymes.

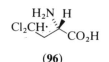

(96)

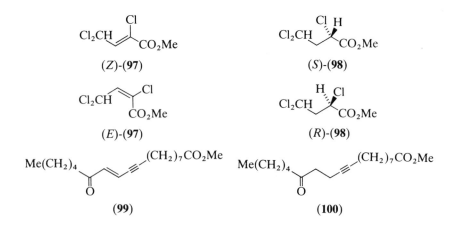

(Z)-(97)   (S)-(98)

(E)-(97)   (R)-(98)

(99)   (100)

Yeast reduction of the enynone (99) gave the alkynic ketone (100) in 74% yield [241]. Reduction of substituted cyclohexene-diones such as (101) has been demonstrated by Leuenberger and his colleagues [237]. Thus compound (101) was biotransformed to give 80% yield of (6R)-2,2,6-trimethyl-1,4-cyclohexanedione after 15 days.

The primary requirement for successful yeast reduction is an electron deficient double bond as demonstrated by Ohta *et al.* [242] who reduced the 2-aryl-1-nitropropenes (102) to furnish the corresponding nitroalkanes (103) in yields of around 50% and enantiomeric excesses of 89–98%. This Japanese group has studied the reduction of nitro-olefins with other microbial systems (*vide infra*).

Ishikawa and Kitazume [243,244] have studied the reduction products of perfluoroalkylated unsaturated alcohols and esters (104). While these reactions were sluggish [seven days for a 68% yield of compound (105) R = $CO_2C_2H_5$] they followed the usual course and gave enantiomeric excesses of

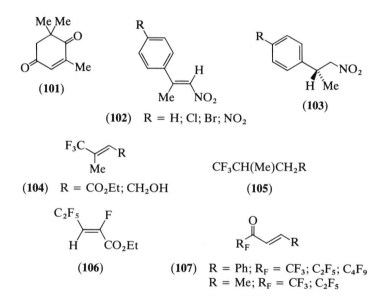

(101)

(102) R = H; Cl; Br; NO_2

(103)

(104) R = CO_2Et; CH_2OH

(105)

(106)

(107) R = Ph; R_F = CF_3; C_2F_5; C_4F_9
       R = Me; R_F = CF_3; C_2F_5

over 60%. As may be expected from the results of experiments on analogous chloroalkenes, the fluorine atom in the ester (106) did not affect the course of the reaction. Reduction of the α,β-unsaturated ketones (107) was also slow, but 10 days was sufficient to produce the corresponding saturated ketones in yields >40% [244]. Small amounts (>9%) of the corresponding alcohols were also produced. Substitution of fluorine or perfluoroalkyl groups on the double bond in this case, however, diverted the usual course of these reactions to produce two series of alcohols. A two-day fermentation involving the compound (108) gave largely the allylic alcohols (109) with enantiomeric excesses of greater than 86%, whereas a 10-day fermentation produced the double bond reduction products (**110**) in yields of >46% and with diastereomeric ratios between 3:1 and 5:1.

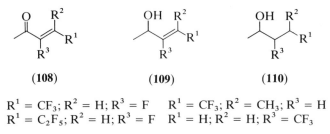

(**108**)   (**109**)   (**110**)

$R^1 = CF_3$; $R^2 = H$; $R^3 = F$ $R^1 = CF_3$; $R^2 = CH_3$; $R^3 = H$
$R^1 = C_2F_5$; $R^2 = H$; $R^3 = F$ $R^1 = H$; $R^2 = H$; $R^3 = CF_3$

As can be seen from this discussion, this reaction presents an extremely efficient and selective method for introducing asymmetry into achiral compounds. As the reactions are extremely easy to perform and the reagent is cheap and readily available, it is to be expected that much more will be heard of this reaction in the future.

3.2.2. Reductions with Enoate Reductases

Due largely to the work of Simon in Germany since the middle 1970s, enoate reductases from *Clostridium kluyveri* (DSM 555) and *Clostridium* sp. La 1 (DSM 1460) are available for the reduction of α,β-unsaturated carboxylates. As with yeast, allylic alcohols and unsaturated aldehydes can be reduced [245], but the majority of studies with these reductases have concentrated on the reduction of unsaturated esters. The responsible enzyme has been isolated and shown to be a conjugated iron–sulphur flavoprotein with a molecular weight of about 450 000 [246]. (In a later review the molecular weight was stated to be 920 000 [247].) It is quite stable (when immobilized [248] its half-life is about 10 days) and sensitive only to the presence of traces of oxygen. The optimum pH for the reduction reaction with the reductase from *Clostridium* La 1 is 6.0, while with the *C. kluyveri* reductase the rate is maximum at pH 7.2. The concentration of sucrose or the ionic strength of the medium are less critical to the activity of the enzyme. The equivalent of hydrogen which the enzyme transfers comes from NADH. In preparative experiments with crude extracts or whole cells, hydrogen gas can help effect the reduction because an active hydrogenase continuously regenerates NADH. The activity ratio of the enoate reductase and hydrogenase in the preparation is dependent upon the growth medium of the *Clostridium* [249]. Glucose as the growth medium induced high levels of hydrogenase, whereas the reductase could be induced by growth on crotonate. Thus the rate-limiting enzyme for the hydrogenation could be either the hydrogenase or the reductase. Of course, isolated enoate reductase does not have the NAD/

hydrogenase complex to provide the necessary NADH, but it has been shown that the radical cation of methylviologen (MV) can act to regenerate NADH or it can indeed replace the NADH by the mechanism shown in Scheme 3.10 [250]. The methylviologen di-cation can be continuously reduced at an electrode to give a greater reaction rate than is observed under the "normal" reaction conditions. Under the "abnormal" conditions, 1 mg of enoate reductase in the presence of 2 mM methylviologen radical cation reduces 5–20 μmol of enoate per minute. Thus, in a remarkable combination of sciences, the electro-enzymic reduction of enoates has been shown to be a viable method for a reduction process on a large scale.

The reductases from each *Clostridium* species show some differences in their rates of reduction of various substrates (for some examples see Table 3.15), but in many cases they are sufficiently similar to be preparatively equivalent [251]. Similar enoate reductases have been demonstrated in 13 other species of *Clostridia* [252], but there are marked differences between them. For example, the substrate specificity of the enoate reductase from *C. sporogenes* is rather narrow compared to that from *Clostridium* La 1 or *C. kluyveri*, and this is true for reductases from other *Clostridia*.

TABLE 3.15

Rate of reduction of some enoates using *Clostridia*

Substrate	Relative rate of reduction	
	C. spec La 1	*C. kluyverii*
(*E*)-3-Methyl-2-butenoate	100	100
(*E*)-2-Butenoate	280	215
(*E*)-3-Methyl-2-pentenoate	11	23
(*Z*)-3-Methyl-2-pentenoate	11	19
(*E*)-4-Methyl-2-pentenoate	130	59
(*E*)-Cinnamate	60	200

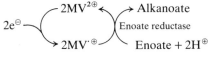

SCHEME 3.10

The stereochemical course of enoate reduction by *Clostridia* species La 1 and *kluyveri* has been studied for (*E*)-crotonate, (*E*)-cinnamate, (*E*)-2-methyl-2-butenoate, and (*Z*)-2-formylaminocinnamate [253]. In each case a *trans* addition of hydrogen (or a hydrogen isotope) to the *si*-faces of the trigonal C-2 and C-3 atoms occurred (Table 3.16). In the case of

(E)-cinnamic acid, the product from a $^2H_2O/C.$ *kluyveri* reduction, $(2S,3R)$-3-phenyl-$[2,3-^2H]$propionic acid, was converted in three steps into $(2R,3S)$- and $(2S,3S)$-phenylalanine. The configuration at C-3 was proved by separate incubations of the product with D- and L-amino acid oxidase. A similar reduction of (E)-2,5-dimethoxycinnamic acid was used in a preparation of $(2S)$-$[^2H]$-2,5-dimethoxyphenylacetic acid [254]. That this mode of reduction is a general phenomenon is illustrated by many other examples provided by Simon in a review [255]. This pattern for the reaction is true even for the enoate reductase derived from *C. sporogenes* which has a different substrate specificity and physical properties [256].

TABLE 3.16

Reduction of enoates using enoate reductase

R^1	R^2	R^3
CH_3	CH_3	H
H	CH_3	H
H	C_2H_5	CH_3
H	CH_3	C_2H_5
H	$CH(CH_3)_2$	H
H	n-$C_{10}H_{23}$	H
H	Ph	H
CH_3	Ph	H
H	p-ClPh	H
H	p-NO_2Ph	H
H	p-MeOPh	H
H	p-Me_2N·Ph	H
H	Ph-σ-OH	H
NHCHO	Ph	H
H		H
H		H
H		H
H		H

These enoate reductases have very broad substrate specificities. Although the rates of reduction vary widely [251], the results cited in Table 3.16 indicate that unsaturated acids substituted with aliphatic, alicyclic or aromatic groups can be reduced with *Clostridium* La 1 to give, where applicable, products of high enantiomeric purity. As has been noted with reductions using bakers' yeast, only the α,β-unsaturated bond of hexa-2,4-dienoic acid is reduced [257]. Reduction of fumarate mono-esters is also possible. This reaction was used in a synthesis of 5-amino-laevulinic acid chirally substituted with deuterium and tritium (Scheme 3.11) [258]. Incubation of monodeuteriofumarate (**111**, R = ^{2}H) with the enoate reductase from *Clostridium* La 1 in water gave, after hydrolysis, (2S)-[2-^{2}H]succinic acid (**112**) which was shown to be configurationally pure. A similar reduction of [2-^{3}H]fumarate (**111**, R = ^{3}H) in deuterium oxide gave the doubly-labelled succinate monoester (**113**).

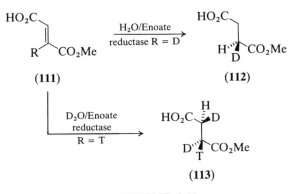

SCHEME 3.11

TABLE 3.17

Some allene carboxylic acids which are reduced by enoate reductase

R^1	R^2	R^3
Et	H	Me
Me	H	Et
Ph	Et	H
Ph	Et	Et

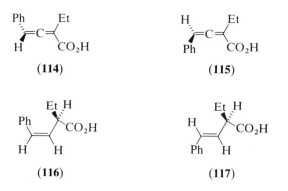

(114)   (115)

(116)   (117)

A reduction which has not been reported with bakers' yeast but is remarkably successful with the enoate reductase from *C. kluyveri* is the reduction of chiral allenes. The enantiomers (114) and (115) gave, in quantitative yield, the products (116) and (117), respectively. Each product had the same stereochemistry at the allylic centre (N.B. the absolute stereochemistry at this centre was not rigorously determined). The reduction of other allenes by *Clostridium* La 1 and hydrogen has been reported (Table 3.17), including a tetrasubstituted allene [257].

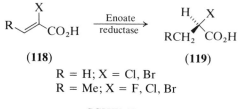

(118)   (119)

R = H; X = Cl, Br
R = Me; X = F, Cl, Br

SCHEME 3.12

Reduction of α-halo-unsaturated acids (118) with the enoate reductase from *C. kluyveri* gave (*R*) halo acids (119) in yields greater than 80% and with high enantiomeric purity [259]. β-Halo-unsaturated acids, however, gave the corresponding saturated, dehalogenated acid [259]. In further work on this reaction, it was demonstrated that, at least for cinnamic acids, this reaction did not proceed through the corresponding 3-halo-saturated acids, but rather that the first hydrogenation step was coupled to the elimination of HX in one enzyme mediated step [260].

The advantages of using an isolated enzyme rather than a whole-cell system are illustrated by the reduction of *p*-nitrocinnamic acid (120). With a whole-cell system both the double bond and the nitro group are reduced, whereas the isolated enzyme reduction resulted only in the reduction of the carbon–carbon bond. There are, however, times when the use of the whole cell is advantageous. Isolated enoate reductase will not reduce allylic alcohols, whereas whole cells of *Clostridium* La 1 or *C. kluyveri* contain an alcohol dehydrogenase which facilitates the reduction of the allylic alcohol by a mechanism similar to that described for yeast [261]. In this case, however, methylviologen exerted an antagonistic effect upon the reduction and ethanol was a far better electron donor than hydrogen [262]. The stereochemistry of this reaction was the same as that observed for the enoate reduction. It follows from this that enoate reductase from *Clostridia* strains will successfully reduce α,β-unsaturated aldehydes [263].

Simon has introduced a series of rules to help in assessing whether a particular unsaturated acid is liable to be reduced. For the general structure (**121**), the group R^1 should not be large, and if R^2 is a phenyl ring, acetamide and ethoxy carbonyl analogues are not reduced while the corresponding formamide and methoxy carbonyl compounds are reduced. If the substituent R^2 is branched in the β-position reduction is inhibited.

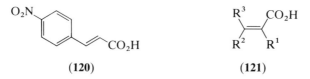

(**120**) (**121**)

Enoate reductases from other sources have been studied in less detail. The reduction of 2-methylfumarate with *Proteus mirabilis* in an atmosphere of hydrogen gave (*S*)-2-methylsuccinate in 95% yield with complete stereospecificity [264]. This, and a similar reduction in deuterium oxide, showed the mechanism to involve *trans* addition to the *re*-face of the C-2 carbon atom and the *si*-face of the C-3 carbon atom (in contrast to the result obtained on reduction with *Clostridia* species). Reduction of (*E*)-cinnamic acid with whole cells of *Peptostreptococcus anaerobius* (DSM 20357) gave a radically different stereochemistry to that observed with Clostridial reduction, in that deuterium was observed to add in a *trans* manner to the *re*-faces of the C-2 and C-3 carbon atoms [265].

3.2.3. Reductions with *Beauveria sulfurescens*

In a study of hydroxylating organisms of use in prostaglandin synthesis, Kergomard and his co-workers incubated 2-methylcyclopentenone with *Beauveria sulfurescens* [266]. This organism was expected to hydroxylate the alicycle, but no such reaction was seen. Instead, the only product isolated, in 90% yield, was (*R*)-2-methylcyclopentanone. A study of this reaction under varying conditions revealed that aeration of the reaction mixture at 300 ml min^{-1} l^{-1} gave a very poor reaction, whereas minimal aeration at 10 ml min^{-1}1^{-1} gave high yields of reduction product. Further studies of this type of process revealed that substitution in the 3-position blocked the reduction and that an ethyl group in the 2-position greatly slowed the reaction. A similar pattern was observed for the reduction of cyclohex-2-enones in that mixtures of saturated ketone and alcohol were obtained. Cyclohex-2-enones substituted with methyl groups in the 4-, 5- or 6-positions were successfully reduced in high yield [267] as were (+)- and (−)-carvone. Reduction of 2,5,5-trimethylcyclohex-2-enone was slow; only 67% reduction occurred after 10-days reaction time. 2,6-Dimethylcyclohex-2-enone and 2,6,6-tri-methylcyclohex-2-enone were not reduced after prolonged reaction times.

Over-reduction in the case of six-membered rings was attributed to differences in ring strain between five- and six-membered rings; a similar pattern is seen in the *Beauveria* reduction of cyclopentanones and cyclohexa-nones wherein the five-membered ring ketone is inert while the six-mem-bered ring ketone undergoes partial reduction to the corresponding alcohol. It was later shown that the reduction to the alcohol occurred after the reduction of the double bond, and that at near neutral pH, ketone reduction did not occur [268].

TABLE 3.18

Reduction of some α,β-unsaturated ketones using *B. sulfurescens*

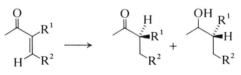

	Ratio of products	
Substrate	Ketone (%)	Alcohol (%)
R^1 = CH$_3$, R^2 = H	70	20
R^1 = H, R^2 = CH$_3$	75	15
R^1 = R^2 = CH$_3$	80	10
R^1 = C$_2$H$_5$, R^2 = CH$_3$	90	0
R^1 = n-C$_4$H$_9$, R^2 = CH$_3$	95	0
R^1 = CH$_3$, R^2 = C$_2$H$_5$	90	5
R^1 = CH$_3$, R^2 = n-C$_3$H$_7$	92	3
R^1 = CH$_3$, R^2 = n-C$_4$H$_9$	—	—
R^1 = CH$_3$, R^2 = n-C$_5$H$_{11}$	—	—

Acyclic unsaturated ketones were also reduced with *Beauveria sulfurescens* (Table 3.18) to mixtures of saturated ketones and alcohols, although significant amounts of alcohol were only obtained with lower molecular weight ketones [266,267]. Reduction of 4-methylpent-3-en-2-one and 3,4-dimethylpent-3-en-2-one gave the corresponding ketones. Two ethyl ketones were also reduced (Scheme 3.13).

Kergomard and his co-workers have also studied the reduction of α,β-unsaturated aldehydes [268] and found, after 48 h fermentation, a mixture of unsaturated and saturated alcohols. In the case of 2-methylpent-2-enal, further studies indicated that after 3 h of fermentation most of the starting material had disappeared and the reaction mixture consisted of 71% unsaturated alcohol and 25% saturated alcohol. Over the next 45 h, the unsaturated alcohol was gradually converted into the saturated alcohol to produce an 82:18 mixture of saturated and unsaturated products. The steric constraints upon the substrate are very similar to those observed for ketones, with the exception that the β-position will tolerate an n-pentyl group (the authors note that this particular reduction gives a higher yield when the reaction is run at a higher temperature). Only alcohols are observed from this reaction, a consequence of the high rate of reduction of the aldehyde group compared to the rate of the reduction of the double bond. A careful study of the rates of each separate step demonstrated a two-fold reaction path (Scheme 3.14). It was also noted that a pH of greater than 7.0 stopped the reaction completely, but at pH values of 7.0 or less there was only a slight change in the ratio of saturated to unsaturated alcohol.

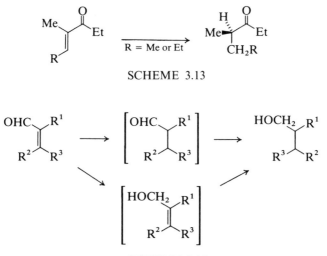

SCHEME 3.13

SCHEME 3.14

The stereochemistry of *B. sulfurescens* reduction has been extensively studied. In all cases in which a cycloalkenone with a 2-substituent was reduced, the resulting asymmetric 2-carbon atom has the (R) configuration [266] and the alcohol moiety produced had the (S) configuration. The reduction of 2-deuteriocycloalkenones was assumed to follow the same course to give (R)-2-[^{2}H]-cycloalkanones [269]. The situation with regard to the reduction of 3-[^{2}H]-cyclopent-2-enone is more clear cut [270]. After 48 h fermentation with *B. sulfurescens*, a 90% yield of (S)-(+)-3-deuteriocyclo-pentanone was obtained: in this case the configuration could be assigned with certainty because it was the enantiomer of the ketone prepared previously by Djerassi *et al.* [271]. A similar reduction of 3-[^{2}H]-cyclohex-2-enone gave a 1:1 mixture of the (3S) ketone and the (3S) deuterio-alcohols. Taken in total, these results indicate that the reduction of cycloalkenones occurs by a *trans* addition of two hydrogen atoms to the *si*-face of C-2 and the *re*-face of C-3.

Reduction of cyclopentenones possessing a remote asymmetric centre indicated a four-fold difference in the rate of reduction between the two enantiomers of the starting enone [267]. For example, reduction of (±)-2,5-dimethyl-2-cyclopent-1-enone (122) (Scheme 3.15) was quite slow, but after 5 days a 30% yield of (2R,5R)-2,5-dimethylcyclopentanone (123) was obtained. Ten-days incubation sufficed to produce a 48% yield of the cyclopentanones (124) and (125) and 40% of recovered starting material (126), which was shown to be the optically pure (5S) isomer [272].

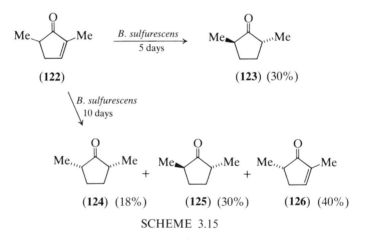

SCHEME 3.15

In contrast to the above results, reduction of acyclic 2-methylalkenones gave optically pure (2S) methyl ketones. The corresponding acyclic 2-ethyl-alkenones gave the saturated product having the (S) configuration [266, 267]. This is explained, assuming a *trans* addition of two hydrogen atoms by classifying the substituents α to the carbonyl in a way analogous to that used in the application of Prelog's rules and then applying this to the *si–re* nomenclature. This results in the familiar 2-*si*–3-*re* addition of the hydrogen atoms as was observed for cycloalkenones.

Reduction of geminally substituted methylene groups is also possible. After 48 h incubation, 2-methylene-cyclohexenone gave a 1:1 mixture of (2*R*)-methylcyclohexanone and (1*S*,2*R*)-methylcyclohexanol in quantitative yield [271]. A similar result was observed for 3-methylene-pentan-2-one, which gave a 65% yield of (3*S*)-methylpentan-2-one.

3.2.4. Conclusion

While bakers' yeast, *Clostridia* enoate reductases and *Beauveria sulfurescens* represent the most extensively studied means of biocatalytic double-bond reduction, other reducing species have been reported. *Penicillium decumbens* [273], various *Streptomyces* strains [274], some *Curvularia* strains [275], *Corynebacterium equi* [276] and *Pseudomonas* sp. NRRL B-3875 [277] are among the many strains which have been reported to possess reducing properties for α,β-unsaturated carbonyl compounds.

A recent paper has described the reduction of various nitropropenes by *Rhodococcus rhodochrous* (IFO 3338) [278]. The products are obtained in good yield, but the low maximum concentration of substrate (<0.6%) tolerated and the necessity to add the substrate in the very early stages of the stationary growth phase of the micro-organism could render this reaction problematic to the novice.

To sum up: enoate reductases, particularly those from *Clostridia* La 1 and *kluyveri*, are very valuable reagents for the stereospecific reduction of unsaturated carbonyl compounds. The isolated enzyme is very specific in its action and certainly possesses advantages over the use of yeast. Brewers' yeast, however, is readily available, and cheap and easy to use, whereas the handling of *Clostridia* species and the isolation of enoate reductase does require some expertise in these matters. Work-up procedures for yeast reactions are sometimes complicated by the difficulty of removing the product from large amounts of cell debris and by the relatively large amounts of solvents which need to be used. In this regard, *Clostridia* reactions are easier to handle in that smaller quantities of solvent are needed and the reactions tend to be easier to work-up. This is particularly true, of course, if the enoate reductase enzyme is isolated and immobilized [248].

Beauveria sulfurescens, like *Clostridia* species, requires some expertise in handling and its utility is applicable only to a limited range of substrates. However, the optical purities of the products make this a valuable preparative method.

REFERENCES

1. For general reviews of the biochemical aspects of dehydrogenases see: P. Boyer (ed.), *The Enzymes*, Vol. 11, 3rd Edition. Academic Press, London, 1975; and P. Boyer, H. Lardy, K. Myrback (eds), Vol. 7, 2nd Edition, 1963. For the use of enzymes from the organic chemist's point of view see: J. B. Jones, C. J. Sih and D. Perlman (eds), *Applications of Biochemical Systems in Organic Synthesis*, in the series *Techniques of Chemistry*, Vol. X, Parts I and II. Wiley, New York, 1976. The latter text covers the following oxidoreductases in considerable detail: yeast alcohol dehydrogenase (YAD), horse liver alcohol dehydrogenase (HLADH), *Curvularia falcata* alcohol dehydrogenase, *Mucor javanicus* alcohol dehydrogenase, pig liver alcohol dehydrogenase and 3α-hydroxysteroid dehydrogenase with comprehensive tables of examples.

2. R. Bentley, *Molecular Asymmetry in Biology*, Vol. II, p. 1. Academic Press, New York, 1970.

3. C. J. Sih and C.-S. Chen, *Angew. Chem. Int. Ed. Engl.* **23**, 570 (1984).

4. H. Simon, J. Bader, H. Gunther, S. Neumann and J. Thanos, *Angew. Chem. Int. Ed. Engl.* **24**, 539 (1985).

5. S. Itoh, M. Mure and Y. Ohshiro, *J. Chem. Soc. Chem. Commun.*, 1580 (1987). B. J. Van Schie, O. H. De Mooy, J. D. Linton, J. P. Van Dijken and J. G. Kuenen, *J. Gen. Microbiol.* **133**, 867 (1987).

6. R. Bowen and S. Pugh, *Chem. Ind. (London)*, 323 (1985). S. W. May and S. R. Padgette, *Biotechnology*, 677 (1983). T. Godfrey and J. Reichet, *Industrial Enzymology*. The Nature Press, New York, 1983.

7. For general reviews containing sections on oxidoreductases see: A. Akiyama, M. Bednarski, M.-J. Kim, E. S. Simon, H. Waldmann and G. M. Whitesides, *Chem. Br.* **23**, 645 (1987). S. Butt and S. M. Roberts, *Chem. Br.* **23**, 127 (1987). S. Butt and S. M. Roberts, *Nat. Prod. Rep.* **3**, 489 (1986). J. B. Jones, *Tetrahedron* **42**, 3351 (1986). J. B. Jones, in *Asymmetric Synthesis* (ed. J. D. Morrison), Vol. 5. Academic Press, Orlando, 1985. A. R. Battersby, *Chem. Br.* **20**, 611 (1984). G. M. Whitesides and C.-H. Wong, *Angew. Chem. Int. Ed. Engl.* **24**, 617 (1985). C. J. Suckling, *Enzyme Chemistry—Impact and Applications*. Chapman-Hall, New York, 1984. F. S. Sariaslani and J. P. N. Rosazza, *Enzyme Microbiol. Technol.* **6**, 242 (1984). C. J. Suckling and H. Wood, *Chem. Br.* **15**, 243 (1979). K. Kieslich, *Microbial Transformations of Non-steroid Cyclic Compounds*. Thieme, Stuttgart, 1976. For steroids see [82,84].

8. F. A. Skinner, S. M. Passmore and R. R. Davenport (eds), *Biology and Activities of Yeast*. Academic Press, New York, 1980.

9. J. N. Strathern, E. W. Jones and J. R. Broach (eds), *The Molecular Biology of the Yeast Saccharomyces: Metabolism and Gene Expression*. Cold Spring Harbor Laboratory Press, Cold Spring Harbor, New York, 1983.

10. A. H. Cook (ed.), *The Chemistry and Biology of Yeast*. Academic Press, New York, 1958.

11. C. Neuberg and A. Lewite, *Biochem. Z.* **91**, 257 (1918). Review: C. Neuberg, *Adv. Carb. Chem.* **4**, 75 (1949). The reduction of furfuraldehyde was observed even earlier by C. J. Lintner and H. H. von Liebig, *Z. Physiol. Chem.* **72**, 449 (1911), who extended the work of W. Windisch, *Wochschr. Brau.* **15**, 189 (1898). See also R. H. A. Plimmer, *The Chemical Changes and Products Resulting from Fermentation*. Longmans, Green and Co., London, 1903.

12. R. U. Lemieux and J. Giguere, *Can. J. Chem.* **29**, 678 (1951). These workers actually used the potassium salts of the β-ketoacids and not the acids as stated in their introduction and elsewhere [34,42].

13. For a review of yeast biotransformations (in Japanese) see: T. Fujisawa, T. Sato and T. Itoh, *Yuki Gosei Kagaku Kyokaishi* **44**, 519 (1986); *Chem. Abs.* **105**, 77444.

14. D. Seebach, M. Sutter, R. H. Weber and M. F. Zuger, *Organic Syntheses* (ed. G. Saucy), **63**, 1 (1985); for useful modifications see J. Ehrler, F. Giovannini, B. Lamatsch and D. Seebach, *Chimia* **40**, 172 (1986).

15. B. Wipf, E. Kupfer, R. Bertazzi and H. G. W. Leuenberger, *Helv. Chim. Acta* **66**, 485 (1983).

16. G. G. Stewart, *BIO/TECHNOLOGY* **3**, 791 (1985).

17. G. I. Georg, H. S. Gill and C. Gerhardt, *Tetrahedron Lett.* **26**, 3903 (1985). G. I. Georg and H. S. Gill, *J. Chem. Soc. Chem. Commun.*, 1433 (1985). G. I. Georg, J. Kant and H. S. Gill, *J. Am. Chem. Soc.* **109**, 1129, 6904 (1987). For the use of (*R*)-3-hydroxybutyric acid in the synthesis of β-lactams see: T. Iimori and M. Shibasaki, *Tetrahedron Lett.* **27**, 2149 (1986); D. M. Tschaen, L. M. Fuentes, J. E. Lynch, W. L. Laswell, R. P. Volante and I. Shinkai, *Tetrahedron Lett.* **29**, 2779 (1988). See also Chapter 4 ref. 278.

18. A. I. Meyers and R. A. Amos, *J. Am. Chem. Soc.* **102**, 870 (1980).

19. K. Mori and K. Tanida, *Tetrahedron* **37**, 3221 (1981); *Heterocycles* **15**, 1171 (1981). K. Mori and M. Ikunaka, *Tetrahedron* **43**, 45 (1987). K. Mori and H. Kisida, *Tetrahedron* **42**, 5281 (1986). K. Mori, *Tetrahedron* **37**, 1341 (1981).

20. K. Mori and S. Kuwahara, *Tetrahedron* **42**, 5539, 5545 (1986). For the use of ethyl (*R*)-3-hydroxy butyrate in the synthesis of pheromones see: K. Mori and T. Ebata, *Tetrahedron* **42**, 4413, 4421 (1986).

21. A. Kramer and H. Pfander, *Helv. Chim. Acta* **65**, 293 (1982).

22. B. Seuring and D. Seebach, *Helv. Chim. Acta* **60**, 1175 (1977).

23. G. Frater, *Helv. Chim. Acta* **62**, 2825, 2829 (1979). M. Larcheveque and S. Henrot, *Tetrahedron Lett.* **28**, 1781 (1987).

24. G. Frater, *Helv. Chim. Acta* **63**, 1383 (1980).

25. G. Frater, U. Muller and W. Gunther, *Tetrahedron* **40**, 1269 (1984).

26. D. Seebach and M. F. Zuger, *Tetrahedron Lett.* **25**, 2747 (1984); *Helv. Chim. Acta* **65**, 495 (1982). Y. Doi, A. Tamaki, M. Kunioka and K. Soga, *J. Chem. Soc. Chem. Commun.*, 1635 (1987). S. Maemoto and K. Mori, *Chem. Lett.*, 109 (1987).

27. M. Hirama, H. Shimizu and M. Iwashita, *J. Chem. Soc. Chem. Commun.*, 599 (1983). M. Hirama, T. Nakamine and S. Ito, *Chem. Lett.*, 1381 (1984).

28. D. Seebach, P. Renaud, W. B. Schweizer and M. F. Zuger, *Helv. Chim. Acta* **67**, 1843 (1984).

29. T. Kitazume and N. Ishikawa, *Chem. Lett.*, 237 (1983). T. Kitazume and J. T. Lin, *J. Fluorine Chem.* **34**, 461 (1987). T. Kitazume and T. Kobayashi, *Synthesis*, 187 (1987). See also [44].

30. D. W. Brooks, R. P. Kellogg and C. S. Cooper, *J. Org. Chem.* **52**, 192 (1987). M. Christen and D. H. G. Crout, *J. Chem. Soc. Chem. Commun.*, 264 (1988).

31. D. Seebach, M. F. Zuger, F. Giovannini, B. Sonnleitner and A. Fiechter, *Angew. Chem. Int. Ed. Engl.* **23**, 151 (1984). D. Seebach and M. Eberle, *Synthesis*, 37, (1986).

32. A. Fauve, M. F. Renard and H. Veschambre, *J. Org. Chem.* **52**, 4893 (1987). J.-C. Gramain, A. Kergomard, M. F. Renard and H. Veschambre, *J. Org. Chem.* **50**, 120 (1985). A. Kergomard, M. F. Renard and H. Veschambre, *J. Org. Chem.* **47**, 792 (1982).

33. C. Fuganti, P. Grasselli, P. Casati and M. Carmeno, *Tetrahedron Lett.* **26**, 101 (1985). C. Fuganti, P. Grasselli, P. F. Seneci and P. Casati, *Tetrahedron Lett.* **27**, 5275 (1986).

34. B.-N. Zhou, A. S. Gopalan, F. Van Middlesworth, W.-R. Shieh and C. J. Sih, *J. Am. Chem. Soc.* **105**, 5925 (1983).

35. B. S. Deol, D. D. Ridley and G. W. Simpson, *Aust. J. Chem.* **29**, 2459 (1976).

36. A. Manzoochi, R. Casati, A. Fiecchi and E. Santaniello, *J. Chem. Soc. Perkin Trans. I*, 2753 (1987).
37. P. Deshong, M.-T. Lin and J. J. Perez, *Tetrahedron Lett.* **27**, 2091 (1986).
38. D. W. Brooks, N. Castro de Lee and R. Peevey, *Tetrahedron Lett.* **25**, 4623 (1984).
39. M. Hirama and M. Uei, *J. Am. Chem. Soc.* **104**, 4251 (1982).
40. H. Hirama, T. Noda and S. Ito, *J. Org. Chem.* **50**, 127 (1985). For (R)-(+)-citronellol see: P. Gramatica, P. Manitto, B. M. Ranzi, A. Delbianco and M. Francavilla, *Experientia* **38**, 775 (1982).
41. M. Akita, H. Matsukura and T. Oishi, *Chem. Pharmacol. Bull.* **34**, 2656 (1986).
42. W.-R. Sheih, A. S. Gopalan and C. J. Sih, *J. Am. Chem. Soc.* **107**, 2993 (1985). C. J. Sih, B.-H. Zhou, A. S. Gopalan, W.-R. Shieh, C. S. Chen, G. Girdaukas and F. Van Middlesworth, in *Enzyme Engineering* (ed. A. J. Laskin), Vol. 7. *Ann. N.Y. Acad. Sci.* **434**, 186 (1984).
43. M. Utaka, H. Higashi and A. Takeda, *J. Chem. Soc. Chem. Commun.*, 1368 (1987).
44. A strain of bakers' yeast from the Oriental Yeast Co., Ltd., was used in this study. In some cases this has been shown to have an opposite enantioselectivity to "Red Star" bakers' yeast. The two other brands of yeast which are often used are Fleischmann's (Standard Brands Inc.) and Fermipan. Although these show some differences, these are often slight (see ref. 55). K. Ushio, K. Inouye, K. Nakamura, S. Oka and A. Ohno, *Tetrahedron Lett.* **27**, 2657 (1986).
45. K. Nakamura, M. Higaki, K. Ushio, S. Oka and A. Ohno, *Tetrahedron Lett.* **26**, 4213 (1985).
46. R. Bernardi, R. Cardillo and D. Ghiringhelli, *J. Chem. Soc. Chem. Commun.*, 460 (1984). D. Ghiringhelli, *Tetrahedron Lett.* **24**, 287 (1983). R. Bernardi, R. Cardillo, D. Ghiringhelli and V. de Pavo, *J. Chem. Soc. Perkin Trans. I*, 1607 (1987). R. Bernardi and D. Ghiringhelli, *J. Org. Chem.* **52**, 5021 (1987).
47. K. Nakamura, K. Ushio, S. Oka and A. Ohno, *Tetrahedron Lett.* **25**, 3979 (1984).
48. For reviews of immobilization of whole cells see: F. Godia, C. Casas and C. Sola, *Process Biochem.* **43** (1987). A. Klibanov, *Science* **219**, 722 (1983). P. G. Rouxhet, J. L. Van Haecht, P. Gerard and M. Briquet, *Enzyme Microbiol. Technol.* **3**, 49 (1981). I. Chibata, T. Tosa and M. Fugimura, *Ann. Rep. Ferm. Proc.* **6**, 1 (1983). V. C. Gekas, *Enzyme Microbiol. Technol.* **8**, 450 (1986). T. Onaka, K. Nakanishi, T. Inoue and S. Kubio, *BIO/TECHNOLOGY* **3**, 467 (1985). M. Kierstan and C. Bucke, *Biotechnol. Bioeng.* **19**, 387 (1977). S. Fukui and A. Tanaka, *Ann. Rev. Microbiol.* **36**, 145 (1982). V. E. Gulaya, S. N. Ananchenko, I. V. Torgov, K. A. Koshcheyenko and G. G. Bychkova, *Bioorg. Khim.* **5**, 768 (1979).
49. K. Nakamura, K. Inoue, K. Ushio, S. Oka and A. Ohno, *Chem. Lett.*, 679 (1987). cf. R. R. Rando, *Biochem. Pharmacol.* **23**, 2328 (1974) and ref. 50.
50. R. P. Lanzilotta, D. G. Bradley and C. C. Beard, *Appl. Microbiol.* **29**, 427 (1975).
51. A. Fauve and H. Veschambre, *Tetrahedron Lett.* **28**, 5037 (1987).
52. For the use of non-fermenting bakers' yeast see: D. Seebach, S. Roggo, T. Maetzke, H. Braunschweiger, J. Cercus and M. Krieger, *Helv. Chim. Acta* **70**, 1605 (1987). See also [57].
53. M. Saita and J. C. Slaughter, *Enzyme Microbiol. Technol.* **6**, 375 (1984). E. E. Egbosimba and J. C. Slaughter, *J. Gen. Microbiol.* **133**, 375 (1987).
54. C. White and G. M. Gadd, *J. Gen. Microbiol.* **133**, 727 (1987).
55. R. Macleod, H. Prosser, L. Fikentscher, J. Lanyi and H. S. Mosher, *Biochemistry* **3**, 838 (1964).
56. R. D. Tanner, L. D. Richmond, C.-J. Wei and J. Woodward, *J. Chem. Techn. Biotechnol.* **31**, 290 (1981).

57. M. Bucciarelli, A. Forni, I. Moretti and G. Torre, *Synthesis*, 897 (1983); *J. Chem. Soc. Chem. Commun.*, 456 (1978).

58. D. W. Brooks, M. Wilson and M. Webb, *J. Org. Chem.* **52**, 2244 (1987).

59. For the use of asymmetric homogenous catalytic hydrogenation to effect this reaction and a general review see: R. Noyori, T. Ohkuma, M. Kitamura, H. Takaya, N. Sayo, H. Kumobayashi and S. Akutagawa, *J. Am. Chem. Soc.* **109**, 5856 (1987). J. M. Brown, *Angew. Chem. Int. Ed. Engl.* **26**, 190 (1987).

60. K. Nakamura, T. Miyai, K. Nozaki, K. Ushio, S. Oka and A. Ohno, *Tetrahedron Lett.* **27**, 3155 (1986). See also [44].

61. R. W. Hoffmann, W. Ladner, K. Steinbach, W. Massa, R. Schmidt and G. Snatzke, *Chem. Ber.* **114**, 2786 (1981).

62. T. Fujisawa, T. Itoh and T. Sato, *Tetrahedron Lett.* **25**, 5083 (1984). See also [44].

63. D. Buisson, C. Sanner, M. Larcheveque and R. Azerad, *Tetrahedron Lett.* **28**, 3939 (1987).

64. T. Oishi and T. Nakata, *Acc. Chem. Res.* **17**, 338 (1984). R.W. Hoffmann, *Angew. Chem. Int. Ed. Engl.* **26**, 489 (1987).

65. D. Buisson, S. Henrot, M. Larcheveque and R. Azerad, *Tetrahedron Lett.* **28**, 5033 (1987).

66. H. Akita, A. Furuichi, H. Koshiji, K. Horikoshi and T. Oishi, *Chem. Pharm. Bull.* **32**, 1333 (1984); *Tetrahedron Lett.* **23**, 4051 (1982). Reduction was successful with other species of yeast.

67. A. Furuichi, H. Akita, H. Koshiji, K. Horikoshi and T. Oishi, *Chem. Pharm. Bull.* **32**, 1619 (1984).

68. H. Akita, H. Koshiji, A. Furuichi, K. Horikoshi and T. Oishi, *Tetrahedron Lett.* **24**, 2009 (1983).

69. S. Inayama, N. Shimizu, T. Ohkura, H. Akita, T. Oishi and Y. Iitaka, *Chem. Pharm. Bull.* **34**, 2660 (1986).

70. R. Chenevert and S. Thiboutot, *Can. J. Chem.* **64**, 1599 (1986). H. Ohta, K. Ozaki and G. Tsuchihashi, *Agric. Biol. Chem.* **50**, 2499 (1986).

71. For a definitive example involving a protease/esterase see: G. Fulling and C. J. Sih, *J. Am. Chem. Soc.* **109**, 2845 (1987).

72. H. Akita, A. Furuichi, H. Koshiji, K. Horikoshi and T. Oishi, *Chem. Pharmacol. Bull.* **31**, 4376 and 4384 (1983). See also [44].

73. M. F. Zuger, F. Giovannini and D. Seebach, *Angew. Chem. Int. Ed. Engl.* **22**, 1012 (1983).

74. C. T. Goodhue and J. R. Schaeffer, *Biotechnol. Bioeng.* **13**, 203 (1971).

75. R. W. Hoffmann, W. Ladner and W. Helbig, *Liebigs. Ann. Chem.*, 1170 (1984). R. W. Hoffmann, W. Helbig and W. Ladner, *Tetrahedron Lett.* **23**, 3479 (1982).

76. D. Seebach and B. Herradon, *Tetrahedron Lett.* **28**, 3791 (1987). D. Seebach, S. Roggo and J. Zimmermann, in *Stereochemistry of Organic and Bioorganic Transformations* (eds W. Bartmann and K. B. Sharpless), *Workshop Conferences Hoechst*, Vol. 17. VCH, Weinham, 1987.

77. A. G. M. Barrett and N. K. Capps, *Tetrahedron Lett.* **27**, 5571 (1986).

78. D. Buisson and R. Azerad, *Tetrahedron Lett.* **27**, 2631 (1986).

79. S. Tsuboi, E. Nishiyama, H. Furutani, M. Utaka and A. Takeda, *J. Org. Chem.* **52**, 1359 (1987). S. Tsuboi, E. Nishiyama, M. Utaka and A. Takeda, *Tetrahedron Lett.* **27**, 1915 (1986). See also [44]. For yeast reduction of α-ketoesters see: K. Nakamura, K. Inoue, K. Ushio, S. Oka and A. Ohno, *J. Org. Chem.* **53**, 2589 (1988).

80. T. Kitahara and K. Mori, *Tetrahedron Lett.* **26**, 451 (1985).

81. H. Kosmol, K. Kieslich, R. Vossing, H.-J. Koch, K. Petzoldt and H. Gibian, *Liebigs Ann. Chem.* **701**, 198 (1967). For the use of *Rhizopus arrhizus* see: P. Bellet, G. Nominé and J. Mathieu, *C.R. Acad. Sci. Ser. C.* **263**, 88 (1966). For later synthetic work see: C. Rufer, H. Kosmol, E. Schröder, K. Kieslich and H. Gibian, *Liebigs Ann. Chem.* **702**, 141 (1967).

82. Reviews: H. Iizuka and A. Naito, *Microbial Transformations of Steroids and Alkaloids.* University of Tokyo Press, Tokyo, 1971; also Springer-Verlag, New York, 1981. W. Charney and H. L. Herzog, *Microbiological Transformations of Steroids.* Academic Press, New York, 1968. A. Capek, O. Hanc and M. Tadra, *Microbial Transformations of Steroids.* Academia, Prague, 1966.

83. W.-M. Dai and W.-S. Zhou, *Tetrahedron* **41**, 4475 (1985).

84. K. Kieslich, in *Biotransformations* (eds H. J. Rehm and G. R. Reed), Vol. 6a. VCH, Weinham, 1984.

85. H. L. Herzog, M. A. Jevnik, P. L. Perlman, A. Nobile and L. B. Hershberg, *J. Am. Chem. Soc.* **75**, 266 (1953).

86. D. W. Brooks, P. G. Grothaus and J. T. Palmer, *Tetrahedron Lett.* **23**, 4187 (1982).

87. D. W. Brooks, P. G. Grothaus and H. Mazdiyasni, *J. Am. Chem. Soc.* **105**, 4472 (1983).

88. D. W. Brooks, H. Mazdiyasni and S. Chakrabarti, *Tetrahedron Lett.* **25**, 1241 (1984).

89. S. Schwarz, G. Truckenbrodt, M. Meyer, R. Zepter, G. Weber, C. Carl, M. Wentzke, H. Schick and H. P. Wentzke, *J. Prak. Chem.* **323**, 729 (1981). D. W. Brooks and K. W. Woods, *J. Org. Chem.* **52**, 2036 (1987). D. W. Brooks, P. G. Grothaus and W. L. Irwin, *J. Org. Chem.* **47**, 2820 (1982). Structures (**3**), (**6**) and (**9**) in this paper have been revised to the enantiomers, see ref. 3 [88].

90. D. W. Brooks, H. Mazdiyasni and P. G. Grothaus, *J. Org. Chem.* **52**, 3223 (1987). K. Mori and H. Mori, *Tetrahedron* **42**, 5531 (1986). R. Ito, R. Masahara and K. Tsukida, *Tetrahedron Lett.* 2767 (1977).

91. D. W. Brooks, H. Mazdiyasni and P. Sallay, *J. Org. Chem.* **50**, 3411 (1985).

92. Y. Takaishi, Y.-L. Yang, D. Di Tullio and C. J. Sih, *Tetrahedron Lett.* **23**, 5489 (1982), improved results were achieved with other microorganisms. T. Fujisawa, E. Kojima, T. Itoh and T. Sato, *Chem. Lett.* **26**, 1751 (1985). C.-Q. Han, D. Di Tullio, Y.-F. Wang and C. J. Sih, *J. Org. Chem.* **51**, 1253 (1986). G. Guanti, L. Banfi and E. Narisano, *Tetrahedron Lett.* **27**, 3547 (1986). G. Guanti, L. Banfi, A. Guaragna and E. Narisano, *J. Chem. Soc. Chem. Commun.*, 138 (1986).

93. S. Takano, *Pure Appl. Chem.* **59**, 353 (1987).

94. Y. Ito and S. Terashima, *Chem. Lett.*, 445 (1987).

95. A. P. Kozikowski, B. B. Mugrage, C. S. Li and L. Felder, *Tetrahedron Lett.* **27**, 4817 (1986).

96. R. L. Crumbie, B. S. Deol, J. E. Nemorin and D. D. Ridley, *Aust. J. Chem.* **31**, 1965 (1978). R. L. Crumbie, D. D. Ridley and G. W. Simpson, *J. Chem. Soc. Chem. Commun.*, 315 (1977).

97. T. Itoh, A. Yoshinaka, T. Sato and T. Fujisawa, *Chem. Lett.* 1679 (1985). T. Fujisawa, T. Itoh, M. Nakai and T. Sato, *Tetrahedron Lett.* **26**, 771 (1985).

98. H. Ohta, Y. Kato and G.-I. Tsuchihashi, *J. Org. Chem.* **52**, 2735 (1987).

99. T. Fujisawa, E. Kojima, T. Itoh and T. Sato, *Tetrahedron Lett.* **26**, 6089 (1985). See also [44].

100. B. M. Trost, J. Lynch, P. Renaut and D. H. Steinman, *J. Am. Chem. Soc.* **108**, 284 (1986).

101. M. M. Midland, in *Asymmetric Synthesis* (ed. J. D. Morrison), Vol. 2A, Chap. 2. E. R. Crandbois, S. I. Haward and J. D. Morrison, *ibid.*, Chap. 3. Academic Press, New York, 1983. M. Nogradi, *Stereoselective Synthesis*, VCH Publishers, New York, 1987.

102. E. J. Corey, R. K. Bakshi and S. Shibata, *J. Am. Chem. Soc.* **109**, 5551 (1987).

103. M. Utaka, H. Watabu and A. Takeda, *J. Org. Chem.* **51**, 5423 (1986). T. Sugimoto, T. Kokubo, Y. Matsumura, J. Miyazaki, S. Tanimoto and M. Okano, *Bioorg. Chem.* **10**, 104 (1981).

104. C. Le Drian and A. E. Greene, *J. Am. Chem. Soc.* **104**, 5473 (1982). A. Belan, J. Bolte, A. Fauve, J. G. Gourey and H. Veschambre, *J. Org. Chem.* **52**, 256 (1987).

105. J. K. Lieser, *Synth. Comm.* **13**, 765 (1983).

106. S. Takano, M. Yanase, Y. Sekiguchi and K. Ogasawara, *Tetrahedron Lett.* **28**, 1783 (1987). D. D. Ridley and M. Stralow, *J. Chem. Soc. Chem. Commun.*, 400 (1975).

107. Nitro and *N*-imido: T. Fujisawa, H. Hayashi and Y. Kishioka, *Chem. Lett.*, 129 (1987). Phenyldimethylsilyl: C. Syldatk, H. Andree, A. Stoffregen, F. Wagner, B. Stumpf, L. Ernst, H. Zilch and R. Tacke, *Appl. Microbiol. Biotechnol.* **27**, 152 (1987).

108. P. A. Levene and A. Walti, *Organic Syntheses Coll. II.* (ed. A. H. Blatt), p. 545. Wiley, New York, 1943.

109. J. Barry and H. B. Kagan, *Synthesis*, 453 (1983).

110. J.-P. Guette and N. Spassky, *Bull. Soc. Chim. Fr.*, 4217 (1972).

111. L. G. Lee and G. M. Whitesides, *J. Org. Chem.* **51**, 25 (1986).

112. D. H. G. Crout and S. M. Morrey, *J. Chem. Soc. Perkin Trans. I*, 2435 (1983). A. F. Drake, G. Siligardi, D. H. G. Crout and D. L. Rathbone, *J. Chem. Soc. Chem. Commun.*, 1835 (1987).

113. H. K. Chenault, M.-J. Kim, A. Akiyama, T. Miyazawa, E. Simon and G. M. Whitesides, *J. Org. Chem.* **52**, 2608 (1987).

114. E. S. Simon, G. M. Whitesides, D. C. Cameron, D. J. Weitz and C. L. Cooney, *J. Org. Chem.* **52**, 4042 (1987).

115. H. Ohta, H. Tetsuki and N. Noto, *J. Org. Chem.* **47**, 2400 (1982).

116. O. Cervinka and L. Hub, *Coll. Czech. Chem. Commun.* **31**, 2615 (1966).

117. M. Utaka, H. Watabu and A. Takeda, *J. Org. Chem.* **52**, 4363 (1987); *Chem. Lett.*, 1475 (1985).

118. F. Moriuchi, H. Muroi and H. Aibe, *Chem. Lett.*, 1141 (1987). See also [44].

119. D. Arigoni and E. L. Eliel, *Topics Stereochem.* **4**, 127 (1969). H. G. Floss and M.-D. Tsai, *Adv. Enzymol.* **50**, 243 (1979).

120. H. S. Mosher, *Tetrahedron* **30**, 1733 (1974). V. E. Althouse, D. M. Feigl, W. A. Sanderson and H. S. Mosher, *J. Am. Chem. Soc.* **88**, 3595 (1966). V. E. Althouse, K. Ueda and H. S. Mosher, *J. Am. Chem. Soc.* **82**, 5938 (1960). K. R. Varma and E. Caspi, *J. Org. Chem.* **34**, 2489 (1969). S. H. Liggero, R. Sustmann and P. von R. Schleyer, *J. Am. Chem. Soc.* **91**, 4571 (1969).

121. R. Lis, W. B. Caldwell, G. A. Hoyer and K. Petzoldt, *Tetrahedron Lett.* **28**, 1487 (1987).

122. M. Imuta, K. I. Kawai and H. Ziffer, *J. Org. Chem.* **45**, 3352 (1980).

123. K. Kabuto, M. Imuta, E. S. Kempner and H. Ziffer, *J. Org. Chem.* **43**, 2357 (1978).

124. M. Imuta and H. Ziffer, *J. Org. Chem.* **43**, 4540 (1978).

125. Y. Tsuda, K.-I. Kawai and S. Nakajima, *Chem. Pharm. Bull.* **33**, 1955 (1985).

126. Y. Yamazaki and H. Maeda, *Agric. Biol. Chem.* **50**, 2621 (1986).

127. D. Buisson, S. El Baba and R. Azerad, *Tetrahedron Lett.* **27**, 4453 (1986). J. Konishi, H. Ohta and G. Tsuchihashi, *Chem. Lett.*, 1111 (1985). M. Imuta and H. Ziffer, *J. Org. Chem.* **43**, 3319 (1978).

128. M. Imuta and H. Ziffer, *J. Org. Chem.* **43**, 3530 (1978). M. R. Uskokovic, R. L. Lewis, J. J. Partridge, C. W. Despreaux and D. L. Pruess, *J. Am. Chem. Soc.* **101**, 6742 (1979).

129. J. Favero, J.-C. Mani and F. Winternitz, *Bioorg. Chem.* **10**, 75 (1981).

130. D. Gröger, P. Schmader and H. Frömmel, *Ger. Offen.* 1, 543, 691 (1966).

131. S. Akamatsu, *Biochem. Z.* **142**, 188 (1923).

132. C. Neuberg, *Biochim. Biophys. Acta* **4**, 170 (1950).

133. W. P. Schneider and H. C. Murray, *J. Org. Chem.* **38**, 397 (1973).

134. M. Miyano, C. R. Dorn, F. B. Cotton and J. Marsheck, *J. Chem. Soc. Chem. Commun.*, 425 (1971).
135. A comprehensive account of 9-, 11- and 15-keto-PG reductions is given by A. Kergomard in [84], pp. 183–193.
136. R. F. Newton, J. Paton, D. P. Reynolds, S. Young and S. M. Roberts, *J. Chem. Soc. Chem. Commun.*, 908 (1979).
137. M. J. Dawson, G. C. Lawrence, G. Lilley, M. Todd, D. Noble, S. M. Green, S. M. Roberts, T. W. Wallace, R. F. Newton, M. C. Carter, P. Hallett, J. Paton, D. P. Reynolds and S. Young, *J. Chem. Soc. Perkin Trans. I*, 2119 (1983). R. F. Newton and S. M. Roberts, *Tetrahedron* **36**, 2163 (1980).
138. G. Lowe and S. Swain, *J. Chem. Soc. Perkin Trans. I*, 391 (1985).
139. G. Hoffmann and R. Wiartalla, *Tetrahedron Lett.* **23**, 3887 (1982).
140. G. Neef, K. Petzoldt, H. Wieglepp and R. Weichert, *Tetrahedron Lett.* **26**, 5033 (1985).
141. I. Bertini, M. Gerber, G. Lanini, C. Luchinat, W. Maret, S. Rawer and M. Zeppezauer, *J. Am. Chem. Soc.* **106**, 1826 (1984). W. Maret, A. K. Shiemke, W. D. Wheeler, T. M. Loehr and J. Sanders-Loehr, *J. Am. Chem. Soc.* **108**, 6351 (1986).
142. W. N. Lipscomb, *Ann. Rev. Biochem.* **52**, 11 (1983).
143. J. P. Klinman, *Crit. Rev. Biochem.* 39, (1981). J. Kuassman and G. Petterson, *Eur. J. Biochem.* **103**, 557, 565 (1980).
144. H. Eklund, B. Nordstrom, E. Zeppezauer, G. Söderlund, I. Ohlsson, T. Boiwe, B.-O. Söderberg, O. Tapia, C.-I. Bränden and A. Akeson, *J. Mol. Biol.* **102**, 27, 61 (1976). E. S. Cedergen-Zeppezauer, I. Andersson and S. Ottonello, *Biochemistry* **24**, 4000 (1985).
145. B. V. Plapp, H. Eklund and C.-I. Bränden, *J. Mol. Biol.* **122**, 23 (1978).
146. J.-F. Biellman, *Acc. Chem. Res.* **19**, 321 (1986). H. Jörnvall, H. von Bahr-Lindström and J. Jeffrey, *Eur. J. Biochem.* **140**, 17 (1984).
147. V. Prelog, *Pure Appl. Chem.* **9**, 119 (1964). *Colloq. Ges. Physiol. Chem.* **14**, 288 (1963).
148. J. B. Jones, *Tetrahedron* **42**, 3351 (1986). See ref. 1; J. B. Jones, in *Organic Synthesis and Interdisciplinary Challenge* (eds J. Streith, H. Prinzbach and G. Schill), p. 179. Blackwell, Oxford, 1985. J. B. Jones, in *Enzymes in Organic Synthesis (Ciba Foundation Symposium 111)* (eds R. Porter and S. Clark), p. 3. Ciba, London, 1985. J. A. Lepoivre, *Janssen Chim-Acta* **2**, 20 (1984). *Chem. Abs.* **103**, 122505.
149. J. B. Jones and I. J. Jakovac, *Can. J. Chem.* **60**, 19 (1982). D. R. Dodds and J. B. Jones, *J. Am. Chem. Soc.* **110**, 577 (1988).
150. C. Hansch and J. P. Bjorkroth, *J. Org. Chem.* **51**, 5461 (1986).
151. M. Nakazaki, H. Chikamatsu, K. Naemura, T. Suzuki, M. Iwasaki, Y. Sasaki and T. Fujii, *J. Org. Chem.* **46**, 2726 (1981). M. Nakazaki, H. Chikamatsu, T. Fujii, Y. Susaki and S. Ao, *J. Org. Chem.* **48**, 4337 (1983).
152. M. Nakazaki, H. Chikamatsu, K. Naemura, Y. Hirose, T. Shimizu and M. Asao, *J. Chem. Soc. Chem. Commun.* 667, 668 (1978). M. Nakazaki, H. Chikamatsu, K. Naemura and M. Asao, *J. Org. Chem.* **45**, 4432 (1980).
153. G. L. Lemière, T. A. Van Osselaer, J. A. Lepoivre and F. C. Alderweireldt, *J. Chem. Soc. Perkin Trans. II*, 1123 (1982).
154. E. Horjales and C.-I. Branden, *J. Biol. Chem.* **260**, 15445 (1985). H. Dutler and C.-I. Branden, *Bioorg. Chem.* **10**, 1 (1981).
155. B. Kaptein, L. Wang-Griffin, G. Barf and R. M. Kellog, *J. Chem. Soc. Chem. Commun.*, 1457 (1987).
156. R. G. Morgan, P. Sartori and V. Reich, *Anal. Biochem.* **138**, 196 (1984). K.-S. You, *Methods Enzymol.* **87**, 101 (1982). See also C.-H. Wong and G. M. Whitesides, *J. Am. Chem. Soc.* **105**, 5012 (1983).
157. A. M. Klibanov and P. Giannousis, *Proc. Natl. Acad. Sci. U.S.A.* **79**, 3462 (1982).

158. W. R. Bowen, S. Y. R. Pugh and N. J. D. Schomburgk, *J. Chem. Technol. Biotechnol.* **36**, 191 (1986).
159. H. Gorisch and M. Schneider, *Biotech. Bioeng.* **26**, 998 (1984).
160. J. Grunwald, B. Wirz, M. P. Scollar and A. M. Klibanov, *J. Am. Chem. Soc.* **108**, 6732 (1986).
161. J. B. Jones and H. M. Schwartz, *Can. J. Chem.* **60**, 335, 1030 (1982).
162. J. B. Jones and D. R. Dodds, *Can. J. Chem.* **57**, 2533 (1979); idem, *J. Chem. Soc., Chem. Commun.* 1080 (1982.)
163. J. Davies and J. B. Jones, *J. Am. Chem. Soc.* **101**, 5405 (1979). L. K. P. Lam, I. A. Gair and J. B. Jones, *J. Org. Chem.* **53**, 1611 (1988).
164. T. A. Van Osselaer, G. L. Lemière, J. A. Lepoivre and F. C. Alderweireldt, *J. Chem. Soc. Perkin Trans. II*, 1181 (1978).
165. G. L. Lemière, T. A. Van Osselaer, J. A. Lepoivre and F. C. Alderweireldt, *Bull. Soc. Chim. Belg.* **88**, 807 (1979).
166. T. A. Van Osselaer, G. L. Lemière J. A. Lepoivre and F. C. Alderweireldt, *Bull. Soc. Chim. Belg.* **87**, 153 (1978).
167. J. B. Jones and H. M. Schwartz, *Can. J. Chem.* **59**, 1574 (1981).
168. G. L. Lemière, T. A. Van Osselaer and F. C. Alderweireldt, *Bull. Soc. Chim. Belg.* **37**, 771 (1978).
169. T. A. Van Osselaer, G. L. Lemière, E. M. Merckx, J. A. Lepoivre and F. C. Alderweireldt, *Bull. Soc. Chim. Belg.* **87**, 799 (1978).
170. J. B. Jones and T. Takemura, *Can. J. Chem.* **62**, 77 (1984).
171. J. B. Jones and T. Takemura, *Can. J. Chem.* **60**, 2950 (1982).
172. T. A. Van Osselaer, G. L. Lemière, J. A. Lepoivre and F. C. Alderweireldt, *Bull. Soc. Chim. Belg.* **89**, 133, 389 (1980).
173. J. J. Willaert, G. L. Lemière, R. A. Dommisse, J. A. Lepoivre and F. C. Alderweireldt, *J. Chem. Res.*, (S)222, (M)2401 (1985).
174. J. Haslegrave and J. B. Jones, *J. Am. Chem. Soc.* **104**, 4666 (1982).
175. T. Takemura and J. B. Jones, *J. Org. Chem.* **48**, 791 (1983).
176. J. J. Van Luppen, J. A. Lepoivre, T. A. Van Osselaer, G. L. Lemière and F. C. Alderweireldt, *Bull. Soc. Chim. Belg.* **88**, 829 (1979). J. Van Luppen, J. Lepoivre, G. Lemière and F. Alderweireldt, *Heterocycles* **22**, 749 (1984).
177. R. W. Fries, D. P. Bohlken and B. V. Plapp, *J. Med. Chem.* **22**, 356 (1979).
178. J. J. Willaert, G. L. Lemière, R. A. Dommisse, J. A. Lepoivre and F. C. Alderweireldt, *Bull. Soc. Chim. Belg.* **93**, 139 (1984).
179. O. Abril and G. M. Whitesides, *J. Am. Chem. Soc.* **104**, 1552 (1982).
180. A. J. Irwin and J. B. Jones, *J. Am. Chem. Soc.* **98**, 8476 (1976); **99**, 556, 1625 (1977).
181. H. G. Davies, T. C. C. Gartenmann, J. Leaver, S. M. Roberts and M. K. Turner, *Tetrahedron Lett.* **27**, 1093 (1986).
182. A. J. Ganzhorn, D. W. Green, A. D. Hershey, R. M. Gould and B. V. Plapp, *J. Biol. Chem.* **262**, 3754 (1987).
183. J. Van Eys and N. O. Kaplan, *J. Am. Chem. Soc.* **79**, 2782 (1957).
184. F. A. Loewus, F. H. Westheimer and B. Vennesland, *J. Am. Chem. Soc.* **75**, 5018 (1953).
185. H. Gunther, F. Biller, M. Kellner and H. Simon, *Angew. Chem. Int. Ed. Engl.* **12**, 146 (1973).
186. E. Hochuli, K. E. Taylor and H. Dutler, *Eur. J. Biochem.* **75**, 433 (1977).
187. H. Dutler, A. Kull and R. Mislin, *Eur. J. Biochem.* **22**, 213 (1971).
188. C.-T. Hsu, N.-Y. Wang, L. H. Latimer and C. J. Sih, *J. Am. Chem. Soc.* **105**, 593 (1983).
189. D. Buisson and R. Azerad, *Tetrahedron Lett.* **25**, 6005 (1984).
190. For other methods of preparation see oxidation, Chapter 4.

191. B. L. Hirschbein and G. M. Whitesides, *J. Am. Chem. Soc.* **104**, 4458 (1982). M.-J. Kim and G. M. Whitesides, *J. Am. Chem. Soc.* **110**, 2959 (1988).
192. H. K. Chenault, M.-J. Kim, A. Akiyama, T. Miyazawa, E. S. Simon and G. M. Whitesides, *J. Org. Chem.* **52**, 2608 (1987).
193. R. DiCosimo, C.-H. Wong, L. Daniels and G. M. Whitesides, *J. Org. Chem.* **46**, 4622 (1981). C.-H. Wong, L. Daniels, W. H. Orme-Johnson and G. M. Whitesides, *J. Am. Chem. Soc.* **103**, 6227 (1981).
194. Z. Shaked, J. J. Barber and G. M. Whitesides, *J. Org. Chem.* **46**, 4100 (1981). Z. Shaked and G. M. Whitesides, *J. Am. Chem. Soc.* **102**, 7105 (1980).
195. C.-H. Wong and G. M. Whitesides, *J. Org. Chem.* **47**, 2816 (1982).
196. C. C. Liu and A. K. Chen, *Process Biochem.*, Sept./Oct., 12 (1982).
197. R. Ruppert, S. Herrman and E. Steckhan, *Tetrahedron Lett.* **28**, 6583 (1987).
198. Y. Yamazaki and H. Maeda, *Agric. Biol. Chem.* **46**, 1571 (1982).
199. J. S. Bindra and R. Bindra, *Prostaglandin Synthesis*. Academic Press, New York, 1977. S. M. Roberts and R. F. Newton, *Prostaglandins and Thromboxanes. An Introductory Text*. Butterworths, London, 1982. S. M. Roberts and F. Scheinmann, *Recent Synthetic Routes to Prostaglandins and Thromboxanes*. Academic Press, London, 1982.
200. K. Laumen and M. P. Schneider, *J. Chem. Soc. Chem. Commun.*, 1298 (1986).
201. S. M. Ali, T. V. Lee and S. M. Roberts, *Synthesis*, 155 (1977).
202. S. Butt, H. G. Davies, M. J. Dawson, G. C. Lawrence, J. Leaver, S. M. Roberts, M. K. Turner, B. J. Wakefield, W. F. Wall and J. A. Winders, *Tetrahedron Lett.* **26**, 5077 (1985).
203. R. F. Newton, D. P. Reynolds, J. Davies, P. B. Kay, S. M. Roberts and T. W. Wallace, *J. Chem. Soc. Perkin Trans. I*, 683 (1983). R. F. Newton, D. P. Reynolds, P. B. Kay, T. W. Wallace, S. M. Roberts, R. C. Glen and P. M. Rust, *J. Chem. Soc. Perkin Trans. I*, 675 (1983).
204. S. M. Roberts, in *Enzymes as Catalysts in Organic Synthesis* (ed. M. P. Schneider), p. 55. Reidel, Dordrecht, 1986. S. M. Roberts, in *Enzymes in Organic Synthesis (Ciba Foundation Symposium* 111) (eds R. Porter and S. Clark), p. 31. Ciba, London, 1985.
205. H. G. Davies, S. M. Roberts, B. J. Wakefield and J. A. Winders, *J. Chem. Soc., Chem. Commun.*, 1166 (1985).
206. S. Butt, H. G. Davies, M. J. Dawson, G. C. Lawrence, J. Leaver, S. M. Roberts, M. K. Turner, B. J. Wakefield, W. F. Wall and J. A. Winders, *J. Chem. Soc. Perkin Trans. I*, 903 (1987).
207. I. H. White and J. Jeffrey, *Biochim. Biophys. Acta* **296**, 604 (1973). J. Kawamura, T. Hayakawa and T. Tanimoto, *Chem. Pharm. Bull.* **28**, 437 (1980). T. Hayakawa, T. Tanimoto, T. Kimura and J. Kawamura, *Chem. Pharm. Bull.* **29**, 456, 476 (1981).
208. C. H. Blomquist, *Arch. Biochim. Biophys.* **159**, 590 (1973). G. Vecchio, P. Pasta, G. Mazzola and G. Carrea, *Biochim. Biophys. Acta.* **914**, 122 (1987).
209. P. Pasta, G. Mazzola and G. Carrea, *Biochemistry* **26**, 1247 (1987). S. Riva, R. Bovara, L. Zetta, P. Pasta, G. Ottolina and G. Carrea, *J. Org. Chem.* **53**, 88 (1988).
210. J. Leaver, T. C. C. Gartenmann, S. M. Roberts and M. K. Turner, in *Biocatalysis in Organic Media* (eds C. Laane, J. Tramper and M. D. Lilly), p. 411. Elsevier, Amsterdam, 1987.
211. D. R. Kelly, S. D. Bull, R. Collier, J. Leaver, J. D. Lewis, F. McKay and M. K. Turner, unpublished results.
212. R. Hilhorst, C. Laane and C. Veeger, *FEBS Lett.* **159**, 225 (1983). R. Hilhorst, R. Spruijt, C. Laane and C. Veeger, *Eur. J. Biochem.* **144**, 459 (1984).
213. M. D. Legoy, V. L. Garde, J. M. Le Moullec, F. Ergan and D. Thomas, *Biochimie* **62**, 341 (1980).
214. C. J. Sih, J. B. Heather, R. Sood, P. Price, G. Peruzzotti, L. F. Hsu Lee and S. S. Lee, *J. Am. Chem. Soc.* **97**, 865 (1975).

215. M. A. Payton, *Trends Biotechnol.* **2**, 153 (1984). T. D. Brock (ed.), *Thermophiles; General Molecular and Applied Microbiology.* Wiley, New York, 1986.
216. D. G. Drueckhammer, C. F. Barbas III, K. Nozaki, C.-H. Wong, C. Y. Wood and M. A. Ciufolini, *J. Org. Chem.* **53**, 1607 (1988).
217. R. J. Lamed, E. Keinan and J. G. Zeikus, *Enzyme Microbiol. Technol.* **3**, 144 (1981). R. J. Lamed and J. G. Zeikus, *Biochem. J.* **195**, 183 (1981); U.S. Pat. 4,352,885 (1982).
218. E. Keinan, E. K. Hafeli, K. M. Seth and R. Lamed, *J. Am. Chem. Soc.* **108**, 162 (1986).
219. K. Otsuka, S. Aono and T. Okura, *Chem. Lett.*, 2089 (1987).
220. E. Keinan, K. K. Seth and R. Lamed, *J. Am. Chem. Soc.* **108**, 3474 (1986).
221. F. Bryant and L. G. Ljungdahl, *Biochem. Biophys. Res. Commun.* **100**, 793 (1981).
222. T. Yamada, K. Kanimura, T. Endo and T. Oshima, *Chem. Lett.*, 1749 (1982).
223. F. G. Fischer and O. Wiedemann, *Annalen* **513**, 260 (1934).
224. F. G. Fischer and O. Wiedemann, *Annalen* **520**, 52 (1935).
225. For a review of this earlier work see: F. G. Fischer, in *Newer Methods of Preparative Organic Chemistry* (ed. W. Foerst), Vol. 1, pp. 159–196. Wiley-Interscience, New York, 1948.
226. F. G. Fischer and O. Wiedemann, *Annalen* **522**, 1 (1936).
227. A. Butenandt, H. Dannenberg and L. Suranyi, *Chem. Ber.* **73**, 818 (1940).
228. L. Mamoli and A. Vercellone, *Chem. Ber.* **70**, 2079 (1937). A. Fauve and A. Kergomard, *Tetrahedron* **37**, 899 (1981).
229. M. Protiva, A. Capek, J. O. Jilek, B. Kakac and M. Tadra, *Coll. Czech. Chem. Commun.* **26**, 1537 (1961).
230. K. Mori, D. M. S. Wheeler, J. O. Jilek, B. Kakac and M. Protiva, *Coll. Czech. Chem. Commun.* **30**, 2236 (1965).
231. C. Fuganti, D. Ghiringhelli and P. Grasselli, *J. Chem. Soc. Chem. Commun.*, 846 (1975). C. Fuganti, *J. Chem. Soc. Chem. Commun.*, 337 (1979).
232. P. Gramatica, B. M. Ranzi and P. Manitto, *Bioorg. Chem.* **10**, 22 (1981).
233. C. Fuganti and P. Grasselli, *J. Chem. Soc. Chem. Commun.*, 995 (1979).
234. P. Gramatica, P. Manitto, B. M. Ranzi, A. Delbianco and M. Francavilla, *Experientia* **38**, 755 (1982). P. Gramatica, P. Manitto, D. Monti and G. Speranza, *Tetrahedron* **42**, 6687 (1986).
235. P. Gramatica, P. Manitto and L. Poli, *J. Org. Chem.* **50**, 4625 (1985).
236. P. Gramatica, G. Giardina, G. Speranza and P. Manitto, *Chem. Lett.*, 1395 (1985).
237. H. G. Leuenberger, W. Boguth, R. Barner, M. Schmid and R. Zell, *Helv. Chim. Acta* **62**, 455 (1979). H. G. W. Leuenberger, W. Boguth, E. Widmer and R. Zell, *Helv. Chim. Acta* **59**, 1832 (1976).
238. P. Ferraboschi, A. Fiecchi, P. Grisenti and E. Santaniello, *J. Chem. Soc., Perkin Trans. I*, 1749 (1987).
239. M. Utaka, S. Konishi and A. Takeda, *Tetrahedron Lett.* **27**, 4737 (1986).
240. M. Utaka, S. Konishi, T. Okubo, S. Tsuboi and A. Takeda, *Tetrahedron Lett.* **28**, 1447 (1987).
241. H. Suemune, N. Hayashi, K. Funakoshi, H. Akita, T. Oishi and K. Sakai, *Chem. Pharm. Bull.* **33**, 2168 (1985).
242. H. Ohta, K. Ozaki and G.-I. Tsuchihashi, *Chem. Lett.*, 191 (1987).
243. T. Kitazume and N. Ishikawa, *Chem. Lett.*, 237 (1983).
244. T. Kitazume and N. Ishikawa, *Chem. Lett.*, 587 (1984).
245. J. Bader, H. Gunther, B. Rambeck and H. Simon, *Hoppe-Seyler's Z. Physiol. Chem.* **359**, 19 (1978) and references therein.
246. W. Tischer, J. Bader and H. Simon, *Eur. J. Biochem.* **97**, 103 (1979). For a simplified method of preparation of the immobilized isolated enzyme see [248].

247. H. Simon and H. Gunther, *Biomimetic Chemistry, Proceedings of the 2nd Kyoto Conference on Newer Aspects of Organic Chemistry*, pp. 207–227. Kodansha, Tokyo, 1983.
248. W. Tischer, W. Tiemeyer and H. Simon, *Biochimie* **62**, 331 (1980)
249. J. Bader and H. Simon, *Arch. Microbiol.* **127**, 279 (1980). For a review on hydrogenase enzymes see: A. I. Krasna, *Enzyme Microbiol. Technol.* **1**, 165 (1979).
250. H. Simon, H. Gunther, J. Bader and W Tischer, *Angew. Chem. Int. Ed. Engl.* **20**, 861 (1981). H. Simon, H. Gunther and J. Thanos, in *Enzymes as Catalysts in Organic Synthesis* (ed. M. P. Schneider), pp. 35–44. Reidel, Dordrecht, 1986. A full description of the apparatus necessary for this reaction is given in [247].
251. M. Buhler and H. Simon, *Hoppe-Seyler's Z. Physiol. Chem.* **363**, 609 (1982).
252. H. Giesel and H. Simon, *Arch. Microbiol.* **135**, 51 (1983).
253. H. Hashimoto, B. Rambeck, H. Gunther, A. Mannschrek and H. Simon, *Hoppe-Seyler's Z. Physiol. Chem.* **356**, 1203 (1975). K. Bartl, C. Cavalar, T. Krebs, E. Ripp, J. Retey, W. E. Hull, H. Gunther and H. Simon, *Eur. J. Biochem.* **72**, 247 (1977).
254. R. Leinberger, W. E. Hull, H. Simon and J. Retey, *Eur. J. Biochem.* **117**, 311 (1981).
255. See [247], pp. 214–215.
256. H. Giesel, G. Machacek, J. Bayeri and H. Simon, *FEBS Lett.* **123**, 107 (1981).
257. Quoted in H. Simon, J. Bader, H. Gunther, S. Neumann and J. Thanos, *Angew. Chem. Int. Ed. Engl.* **24**, 539 (1985).
258. A. R. Battersby, A. I. Gutman, C. J. R. Fookes, H. Gunther and H. Simon, *J. Chem. Soc. Chem. Commun.*, 645 (1981).
259. H. Hashimoto and H. Simon, *Angew. Chem. Int. Ed. Engl.* **14**, 106 (1975).
260. H. Sedlmaier, W. Tischer, P. Rauschenbach and H. Simon, *FEBS Lett.* **100**, 129 (1979).
261. J. Bader, M.-A. Kim and H. Simon, *Hoppe-Seyler's Z. Physiol. Chem.* **362**, 809 (1981).
262. L. Angermaier, J. Bader and H. Simon, *Hoppe-Seyler's Z. Physiol. Chem.* **362**, 33 (1981).
263. H. Simon, *Nachr. Chem. Technol.* **23**, 66 (1975).
264. E. Krezdorn, S. Hocherl and H. Simon, *Hoppe-Seyler's Z. Physiol. Chem.* **358**, 945 (1977).
265. H. Giesel, G. Machacek, J. Bayerl and H. Simon, *FEBS Lett.* **123**, 107 (1981).
266. A. Kergomard, M. F. Renard and H. Veschambre, *Tetrahedron Lett.* 5197 (1978).
267. A. Kergomard, M. F. Renard and H. Veschambre, *J. Org. Chem.* **47**, 792 (1982).
268. M. Desrut, A. Kergomard, M. F. Renard and H. Veschambre, *Tetrahedron* **37**, 3825 (1981).
269. G. Dauphin, J. C. Gramain, A. Kergomard, M. F. Renard and H. Veschambre, *Tetrahedron Lett.* **21**, 4275 (1980). Reduction of 2-[2]H-cyclohexenone was reported by J. M. Schwab, *J. Am. Chem. Soc.* **103**, 1876 (1981).
270. G. Dauphin, J. C. Gramain, A. Kergomard, M. F. Renard and H. Veschambre, *J. Chem. Soc. Chem. Commun.*, 318 (1980).
271. J. W. Simek, D. L. Mattern and C. Djerassi, *Tetrahedron Lett.* 3671 (1975).
272. J.-C. Gramain, A. Kergomard, M. F. Renard and H. Veschambre, *J. Org. Chem.* **50**, 120 (1985).
273. T. L. Miller and E. J. Hessler, *Biochim. Biophys. Acta* **202**, 354 (1970).
274. M. Desrut, A. Kergomard, M. F. Renard and H. Veschambre, *Biochem. Biophys. Res. Commun.* **110**, 908 (1983).
275. A. Kergomard, M. F. Renard and H. Veschambre, *Agric. Biol. Chem.* **46**, 97 (1982).
276. H. Ohta, J. Konishi and G.-I. Tsuchihashi, *Agric. Biol. Chem.* **49**, 665 (1985).
277. M. Miyano, C. R. Dorn, F. B. Colton and W. J. Marsheck, *J. Chem. Soc. Chem. Commun.*, 425 (1971).
278. K. Sakai, A. Nakazawa, K. Kondo and H. Ohta, *Agric. Biol. Chem.* **49**, 2331 (1985).

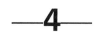

—4—

Oxidation Reactions

4.1. THE OXIDATION OF ALCOHOLS AND ALDEHYDES

4.1.1. Oxidation of Alcohols

Compared with the reduction of ketones, there have been comparatively few reports of the enzymic oxidations of alcohols and even fewer studies involving the conversion of aldehydes into carboxylic acids [1]. In general, the oxidation of alcohols is thermodynamically unfavourable and many oxidations are inhibited by the products formed (e.g. acetaldehyde inhibits yeast alcohol dehydrogenase). This is because a ketone or aldehyde is often bound better by a hydrophobic active site than is an alcohol [2]. Enzymic oxidations work best at high pH (8–9) and this may be unsuitable for some substrates and cofactor reoxidants.

In synthetic terms, oxidation of an alcohol may result in loss of a valuable centre of chirality. The majority of reported examples involve either the kinetic resolution of racemic secondary alcohols by enantioselective oxidation or the oxidation of prochiral diols to chiral lactols and lactones. It must be stressed that any of the enzymes used for reductions can, in principle, be used for oxidations and that the name of an enzyme is not necessarily a guide to its suitability for oxidation over reduction or *vice versa*.

4.1.1.1. Kinetic Resolution of Chiral Alcohols by Enantioselective
Oxidations and Related Reactions

Of the organisms that can effect oxidation of organic alcohols, there has been much interest in methylotrophs; that is microbes that can grow exclusively on methanol [3]. Many of these organisms produce useful dehydrogenases that are specific for primary alcohols [4], benzylic alcohols [5], D-series secondary alcohols [6], methyl carbinols [7] and butan-2,3-diol [8]. A dehydrogenase highly selective for L-methyl carbinols was isolated from *Comamonas terrigena*, a soil bacterium [9].

The detergent degrading bacterium *Pseudomonas* C12B normally selectively oxidizes D alkan-2-ols. However, when it is grown on a medium containing secondary alcohols, an L alkan-2-ol dehydrogenase and a dehydrogenase selective for symmetrical ketones are induced [10]. A methanol dehydrogenase showing high activity for 2-chloroethanol oxidation was induced in *Xanthobacter autotrophicus* GJ10 [11].

Corynebacterium equi IFO 3730 oxidizes (S)-(phenylsulphenyl)-propan-2-ol to the corresponding ketone (1) at pH 8. The (R)-configuration alcohol is unaffected. When the corresponding sulphoxide was subjected to these conditions only the (R) hydroxysulphone was isolated [12]. At pH 6.5 the

MeCOCH$_2$SPh

(1)

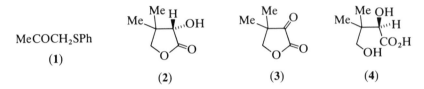

(2)   (3)   (4)

reduction process predominates. Washed cells of *Rhodococcus erythropilis* IFO 12540 oxidize L-pantoyl lactone (2) to the α-keto lactone (3) which hydrolyses and is then reduced to give D-pantoic acid (4) which can be isolated [13].

Yeast alcohol dehydrogenase (YAD) [1,14] oxidizes aliphatic primary alcohols, secondary alcohols of the L series [(S) configuration] and D-lactic acid [15]. The fastest rate is observed with ethanol [16] from which the pro-(R) proton is removed [17]. The rate of oxidation is much slower for secondary alcohols, and for a series of methyl n-alkyl carbinols the rate decreases as the chain length is increased [15]. The enzyme shows little or no activity with alicyclic secondary alcohols.

There are several enzymes available that catalyse the oxidation of specific hydroxyl groups in polyols. These are usually derived from sugar or glycerol metabolism. Glycerol dehydrogenase is specific for (R) alcohols and oxidizes (R)-butan-1,2-diol to 1-hydroxybutan-2-one. (S)-Butan-1,2-diol can be recovered unchanged, although it has poor optical purity (e.e. = 36%). Similarly, *cis*-cyclohexane 1,2-diol is oxidized to (S)-α-hydroxy-cyclohexanone (*vide infra*) [18]. HLAD catalyses the oxidation of the pro-(S) hydroxyl group of glycerol to give (S)-glyceraldehyde [19]. The dihydroxyacetone reductase from *Mucor javanicus* [1] is highly selective for (R) alcohols [20], whereas 20β-hydroxysteroid dehydrogenase (E.C. 1.1.1.53) (which oxidizes a similar grouping in 20-dihydrocortisone) is (S) selective [21].

D-Galactose oxidase (E.C. 1.1.3.9) catalyses the oxidation of the terminal hydroxyl group of galactose to an aldehyde. Polyols that have the same absolute configuration at the two adjacent hydroxyl positions are also oxidized: for example, xylitol (5) is converted into L-xylose (6) [22]. Glycerol is oxidized to (S)-glyceraldehyde, and (R)-3-chloro- and (R)-3-bromo-propane-1,2-diols are oxidized more quickly than their enantiomers [23]. L-Fucitol (6-desoxy-L-galactitol) is oxidized by *Acetobacter suboxydans* to L-fuco-4-ketose (7) [24]. Similar selectivity may be possible with the two widely available enzymes glucose oxidase (E.C. 1.1.3.4) and glucose dehydrogenase (E.C. 1.1.1.47).

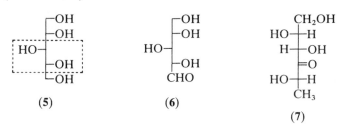

(5)   (6)   (7)

   Some enzymes are capable of catalysing more than one type of reaction. For example, the hydrocarbon monooxygenase system from *Pseudomonas oleovorans* is capable of hydroxylation of non-activated carbon centres; it also epoxidizes double bonds (see Section 4.5) and catalyses the oxidation of primary alcohols to aldehydes [25]. Chloroperoxidase from *Caldariomyces fumago* catalyses the oxidation of allylic, benzylic and propargylic alcohols to unsaturated aldehydes with a selectivity reminiscent of manganese dioxide [26]. Cultured cells of *Nicotiana tabacum* which are able to hydroxylate unfunctionalized carbon centres also contain an alcohol dehydrogenase similar to HLAD which enantioselectively oxidizes several cyclic terpenes [27]. Horse liver alcohol dehydrogenase (HLAD) oxidizes allylic alcohols without affecting the double bond [28], whereas cell-free extracts of the alkene-epoxidizing micro-organism *Pseudomonas oleovorans* yield saturated ketones [29].

   In addition, HLAD oxidizes a wide range of 1,2-diols and α-amino primary alcohols to L hydroxy (or L amino) aldehydes [30]. YAD shows similar activity with 1,2-diols, but does not oxidize α-amino primary alcohols [30,31]. HLAD displays a much lower enantioselectivity than YAD in the oxidation of aliphatic secondary alcohols [6], and is much better suited for the oxidation of alicyclic secondary alcohols. Indeed, cyclohexanol is often used as a standard for evaluating new reaction conditions [32,33], and NAD^+ regenerating systems [33,34].

   Regeneration of the requisite cofactor NAD^+ is, in fact, the major difficulty when using enzymes such as HLAD in the oxidative mode. Jones and Taylor [32] introduced a flavin mononucleotide (FMN) system in which dioxygen is the ultimate oxidant (in the ensuing process dioxygen is reduced to hydrogen peroxide). The rate of transfer of hydrogen from NAD(P)H to FMN was improved by adding FMN reductase from *Photobacterium fisherei* (ATCC 7744) and the hydrogen peroxide produced was reduced with catalase [35]. Up to 50-fold regeneration of NAD^+ was possible with the original system. The catalysed system is approximately 25-fold faster*. Regeneration rates of 500–1000 times have been achieved with 2-oxoglutarate and glutamate dehydrogenase [33] or with acetaldehyde as a coupled oxidant [2]. Regeneration rates of up to 3000 times are possible with electrochemical regeneration [36] and a system using diaphorase is stable for at least 20 days [37].

The rate of oxidation of alicyclic secondary alcohols by HLAD is limited by the rate of hydride transfer, whereas for simple aliphatic alcohols the NAD^+ dissociation from the enzyme is rate limiting [38]. This rate difference enables the oxidation of primary alcohols in the presence of secondary alcohols. The racemic cyclopentanediol (8) was oxidized to the lactol (9) and then the lactone (10) (49% optical purity); the recovered starting material was of 23% optical purity[†]. The enantioselectivity of the second oxidation has been shown to be the same as that of the first by subjecting racemic lactol to the same conditions [39]. Thus, the enantiomer (9) was preferentially oxidized. Similar results were obtained with cyclohexane derivatives [40], but, in general, the enantioselectivity in these oxidations is very low: in several cases, e.g. (11) and (12), essentially racemic products were obtained [39,40].

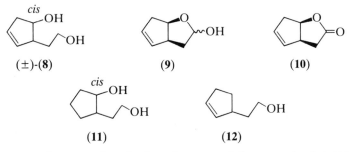

In contrast, the secondary hydroxyl group of the racemic diol (13) was oxidized in preference to the primary one to give the hydroxy ketone (14) and the lactone (15) as well as recovered starting material (Scheme 4.1). The enantioselectivity was predicted using the diamond lattice model [39], which also predicted the oxidation of (1S)-*exo*-bicyclo[2.2.1]heptanol (16) in preference to its enantiomer [38,39].

* The rate of regeneration is as important as the fidelity of the NADH → NAD^+ transformation, e.g. if this reaction is only 99% efficient, after 100 cycles only 37% of the NAD^+ will remain. Efficiencies of 99.97%, 99.99% and 99.999% are required for similar recoveries after 3000, 10 000 and 100 000 cycles, respectively [36].

† In most cases the optical purity may be assumed to be equivalent to the enantiomeric excess.

As stated in other sections, there has been considerable interest in enzymes from thermophilic bacteria [41]. The alcohol dehydrogenase from *Thermoanaerobium brockii* [42], which is selective for the oxidation of secondary alcohols over primary alcohols [43], has been investigated in some detail [42–44]. In the reduction mode, short-chain ketones gave (*R*)-configuration alcohols, whereas long-chain ketones gave (*S*)-configuration alcohols; a similar selectivity should be exhibited in the oxidation mode. Regeneration of the cofactor NADP$^+$ is still problematic and a system is not yet available for routine use by organic chemists. The alcohol dehydrogenase from *Thermoanaerobacter ethanolicus* is also selective for the oxidation of secondary alcohols, but the enantioselectivity has not been investigated [45].

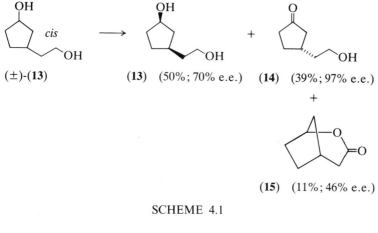

SCHEME 4.1

(16)

There is a huge range of dehydrogenases available for selective oxidation of hydroxyl groups in steroidal polyols. 3- [46–48], 7- [46,47], 11-, 12- [47,49], 17- and 20- [50] dehydrogenases [51] are readily available with a choice of α/β selectivity [52].

4.1.1.2. The Oxidation of Prochiral Diols

The selective modification of one group in a molecule containing a prochiral centre (17) or having *meso* symmetry (18) will result in a chiral product (19) or (20) (Scheme 4.2). In contrast to a kinetic resolution, all of the material

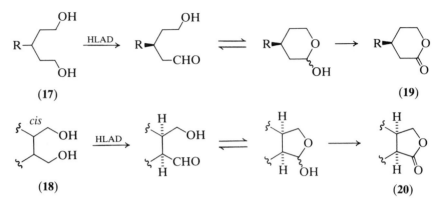

(17)   **(19)**

(18)   **(20)**

SCHEME 4.2

may be utilized and the operational difficulties involved in stopping a reaction at 50% conversion are avoided. These advantages have stimulated a number of investigations concerned with the oxidation of diols to lactones by HLAD. In virtually all cases, it is the pro-(*S*) hydroxyl group that is oxidized, irrespective of the other groups present. Thus 3-substituted acyclic 1,5-diols (**17**) are oxidized with high selectivity to the lactones (**19**) when small groups such as methyl, ethyl and propyl are present at the 3-position. However, as the size of the 3-substituent is increased, the rate of reaction and the optical purities decrease. The lactol/lactone mixtures which are isolated can be oxidized to the lactone with silver oxide [53]. The optical purities of these lactols are less than those of the lactones and so the selectivities of the two steps are directed towards the same enantiomeric series [i.e. pro-(*S*), then (*S*)] [54]. Larger scale preparative reactions can be run reasonably efficiently with the FMN-reductase NAD$^+$ recycling system [35].

Some *Gluconobacter* spp. oxidize the pro-(*R*) hydroxyl group of 3-substituted acyclic-1,5-diols (**17**) to provide the (*R*) lactones. 2-Substituted 1,3-diols gave the corresponding (*R*) acids, such as (*R*)-β-hydroxyisobutyric acid (**21**) (e.e. = 83%) [55].

(21)

An important potential application of enantiospecific alicyclic diol oxidations is the provision of the valuable natural product (*R*)-mevalonolactone (**23**) from the diol (**22**, R = H). Micro-organisms such as *Gluconobacter* sp. [55] and *Flavobacterium oxydans* [56] oxidize the diol to give the unnatural (*S*) enantiomer [the natural (*R*) enantiomer may be metabolized].

HLAD also gives the (S) enantiomer of low optical purity, so, at present, hydrolysis of the prochiral ester (22, R = COMe) is a more efficient process [56]. As illustrated in Scheme 4.2, *meso*-diols ($_{18}$) are usually oxidized with pro-[(S),(S)] selectivity to give the lactones (20). However, oxidation of the diols (24) and (27) afforded the lactones (25) and (28) and lactols (26) and (29) all showing very high optical purity. The initial oxidation is non-selective, while the lactol oxidation is highly (S) selective. The result was confirmed by treating the lactol [(29) and its enantiomer] with HLAD. The same mixture of lactone (28) and lactol (29) was isolated [57].

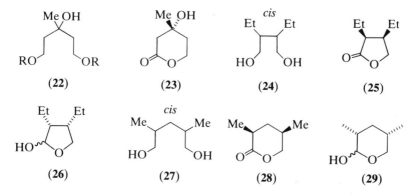

(22)   (23)   (24)   (25)

(26)   (27)   (28)   (29)

Preparation 4.1   Preparation of (+)-(1R,6S)-8-oxabicyclo[4.3.0]nonan-7-one (31; n = 4)

In a 1-l Erlenmeyer flask were placed 475 ml of distilled water and 3.75 g (0.05 mol) of reagent-grade glycine, and the pH was adjusted to 9 by the careful addition of aqueous 10% sodium hydroxide. In the buffer solution thus obtained were dissolved 2.00 g (13.87 mmol) of *cis*-1,2-bis(hydroxymethyl)cyclohexane, 0.58 g (0.852 mmol) of β-NAD, and 7.8 g (16.2 mmol) of FMN. To the clear orange solution obtained was added 80 units of horse liver alcohol dehydrogenase (HLAD). The solution was gently stirred for 1 min, the pH was readjusted to 9 and the mixture was kept at room temperature with the mouth of the flask loosely covered by a watchglass. After a few minutes the colour of the solution began to darken and after several hours became an opaque green-brown. The pH was readjusted to 9 after 6, 12, 24, 48 and 72 h by the careful addition of aqueous 10% sodium hydroxide (the pH of the mixture decreases progressively as the reaction proceeds). After 4 days the mixture was brought to a pH of ca. 13.3 by the addition of 20 ml of aqueous 50% sodium hydroxide solution. After 1 h the mixture was continuously extracted with chloroform for 10 h. The chloroform extract was discarded. The aqueous layer was acidified to pH 3 with concentrated hydrochloric acid and again extracted continuously for 15 h with chloroform. To the green-orange solution were added charcoal (0.5 g) and magnesium sulphate. The dried and partially decolourized mixture was filtered through a bed of Celite, and the chloroform was removed under reduced pressure using a rotatory evaporator. The residual orange-green oil was distilled in a Kugelrohr to give 1.4–1.5 g (72–77% yield, (+)-(1R,6S)-8-oxabicyclo[4.3.0]nonan-7-one (e.e. >97%), as a colourless oil, b.p. 85–100°C (0.1–0.05 mm), m.p. 26–29°C, [α]$_D^{22}$ + 51.3° (CHCl$_3$, c = 1.1).

The oxidation of *cis* cyclic primary diols (**30**, $n = 1$–4) by HLAD is very fast and is completely enantioselective for oxidation of the pro-(S) hydroxyl group to give the corresponding lactones (**31**, $n = 1$–4) (e.e. = 100%). In contrast, kinetic resolution by oxidation (to 50% conversion) of the corresponding *trans* compounds gave essentially racemic products [60]. The cyclobutyl lactone (**31**, $n = 2$) was used in a synthesis of grandisol [58] and the cyclopentyl lactone (**31**, $n = 3$) was used in a synthesis of multifidene and viridiene [59]. The oxidation of the cyclohexanediol (**30**, $n = 4$) is particularly interesting because it is not a *meso* compound but an equilibrating pair of enantiomers, so, strictly, the oxidation is a second-order kinetic resolution [60]. Nevertheless, pro-(S) selectivity is observed as usual. This is a highly reproducible process and is described in *Organic Syntheses* [61]. The biotransformation has been further improved by effecting cofactor regeneration using glutamate dehydrogenase [33,62]. Similarly, the series of diols (**32**) [63] and (**34**) [64] (as well as several closely related analogues) gave excellent yields of the corresponding lactones (**33**) and (**35**) with complete enantioselectivity.

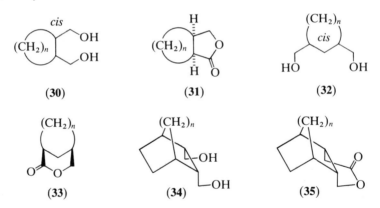

Cyclic 1,2-diols are not oxidized by HLAD, but it is noteworthy that cyclohexane-1,2-diol has been converted into (2S)-hydroxycyclohexanone using immobilized glycerol dehydrogenase [18,65].

Preparation 4.2 Preparation of (2S)-hydroxycyclohexanone by oxidation of 1,2-cyclohexanediol

A 500-ml round-bottomed flask was charged with 1,2-cyclohexanediol (1 g, 8.6 mmol), NAD (0.1 g, 0.13 mmol), methylene blue (0.035 g, 0.09 mmol), pH 9.0 glycine buffer (80 ml), and immobilized GDH (*Cellulomonas* sp. 145 U) and diaphorase (140 U). The reaction mixture was stirred while open to the atmosphere; the colour was dark blue. The reaction progress was followed by GLC analysis. After three days, the reaction appeared to be 30% complete and no further progress was observed. The immobilized enzymes were separated by centrifugation. The recovered activities of the enzymes were 83% for GDH and 58% for diaphorase. The aqueous solution was extracted with two 200-ml portions of methylene chloride. The combined organic extracts were concentrated, redissolved in ether (20 ml), filtered through charcoal and Celite, and concentrated to give the title compound as a colourless oil (0.2 g), which crystallized upon standing: $[\alpha]_D$ -0.07 ($c = 0.53$).

4.1.2. Oxidation of Aldehydes and Carboxylic Acids

Many alcohol dehydrogenases are also able to oxidize aldehydes; however, there are several fairly specific aldehyde dehydrogenases (E.C. 1.2.X.Y) which utilize $NAD(P)^+$ as cofactor. It is these enzymes which are responsible for alcohol tolerance in man [66]. The most readily and commercially available enzyme of this type is yeast aldehyde dehydrogenase (E.C. 1.2.1.5) [67] which is activated by potassium ions, stabilized by polyhydric alcohols [68], and inhibited by cyclopropanone hydrate [66]. Procedures have been developed for the purification of the enzyme from bovine [69], rat [70] and mouse [66] liver *inter alia*, and for immobilization of the protein [71,72]. The yeast enzyme has been mostly used for the oxidation of simple aliphatic [67–69] and aromatic [70] aldehydes. In combination with HLAD it oxidizes 1,2-diols, α-amino primary alcohols, and L α-hydroxy and L α-amino acids [30,31]. The same system has been used for the regeneration of NADH [72]. Yeast and liver aldehyde dehydrogenases oxidize aromatic aldehydes [70,73]. The rate of oxidation decreases in the presence of a large *ortho*-halo substituent, but the reaction is not affected by a *para*-halo substituent [70].

Formate dehydrogenase (FDH) catalyses the oxidation of formate to carbon dioxide. Several forms of the enzyme exist, most of which are sensitive to inhibition by dioxygen. Some of these isoenzymes use NAD^+ [74] as a hydrogen acceptor, whilst others use dyes such as benzylviologen [75] and FMN. Two dye-linked forms were isolated from *Methanococcus vannielii* [76], one of which contained molybdenum and iron, whilst the other contained molybdenum or tungsten iron–sulphur complexes [77] and selenium as selenocysteine [78]. The gene for FDH in *Escherichia coli* has been identified and cloned [79] and methods for large-scale extraction from *Candida biodinii* have been developed [80].

The major interest in FDH has been for the regeneration of NADH from NAD^+. It can be used alone with formate [81] or in combination with YAD and aldehyde dehydrogenase with methanol [72].

Pilot-scale reduction [82] and reductive animation [83] of 2-ketoacids was achieved using FDH for the recycling of polyethylene glycol linked NADH retained by an ultrafiltration membrane. FDH is also active in the presence of polymeric pyridinium salts [84].

Resting cells of *Proteus mirabilis* catalyse the disproportionation of α-ketoacids to (2R) hydroxy acids, the noracid and CO_2 (Scheme 4.3) [85]. The presence of hydrogen and/or formate suppresses the oxidative decarboxylation and the (2R) hydroxy compound becomes the major product [86].

$$2\,R\overset{O}{\underset{}{\bigwedge}}CO_2^{\ominus} + H_2O \quad\xrightarrow{\textit{P. mirabilis}}\quad RCO_2^{\ominus} + \;\underset{R}{\overset{H\;\;OH}{\bigwedge}}CO_2^{\ominus} + CO_2$$

SCHEME 4.3

4.2. BAEYER–VILLIGER TYPE OXIDATION OF A CARBONYL COMPOUND

Enzyme catalysed Baeyer–Villiger type reactions were first demonstrated in 1953, when *Fusarium*, *Penicillium*, *Cylindrocarpon*, *Aspergillus* and *Gliocladium* species were shown to cleave off steroidal C-17 side chains and insert an oxygen atom into the D ring [87]. In a study of the *Penicillium lilacinum* catalysed conversion of testosterone (**36**) into testololactone (**37**), Prairie and Talalay [88] established the involvement of two enzymes in the transformation. Initially a dehydrogenase converted testosterone into 4-androsterone-3,17-dione and, in a separate slower step, this was converted into testololactone by an NADH-dependent enzyme requiring the presence of oxygen. The requirement for molecular oxygen was demonstrated by $^{18}O_2$-labelling studies, the $^{18}O_2$ being incorporated as the ring oxygen atom of testololactone. The cleavage of the 17-acetyl group of progesterone by an oxygenase from *Cylindrocarpon radicola* was studied by Rahim and Sih [89] and shown to involve four steps, first a Baeyer–Villiger type cleavage to give testosterone acetate, secondly esterase hydrolysis, thirdly oxidation to the 3,17-dione, and finally Baeyer–Villiger cleavage to afford testololactone. Both of the Baeyer–Villiger type reactions were shown to be catalysed by the same mono-oxygenase enzyme. The A ring of some steroids can also be cleaved in Baeyer–Villiger style using the micro-organism *Glomerella fusarioides* [90].

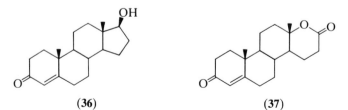

(36)   (37)

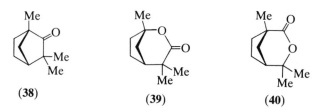

(38)   (39)   (40)

Similar transformations have been reported for non-steroidal substrates. Aliphatic methyl ketones, such as tridecan-2-one, are converted into the corresponding alcohol via the acetate by *Penicillium multivorans* and *Pseudomonas aeruginosa* [91], and studies on simple alicyclic ketones with *Pseudomonas* species and a cyclohexanone oxygenase from *Acinetobacter* have been reported [92]. These studies have shown that, with unsymmetrical ketones, the group which migrates generally follows the sequence of tertiary > secondary > primary carbon atom as is seen in the classical Baeyer–Villiger reaction. The monocyclic terpene fenchone (38), upon oxidation with *Corynebacterium* sp., strain T1, gives a mixture of 1,2- and 3,4-fencholides (39) and (40) in a ratio of 9 : 1 [93]. In contrast, peracetic acid oxidation of fenchone gives a 4 : 6 ratio of the lactones (39) and (40). Peracetic acid oxidation of camphor gives the expected lactone [94], but microbial oxidation with *Pseudomonas* species follows a different course (Scheme 4.4) [95].

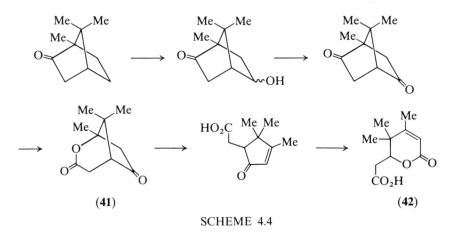

(41)   (42)

SCHEME 4.4

Initial hydroxylation at C-5 is followed by oxidation and subsequent Baeyer–Villiger reaction to give the keto-lactone (41). This compound undergoes lactone-ring fission, dehydration, and a further Baeyer–Villiger reaction to give the lactone acid (42). Transformations of the homologous series of cyclopentylketones (43) with *Acremonium roseum* have been studied by Nespiak and co-workers [96]. For the methyl and ethyl ketones (**43**, R = CH$_3$ or C$_2$H$_5$), the hydrolysed Baeyer–Villiger product (**44**) was the only compound isolated after two days' incubation; longer reaction times merely resulted in further oxidation of this alcohol to the corresponding acid (**45**). The n-propyl and i-propyl ketones (**43**, R = n- or i-C$_3$ H$_7$), in contrast, gave no more than traces of (**44**) and (**45**), but gave low yields of the corresponding allylic alcohol (**46**). The reaction path changed yet again with the n-butyl ketone (**43**, R = n-C$_4$H$_9$) which gave a 5:1 mixture of Baeyer–Villiger and allylic oxidation products. The Baeyer–Villiger reaction in each case was completely regioselective. Retention of configuration in an enzymic Baeyer–Villiger reaction was demonstrated by Schwab who studied the oxidation of (2R)-[2-^{2}H]-cyclohexanone, a sterically unbiased ketone, with cyclohexanone oxygenase from *Acinetobacter* [97].

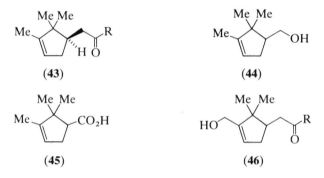

It is known that *m*-chloroperoxybenzoic acid oxidation of the ketol (**47**) gives the hydroxy-lactone (**49**), presumably by a spontaneous rearrangement of the intermediate lactone (**48**) [98]. This reaction also occurs when the (S) ketol (**47**) is subjected to oxidation with a well-aerated culture of *Curvularia*

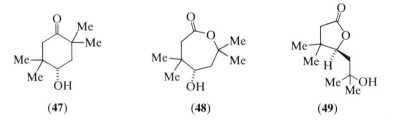

lunata if the reducing medium (glucose) is omitted [99]. Similar oxidation of a racemic mixture of the ketol (**47**) gave the lactone (**48**) and unchanged (R) ketol with an enantiomeric excess of 97%, an interesting example of remote stereoselection.

Detailed studies of the mechanism of the enzyme catalysed Baeyer–

Villiger type reaction with cyclohexanone oxygenase indicate no difference in mechanism between this and the normal peracid catalysed reaction [97,100]. The migratory aptitudes are in a similar order, the configuration of the migrating group is retained, and from the results of a number of experiments on cyclic ketones (see above), the inserted oxygen atom is known to appear in the ring. Operationally these enzyme catalysed reactions can present problems for the following reasons.

(1) Microbial mono-oxygenases tend to have short lifetimes and the preparation of the requisite enzyme systems, crude or relatively pure, requires equipment not routinely available in a chemistry laboratory.

(2) The multitude of whole-cell systems which can catalyse this reaction makes the preliminary selection of a suitable bacterium or fungus confusing for a novice in the field.

For these reasons there seems to be little advantage at the present time in using an enzyme-based system when the usual chemical procedure is simpler to operate and is usually faster. The only exception to this may be in cases where optically active lactones are required from prochiral ketones, or where an enantioselective oxidation is helpful.

4.3. HYDROXYLATION

The hydroxylation of "unactivated" centres in hydrocarbons is potentially the most useful of all biotransformations. This is because the process has so few counterparts in "traditional" organic synthesis, although parallels can be drawn with the Barton reaction of steroidal 20-nitrites [101] or Breslow's remote functionalization templates [102]. In some cases the oxidation of methine carbons with ozone [103] and peracids [104] has a similar regioselectivity to that obtained with micro-organisms.

General texts [105–109] and reviews [110,111] have appeared on the subject and compilations of data of the hydroxylation of steroids [112–117], alkaloids [116–119], terpenes [120], and hydrocarbons [121,122] have been published.

The majority of hydroxylations [123] are performed with whole micro-organisms [121–126] because the enzymes involved are usually membrane bound and hence cannot easily be purified. However, two representative free enzymes, cytochrome $P450_{cam}$ (Cyt-$P450_{cam}$*, E.C. 1.14.15.1) [127–131] and horseradish peroxidase (HRP, E.C. 1.11.1.7), *are* readily available. Both are haemoproteins (similar to myoglobin or haemoglobin) and both can use

---

* 450 refers to the wavelength of light (in nm) adsorbed by the reduced enzyme Fe(II)/carbon monoxide complex. Cam is a contraction of D(+)-camphor, the substrate which is converted by this enzyme into 5-*exo*-hydroxycamphor.

hydrogen peroxide as an oxygen source [132]; in addition, Cyt-P450$_{cam}$ is able to utilize dioxygen [130,131,133–141].

Cyt-P450 is a generic term for a wide range of haemoproteins capable of oxidase activity found in virtually all organisms [133,142]. The various forms are usually distinguished by a suffix subscript. Detoxification organs, such as the liver, as well as some micro-organisms, can be induced to produce individual (or several) Cyt-P450 isozymes by the presence of a particular substrate [143]. Phenobarbitol [144], β-naphthoflavone [145] and ethanol [146] induced Cyt-P450s have been purified from rabbit, rat and human livers. However, it should be emphasized that, while a Cyt-P450 may be induced by a particular substrate, most enzymes in this category are capable of oxidizing a wide range of substrates.

The most intensively investigated enzyme of this type is Cyt-P450$_{cam}$ from *Pseudomonas putida*; this enzyme is unique in that it is both soluble and crystalline. The amino acid sequence [147] and crystal structures in the presence [148] and absence of camphor [149] have been determined. The gene has been identified and expressed in *E. coli* [150].

Cyt-P450 is a mono-oxygenase, i.e. only one atom of oxygen from dioxygen is incorporated into the organic substrate [135,139,151]; the other oxygen atom is released as water. A hydrogen atom is supplied by NADPH, a process mediated through a flavoprotein, namely NADPH cytochrome P450 reductase [136,152]. If no suitable substrate is available dioxygen is reduced to two molecules of water [141,153].

A proposed mechanism [127,134] for Cyt-P450 hydroxylation is shown in Figure 4.1. The iron species is coordinated equatorially by protoporphyrin IX (haem) and axially by the sulphur of a cysteine residue. The bound hydrocarbon residue is enclosed in a deep hydrophobic pocket [127,148, 149,154] on the opposite face of the porphyrin molecule adjacent to the oxygen binding site [140,141,155]. At some stage the hydrocarbon residue is converted into a short-lived ($< 10^{-9}$ s) free radical [156]. The hydroxylation of many steroids occurs with retention of configuration [157]; in other cases stereochemical integrity is lost [153]. In addition, hydroxylation adjacent to a double bond often takes place with allylic scrambling [158].

The hydroxylation of unactivated centres is often the predominant reaction observed but, of course, the product(s) may not be stable or other processes may be preferred. Oxidation of tertiary amines [126], halogens, thioethers and phosphines gives the corresponding oxides, whereas oxidation in the α-position to such a group results in loss of the heteroatom (e.g. bromocamphor [134], ethers and esters [159] (Figure 4.2)).

Virtually any class of compound can be hydroxylated by Cyt-P450, but primary amines [160], hydrazines, terminal alkenes or acetylenes [293] and powerful alkylating reagents [162] act as "suicide inhibitors".

Overall reaction: $RH + NADPH + H^+ + O_2 \longrightarrow ROH + H_2O + NADP$

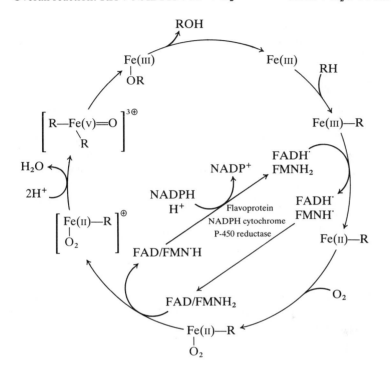

FIG. 4.1. Oxidation of compound RH to the alcohol ROH using Cyt-P450.

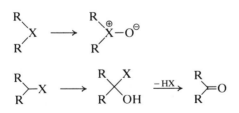

FIG. 4.2. Enzyme catalysed oxidation at, or adjacent to, a heteroatom.

Peroxidases, such as horseradish peroxidase, catalyse the oxidation [163] of a range of organic compounds and inorganic ions (e.g. iodide ion; see Section 5.4). Aromatic rings may be hydroxylated [164] but unactivated saturated carbon atoms are not. There is no crystal structure available for horseradish peroxidase (HRP), but two such structures have been published for cytochrome-C peroxidase. Functionally, cytochrome-C peroxidase is closely related to membrane-bound cytochrome oxidase, but it is also capable of similar reactions to HRP. The resting enzyme consists of two clearly defined domains between which lies a crevice occupied by a non-covalently bound high-spin ferric haem group ligated to histidine and water in the axial positions [165,167]. The water molecule is displaced by hydrogen peroxide when the latter is bound [166].

When solutions of peroxidases (normally coloured brown) are treated with hydrogen peroxide in the absence of electron donors, a bright-green form of the enzyme is produced (compound I) [167] which can be transformed sequentially to a red form (compound II) [167] and the resting enzyme by the addition of electron donors [168]. Compound I is the fully oxidized form of the enzyme and compound II has been reduced by one electron. One-electron reduction of compound II yields unoxidized enzyme. Both compounds I and II take up one oxygen atom from hydrogen peroxide, the other oxygen atom from the peroxide being released as water.

HRP requires the binding of two calcium(II) ions per molecule of enzyme for catalytic activity [169]. It is active in water/DMF mixtures down to $-65°C$ [170] and in non-polar organic solvents when precipitated on glass powder [171].

As stated previously, the majority of hydroxylations are performed with whole micro-organisms. This strategy has the disadvantage that the organisms must be maintained under sterile conditions and supplied with nutrients to prevent cell death and/or mutation. However, many fungi produce spores which can be isolated from the fungal mycelium and stored under much less rigorous conditions which, nevertheless, maintain dormancy. Thus the fungal spores provide a useful "off the shelf" alternative to micro-organisms [172].

One major problem in this whole area involving the hydroxylation of ostensibly non-activated carbon centres using whole cells or isolated enzymes is that the predictability of the point and regioselectivity of oxidation is poor. Only a few attempts have been made to rationalize experimental results in order to provide an active-site model that may be useful when considering the oxidation of a novel substrate. Not unnaturally, active-site models have emerged from studies on closely related sets of compounds. For example, Jones [125,196] explained the 12β,15α-dihydroxylation of 3- or 4-ketosteroids by *Calonectria decora* with an active-site model involving coordination of the keto-group, and oxygen incorporation at one or two sites situated 7 Å apart. The two hydroxylating sites would be separated by 4 Å. In an alternative picture, the steroid is bound consecutively on one face and then the other ("normal" or "capsized binding") to undergo sequential hydroxylation. The model nicely explains the 1β,6α-dihydroxylation of 16- and 17-ketosteroids.

Fonken and co-workers [173] proposed an active-site model for *Sporotrichum sulfurescens* (modified by Furstoss *et al.* [174]) in which hydroxylation occurs approximately 5.5 Å from the coordination site; however, one must remain circumspect about this rather crude model. Indeed the French workers indicate that hydroxylation can occur at a distance of 3.3–6.2 Å from the directing group.

In summary, it is extremely difficult to predict the likely site of oxidation for any novel substrate using micro-organism derived mono-oxygenases. In the absence of clear guidelines, there remain three main strategies for achieving regioselectivity and/or stereoselectivity in biocatalytic hydroxylation procedures: (a) variation of the culture; (b) broad screening (including mutation of micro-organisms); and (c) judicious modification of the substrate. The first two avenues require a substantial commitment by a microbiologist and the more practical approach for the novice is to find a "lead" micro-organism and then modify the substrate before fine tuning the culture conditions.

Details of the hydroxylation of various classes of substrates are given in the following Sections.

4.3.1. Hydroxylation of Steroids

The 7-hydroxylation of cholesterol by *Proactinomyces roseus* in 1948 is the earliest record of a microbial hydroxylation [175]* and at much the same time Hechter *et al.* used perfused bovine adrenal glands to transform cortexone (**50**) to corticosterone (**51**) [176]. Workers from the Upjohn Company announced that aerated *Rhizopus arrhizus* was capable of the 11α-hydroxylation of progesterone (**52**) [177] and, within a few months, the Squibb Insititute reported the same reaction with *Aspergillus niger* [178]. The 11β-hydroxy configuration is required for the optimum biological activity of hydrocortisone and the corticosteroids, but the necessary inversion of configuration of the hydroxyl group is readily achieved by oxidation and reduction. The biocatalytic process removed 10–12 steps from the previous synthesis and made hydrocortisone available for therapy at reasonable cost for the first time. 11β-Hydroxylation of Reichstein's substance S (**53**) was achieved at Pfizer using *Curvularia lunata* [179]. This transformation continues to attract attention [107] especially using immobilized cell systems [112,157,180,181]. 9α- [182] and 16α-Hydroxylation processes [112,183] are used industrially in the production of hydrocortisone analogues [107].

---

* This result has been disputed; the alternative explanation is that the substrate may simply have been oxidized by air.

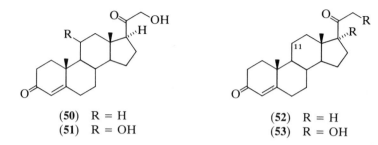

(50) R = H
(51) R = OH

(52) R = H
(53) R = OH

A similar problem was presented by ursodeoxycholic acid (54) and chenodeoxycholic acid (55) which dissolve the cholesterol deposited in gallstones [184]. It was perceived that it would be extremely useful if both compounds could be prepared by 7-hydroxylation of the more readily available lithiocholic acid (56). Selective 7β-hydroxylation was achieved using *Fusarium equiseti* after eight culture collection strains and 609 soil fungal strains were screened [185]. Other micro-organisms gave mixtures of 15-hydroxylated products [186].

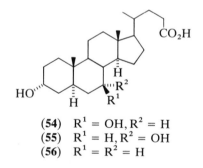

(54) R^1 = OH, R^2 = H
(55) R^1 = H, R^2 = OH
(56) R^1 = R^2 = H

Nowadays, virtually any centre in the steroid nucleus can be hydroxylated stereospecifically [112–117] (Table 4.1) using a range of micro-organisms. The immense industrial importance of this area elicited a great deal of research activity, notably from Kieslich [187–189], Jones [190–196], Holland [158, 197–201] and Crabb [202] and their co-workers. Despite this volume of excellent work there is still a considerable gulf in our knowledge between the precise understanding of the molecular structure of mono-oxygenase enzymes (such as Cyt-P450$_{cam}$) and the vast majority of biological hydroxylations catalysed by whole micro-organisms. Only about a dozen cell-free steroid hydroxylation systems have been prepared and utilized and most of these systems are no more selective than the parent micro-organisms [107].

TABLE 4.1

References to selective hydroxylation of steroids using micro-organisms

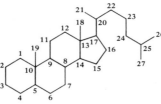

Position of hydroxylation	Stereochemistry of incoming hydroxyl group	Ref.
1	α	190
1	β	184, 187, 190, 202
2	α	190, 209
2	β	190, 197, 198
3	α	190, 199
3	β	190
4	α	190, 209
4	β	114, 115
5	α	190
6	α	190, 202, 210
6	β	158, 187, 190, 197, 198, 199, 202, 211
7	α	190, 198, 202, 222, 376
7	β	158, 185, 187, 190, 198, 202, 210, 211
9	α	158, 182, 187, 189, 190, 202, 377
10	β	190, 198 (nor-methyl)
11	α	112, 157, 158, 177, 178, 180, 182, 187, 190, 202, 210, 211, 212
11	β	112, 157, 176, 179, 180, 182, 184, 187, 202, 212, 213, 214, 377
12	α	190, 202
12	β	184, 187, 190, 202, 215
13	α	216 (Baeyer–Villiger type oxidation)

TABLE 4.1 (continued)

Position of hydroxylation	Stereochemistry of incoming hydroxyl group	Ref.
14	α	182, 188, 189, 190, 202, 222, 377
15	α	184, 187, 188, 190
15	β	184, 186, 187, 215, 217
16	α	112, 180, 183, 187, 190
16	β	190, 202
17	α	112, 187, 190, 204, 210
17	β	190, 212, 214
18		114, 115, 186
19		180, 190, 206, 218
20		203
21		200, 214, 219
24		220
25		221
32	α	205, 222

A methodology employed to prevent an unwanted hydroxylation of a steroid is to block one face of the molecule by a large atom or group which can subsequently be removed. Thus, 11β-hydroxylation of the fluorosteroid (57) by *Curvularia lunata* is complicated by concomitant 9α- and 14α-hydroxylation. This undesirable oxidation is not a problem for the bromofluorosteroid (58) because the axial 5α-bromine atom blocks the bottom face [189]. On the other hand, if the 11β-position is blocked by an 18-methyl group (i.e. by an ethyl group at C-13), 14α-hydroxylated products are then obtained [107,188]. Since many micro-organisms have dehydrogenase activity, it is often worthwhile protecting ketone groups as ketals and hydroxyl groups as acetate moieties, although the latter group may be cleaved by native esterases [159,189].

In the steroid system (and in other cases) when an alkene double bond is present at a centre which is normally hydroxylated by a given micro-organism, it will often be epoxidized [137]. Moreover, hydroxylation adjacent to carbonyl groups and in the γ-position of α,β-unsaturated ketones (e.g. at C_6 of Δ^4-3-ketosteroids) [199] may also proceed via the epoxidation of enols and dienols (Scheme 4.5) [198].

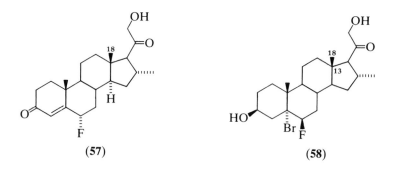

(57)   (58)

Hydroxylation of the steroid nucleus may be accompanied by loss of portions of the carbon skeleton. For example, cleavage of the side chain in cholesterol (59) yields pregnenolone (60) [203]: hydroxylation at the 17-position in progesterone (61) is followed by cleavage to afford androstenedione (62) [204]. 14-Demethylation of lanosterol [205], bile acid side-chain cleavage [184], as well as 19-demethylation of androstene dione (62) to yield the A-ring aromatic steroid estrone (63) [206], all result from Cyt-P450 hydroxylations followed by oxidative carbon–carbon bond cleavage. These transformations, of course, are all steps in "normal" steroid biosynthesis and hence they are more likely to be catalysed by mammalian enzyme preparations than by micro-organisms [114,184,207].

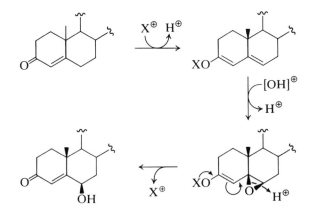

SCHEME 4.5

It should be noted that the identification of hydroxylated steroids is not a trivial undertaking; the Jones group characterized over 1000 hydroxylated steroids, mostly by ^{1}H-NMR [191,196] and IR spectroscopy [192], of the derived ketones. More recently, ^{13}C-NMR spectroscopy [201,208] has played a key role in such structure determination work.

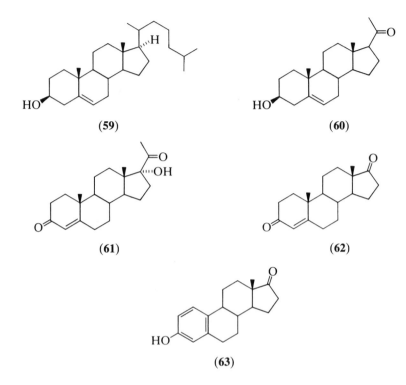

(59)   (60)

(61)   (62)

(63)

4.3.2. Hydroxylation of Alicyclic Compounds (Other than Steroids) and Saturated Heterocyclic Compounds

Hydroxylation of alicyclic compounds other than steroids has been studied extensively over the past 20 years and only the major compilations [107,108] can give full justice to the volume of the work done in this area. The majority of the mechanistic work has employed Cyt-P450$_{cam}$ from *Pseudomonas putida* which converts D-camphor (**64**) into 5-*exo*-hydroxycamphor (**65**). Surprisingly, the hydrogen abstraction step is not stereoselective [223].

In contrast, phenobarbitol induced Cyt-P450$_{LM2}$ obtained from rabbit liver is totally non-selective and the product ratios obtained are similar to those obtained by chemical hydroxylation [224]. Cyclic alkenes give a mixture of allylically hydroxylated products, epoxides [225] and on some occasions *cis* diols [226,227]. Oxidizing agents other than oxygen (e.g. iodosylbenzene) may be used in these reactions, but an attempt to use an iminophenyliodinane to transfer nitrogen in place of oxygen resulted in hydroxylation rather than amination [228].

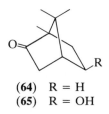

(64) R = H
(65) R = OH

As observed in the steroid series, electron-rich functional groups seem to promote hydroxylation. At least in a few cases this may be because the polar group renders the molecule more soluble in water and thereby less volatile. Volatility is a particular problem in reaction mixtures that require continuous aeration. Amides and, to a lesser extent, hydroxyl and ketone groups seem to be the best directing groups. It is often worthwhile converting an alcohol to the corresponding urethane to increase regioselectivity in the hydroxylation process [174].

Virtually all small- and medium-size ring cycloalkanes have been hydroxylated and mixtures of products are frequently observed [107]. A useful prostaglandin synthon was prepared by oxidation of the cyclopentenone **(66)** to furnish the alcohol **(67)** (67% yield) using *Aspergillus niger* [229]. The procedure was adapted from the known (and less selective) hydroxylation of cinerone **(68)** [230].

Sixty-one cultures were screened for their capacity to hydroxylate cyclohexylcyclohexane **(69)**. Two species (*Cunninghamella blakesleena* and *Geotrichum lacrispora*) gave the diequatorial diol **(70)** as the major product; the latter compound was used in the synthesis of a leukotriene-B_3 analogue; 3,4'- and 3,3'-diequatorial diols were significant by-products [231]. Fonken had previously undertaken a massive study of the hydroxylation of substituted cyclohexylcyclohexanes and phenylcyclohexyl substrates [106,107,232–237].

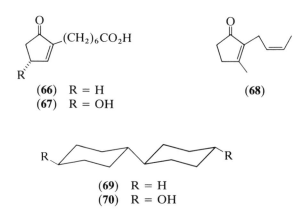

(66) R = H
(67) R = OH

(68)

(69) R = H
(70) R = OH

Cyclododecanol is oxidized to a mixture of alcohols by *Beauveria sulfurescens*. The mixture was chemically oxidized whereupon 1,5- 1,6- and 1,7-cyclododecandiones were isolated. Cyclotridecanol gave mostly the 1,7-dione [232]. Ashton *et al.* [193] observed similar results with cyclododecanone and cyclopentadecane and with larger rings annulated to cyclopentanones and cyclohexanones [194,195]. When cyclododecylamine was treated with *B. sulfurescens*, hydroxycycloalkyl acetamides were isolated. This unexpected acetylation proved to be a guide for further work, because micro-organisms which failed to hydroxylate or acetylate cyclododecylamine were shown to be able to hydroxylate *N*-cyclododecylacetamide. The hydroxylation of acetamides, benzamides and *p*-toluenesulphonamides was found to be a general and facile reaction [233]. A wide range of substrates (71) [233], (72) [234], (73) [235] and (74) [236], gave fair to good yields of the product hydroxylated in the position indicated by the arrows [173,233–237].

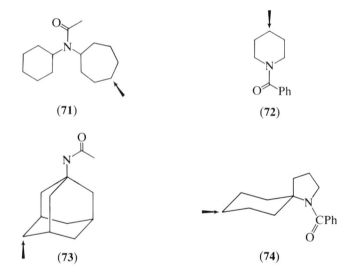

(71)   (72)

(73)   (74)

In cases where a direct competition between different sized rings was possible [e.g. compound (71)], hydroxylation was preferred in the sequence cycloheptyl > cyclohexyl > cyclopentyl. The major products obtained were those in which hydroxylation had occurred 5.5 Å from the carbonyl oxygen atom and on the face *trans* to the amide function [125,173]. Furstoss *et al.* [174] have extended the studies on the hydroxylations with this strain of micro-organism, utilizing a range of polycyclic amides. They conclude that the point of hydroxylation can vary from 3.3 to 6.2 Å for the amide moiety [174]. For example, the two amides (75) and (76) both give a single hydroxylated product, (77) and (78), respectively, in approximately 50%

yield: the incorporated hydroxyl group has the same relative position
[238]. All four pinane stereoisomers [(±)-(79)] gave hydroxylated products
[(±)-(80)] [239].

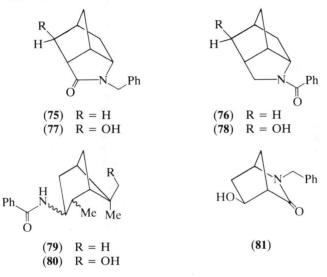

(75) R = H
(77) R = OH

(76) R = H
(78) R = OH

(79) R = H
(80) R = OH

(81)

Hydroxylation of *N*-benzyl-2-aza-bicyclo[2.2.1]heptan-3-one gave the
hydroxylated compound **(81)**, a possible synthon for 2-deoxy-*C*-nucleosides
[240].

The ability of micro-organisms to hydroxylate terpenes [107,108,120]
increases the latter compounds' value as chiral starting materials for synthesis and may provide new materials for the fragrance and aroma industries
[111]. 1,4-Cineole **(82)** gives 8-hydroxy cineole **(83)** as the major product
on incubation with *Streptomyces griseus*; 2-*exo*- and 2-*endo*-cineole **(84)** were
obtained as minor products. 2-*endo*-Cineole was the major product formed
on incubation of 1,4-cineole with several other micro-organisms [241]. The
norbornane **(85)** was hydroxylated with *Aspergillus awamori* to give the
endo-alcohol **(86)** in 57% yield (e.e. = 84.7%) [242].

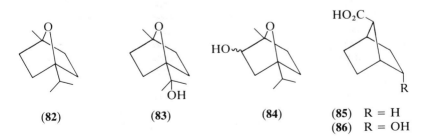

(82)

(83)

(84)

(85) R = H
(86) R = OH

A wide range of cyclic and monocyclic unsaturated terpenes has been vicinally *trans* dihydroxylated (—CH=CH → —CHOH—CHOH), presumably via the corresponding epoxide. Allylic hydroxylation is often a significant side reaction [227].

Nicotiana tabacum ("Bright Yellow") preferentially hydroxylates vinylic methyl groups [243]. Linalool gave predominantly the *trans* hydroxymethyl product [244]. Unfortunately, as intimated above, *trans* diols (formed by epoxidation of the double bond and hydrolysis) are significant by-products or indeed the major products formed when the substrates are monocyclic terpenes [245].

Compactin (**87**) is transformed into 3 α-hydroxy-compactin (**88**) *in vivo* and by *Mucor hiemalis* as well as several other species [246]. The methylated analogue, monacolin K (mevinolin) (**89**) is also hydroxylated, but the stereochemistry of the product (**90**) was not determined. The alcohols (**88**) and (**90**) underwent allylic rearrangement to afford the corresponding 6α-hydroxyl compounds (**91**) and (**92**) [247]. Several new micro-organisms used in this study were identified as *Nocardia* sp. [248].

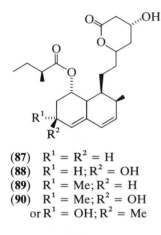

(**87**)   R¹ = R² = H
(**88**)   R¹ = H; R² = OH
(**89**)   R¹ = Me; R² = H
(**90**)   R¹ = Me; R² = OH
   or R¹ = OH; R² = Me

(**91**)   R = H
(**92**)   R = Me

The hydroxy decalin (**93**) is converted into the diequatorial diol (**94**) by a rat liver homogenate [249]. A similar regioselectivity is observed in the hydroxylation of the ketoester (**95**) by *Calonectria decora* to give the ketol [96,251]. Cinnamolide (**97**) is hydroxylated at the 3β-position by *Aspergillus niger* in 46% yield to give the compound (**98**), whereas the triol (**100**) can be prepared from the diol (**99**) in a remarkable 70–80% yield using *A. niger* or *Cunninghamella elegans* [250].

1,9-Dideoxyforskolin (**101**) occurs in almost equal amounts to forskolin (**102**) in the Indian herb, *Coleus forskohlii*. This herb is the sole source of forskolin. Of 263 fungal isolates screened for the conversion of 1,9-dideoxy

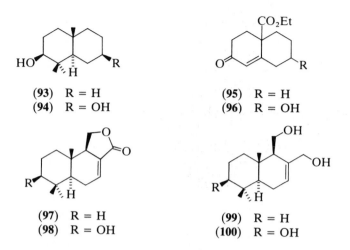

(93) R = H
(94) R = OH

(95) R = H
(96) R = OH

(97) R = H
(98) R = OH

(99) R = H
(100) R = OH

deacetylforskolin (103) into deacetylforskolin (104), one was active! The yield was 0.76%, which the authors noted should be optimized to develop a viable process [252].

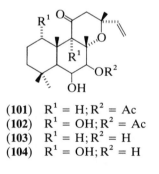

(101) R^1 = H; R^2 = Ac
(102) R^1 = OH; R^2 = Ac
(103) R^1 = H; R^2 = H
(104) R^1 = OH; R^2 = H

Aphidicolin (105) was hydroxylated at the 6β-position in 54% yield by *Streptomyces punipalus* to give the poly-hydroxy compound (106) [253]. Other micro-organisms gave simple redox products (3-ketone and/or 18-

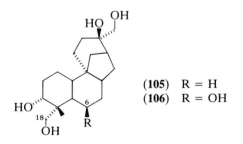

(105) R = H
(106) R = OH

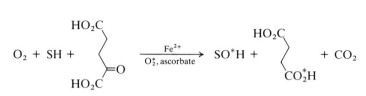

$$O_2 + SH + \underset{HO_2C}{\overset{HO_2C}{\diagdown}} =O \quad \xrightarrow[O_2^*, \text{ascorbate}]{Fe^{2+}} \quad SO^*H + \underset{CO_2^*H}{\overset{HO_2C}{\diagdown}} + CO_2$$

FIG. 4.3.   Stoichiometry of reactions involving 2-oxoglutarate coupled oxidases.
SH, substrate; O*, oxygen isotope.

carboxylic acid) plus acetylated products (at C-13 and C-18) reminiscent of
the amine acetylation observed with *B. sulfurescens* [233]. This range of
activities is common with whole micro-organisms. Although it may be
frustrating for the synthetic organic chemist, it is obviously a bonus for
pharmaceutical companies involved in drug-screening programmes.

Oxidases that are coupled to 2-oxo-glutarate (Figure 4.3) are widespread
in nature and include such diverse enzymes as several gibberellin hydroxylases [enzymes involved in the biosynthesis of compounds such as gibberellin
A_8 (**107**)] [254], peptidyl proline hydroxylase, flavanone-3-hydroxylase,
thymine-7-hydroxylase, *p*-hydroxyphenylpyruvate hydroxylase and γ-butyro-
betaine-3-hydroxylase [255] (see also Section 4.3.3 on aliphatic hydroxylations). A notable bio-hydroxylation procedure involves the conversion of
2,2-dimethylgibberellin A_4 (**108**) into 2,2-dimethylgibberellin A_1 (**109**) by
cultures of *Gibberella fujikuroi*, a process that can be carried out on a
multimilligram scale [256].

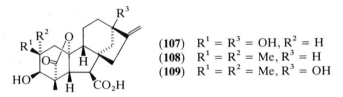

(**107**) $R^1 = R^3 = OH, R^2 = H$
(**108**) $R^1 = R^2 = Me, R^3 = H$
(**109**) $R^1 = R^2 = Me, R^3 = OH$

Streptomyces griseoviridus hydroxylates free L-proline to give *trans*-4-
hydroxy-L-proline [257], whereas the equivalent mammalian enzyme only
accepts the peptide triplet sequence X-Pro-Gly [258]. Some other saturated
nitrogen heterocyclic compounds (such as agroclavine [259], quinidine [260]
and 3-exomethylene cephalosporin C [261]) have been shown to undergo
enzyme-catalysed hydroxylation.

7,18-Diacetylpusillatriol (**110**) (an *ent*-beyerene) gave on incubation with
Rhizopus nigricans the product of *ent*-3β-hydroxylation (**111**) and *ent*-15,
16α-epoxidation [262], whereas several other similar substrates gave mixtures
of 1- and 7-hydroxylated *ent*-kauranes [263] or products arising from *ent*-
16β-hydroxylation [264]. Parallels can be drawn between the hydroxylation
of the above substrates and the hydroxylation pattern of steroids.

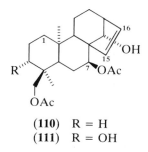

(110) R = H
(111) R = OH

4.3.3. Hydroxylation of Aliphatic Compounds

Many micro-organisms are able to metabolize alkanes possessing carbon chains longer than C_9. Rather few are able to metabolize shorter hydrocarbon chains (C_2 to C_8). There is a huge number of micro-organisms which oxidize methane (or other C_1 compounds such as methanol or methylamine) to CO_2. Some of these grow exclusively on C_1 compounds, whilst others utilize a range of organic materials [121,122]. The soil is a particularly rich source of these micro-organisms, simply because such organisms are able to assimilate the methane produced by the anaerobic degradation of organic material (the process of rotting).

The metabolism of alkenes by micro-organisms can involve hydroxylation, dehydrogenation to dienes or epoxidation [137]. Hydroxylation and epoxidation may be mediated by the same enzyme system [265]; α,ω-dihydroxylation is often observed with yeast, but only a few bacteria behave in this way [266]. At least two major types of hydroxylation system have been identified as active in these processes [107], a Cyt-P450 system [127] and one based on rubredoxin [265,267] (an iron–sulphur protein), but other systems have been investigated [121,268].

Hydroxylation of octane with rat liver microsomal Cyt-P450 proceeded with retention of configuration at C_1 [269]. On the other hand, phenobarbitol-induced Cyt-P450 gave a mixture of 1-, 2- and 3-octanol. A large kinetic-isotope effect of 9.8 was observed for a reaction employing [1,1,1-$^2H_3$]-n-octane [270]. γ-Butyrobetaine (112) was converted into L-carnitine (113) by γ-butyrobetaine hydroxylase (E.C. 1.14.11.1) [255]. This enzyme is a 2-oxoglutarate linked oxidase.

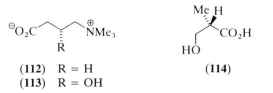

(112) R = H
(113) R = OH

(114)

The main synthetic interest in this area has been the β-hydroxylation of carboxylic acids and esters. (S)-(+)-β-Hydroxy-isobutyric acid (114) is prepared by oxidation of isobutyric acid with *Pseudomonas putida* in 48% yield [271]. The process is operated on a commercial scale in Japan. A patent application has been made for the preparation of the (R) enantiomer using *Candida* sp. [272]. These useful chiral synthons have been used *inter alia*, in the synthesis of muscone [273] and α-tocopherol [274].

Straight-chain carboxylic acids can also be hydroxylated. Both the (R) and (S) enantiomers of ethyl 3-hydroxypentanoate were produced from ethylpentanoate by different strains of *Candida rugosa*; the enantiomeric excesses were either fair or not reported [275]. These synthons have been used in the syntheses of lardolure [276], cladospolide-A [277] and carbapenems [278]. The same compounds may also be produced by yeast reduction of β-ketoesters and this is probably the method of choice at the present time.

In an unusual reaction, the enolic form of isobutyraldehyde was converted into formic acid and triplet-state acetone by horseradish peroxidase compound I. The reaction is sufficiently fast that it can be used to measure the rate of keto–enol tautomerism [279].

4.3.4. Hydroxylation of Benzylic Positions and other Aromatic and Heteroaromatic Side Chains

The hydroxylation of ethylbenzene by rat-liver microsomes (a Cyt-P450 preparation) gives 80% (R)-(+)- and 20% (S)-(−)-phenylethanol. ^{18}O from $^{18}O_2$ was incorporated into the product, but not deuterium from D_2O. (+)-α-[D]-Ethylbenzene (115) was hydroxylated with retention of configuration and virtually all the deuterium was retained indicating a significant primary isotope effect [280]. Using homogeneous Cyt-P450$_{LM2}$, essentially racemic 2-phenylethanol was produced from ethyl benzene. With specifically labelled α-[D$_1$]ethyl benzenes, pro-(R) abstraction resulted in 25% inversion, whilst for pro-(S) abstraction 40% inversion took place [281]. These results may require revision to take account of secondary isotope effects; nevertheless they accord with the general observation that Cyt-P450$_{LM2}$ (a phenobarbitol induced cytochrome) is generally less stereoselective than other cytochrome preparations [280]. Toluene-α-[D$_1$] and toluene-α,α-[D$_2$] also showed substantial retention of deuterium upon hydroxylation with several types of liver microsome [282].

The fungus *Mortierella isabellina* hydroxylates (R)- and (S)-α-[D$_1$]ethylbenzene by stereospecific abstraction of the pro-(R) hydrogen atom apparently without the usual large primary isotope effect. However, the oxygen insertion is only partially stereoselective (e.e. = 33%) [283]. A wide range of

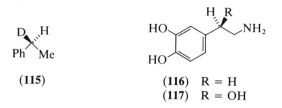

(115)

(116)   R = H
(117)   R = OH

substituted ethylbenzenes and analogous compounds (e.g. tetralins) was also incubated with *M. isabellina*, a *Helminthosporum* sp. and *Cunninghamella echinulata* var. *elegans*. The yields and enantiomeric excesses obtained were mostly poor, but in a few cases good to excellent e.e. values were observed. Unfortunately, this was not due to the hydroxylating enzyme but to an enantioselective oxidoreductase which preferentially oxidized the (*S*)-configuration alcohols [284].

Dopamine β-hydroxylase (E.C. 1.14.17.1) (a copper containing mono-oxygenase [285]) hydroxylates dopamine (116) at the benzylic position. The pro-(*R*) proton is removed and oxygen insertion occurs with retention of configuration to give (−)-noradrenaline (117) [286]. An identical stereochemical result is obtained for the tyramine residue in the conversion of *O*-methylnorbelladine to haemanthramine [287], but the benzylic hydroxylation of *N*-acetyldopamine by insect cuticle gives racemic *N*-acetylnoradrenaline [288].

Cunninghamella elegans hydroxylates indolines, 1,2,3,4-tetrahydroquinolines, 1,2,3,4-tetrahydroisoquinolines and a 2,3,4,5-tetrahydro-1*H*-1-benzazepine at the benzylic position within the saturated-ring system. If an aromatic methyl group was also present, this was hydroxylated concurrently. Concomitant oxidation of 4-hydroxy-1,2,3,4-tetrahydroquinoline to the corresponding ketone also occurred [289]. Oxidation to both the aldehyde and carboxylic acid group was observed in the oxidation of the benzylic methyl moiety of lucanthone by *Aspergillus sclerotiorum* [290].

2,6-Dimethylpyridine is hydroxylated at one methyl group by *Curvularia lunata*, *C. geniculata*, *Cunninghamella blakesleena* and a *Pseudomonas* sp., whereas dihydroxylation occurred with *Beauveria sulfurescens* [291].

Liver microsomal Cyt-P450 does not hydroxylate cumene (118) on the benzylic carbon atom as might be expected, but selectively at the pro-(*R*) methyl group (e.e. = 28%) to give the product (119) [161]. A patent application has been made claiming the preparation of 2-arylpropionic acids by hydroxylation and oxidation of cumenes with *Cordyceps* sp. For example, the naphthalene derivative (120) was transformed by *C. militaris* into the (*S*)enantiomer of the anti-inflammatory drug naproxen (121) (e.e. = 97.8%) [292]. Cyt-P450 transforms phenylacetylenes into arylacetic acids, but this

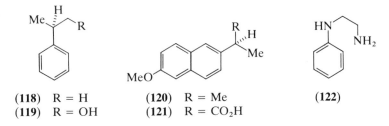

(118) R = H	**(120)** R = Me	**(122)**
(119) R = OH	**(121)** R = CO_2H	

is accompanied by suicide inactivation of the enzyme by alkylation of a porphyrin nitrogen atom [293].

Ethyl 4-alkyl-β-carboline-3-carboxylates are hydroxylated at the homo-benzylic position of the 4-alkyl group by *Streptomyces lavendulae* and *S. griseus*. Spontaneous ring closure to the δ-lactone then takes place [294]. Dopamine β-mono-oxygenase cleaves phenylalkylamines such as **(122)** by hydroxylation α to the nitrogen atom adjacent to the aromatic ring and breakdown of the resultant aminol to aniline and α-amino acetaldehyde [285]. When the amine function forms part of a ring system, the aminol can sometimes be isolated (e.g. in the conversion of leurosine into 5-hydroxy-leurosine [295]). Alternatively, the α-aminol may give rise to an electrophilic iminium ion [296] or possibly to an enamine [297]. Cytochrome $P450_{LM2}$ catalyses the dealkylation of *N,N*-dimethylaniline in association with O_2, NADPH and NADPH-Cyt-P450 reductase: the same catalyst promotes dealkylation of *N,N*-dimethylamine-*N*-oxide without the need for an exogenous oxidant [298]. *B. sulfurescens* is a useful organism for the cleavage of *N*-methyl groups from urethanes [299].

Cytochrome P450s are also able to dealkylate phenylalkyl ethers by geminal hydroxylation and hydrolysis [300,301]. 7-Ethoxycoumarin **(123)** is the standard assay substrate for this activity; it is converted into 7-hydroxycoumarin **(124)** and acetaldehyde. However, when the ethyl group in this compound **(123)** is replaced by an [α,α-D_2]ethyl group **(125)**, a new product **(126)** is obtained resulting from hydroxylation of the aromatic ring (see Section 4.3.5).

Finally, it is interesting to note that novobiocin **(127)** is selectively oxidized at a terminal methyl group by a soil actinomycete to give the hydroxy derivative **(128)**, despite the presence of a very highly activated benzylic position (i.e. C-7) [302].

4.3.5. Hydroxylation of Homocyclic and Heterocyclic Aromatic Compounds

The general aspects of reactions involving the hydroxylation of homocyclic

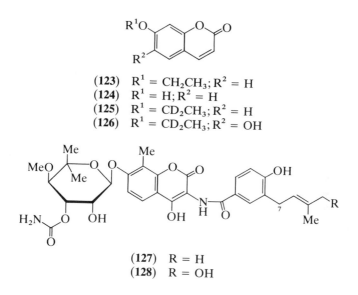

(123) $R^1 = CH_2CH_3; R^2 = H$
(124) $R^1 = H; R^2 = H$
(125) $R^1 = CD_2CH_3; R^2 = H$
(126) $R^1 = CD_2CH_3; R^2 = OH$

(127) R = H
(128) R = OH

and heterocyclic compounds have been reviewed [105,107,123,125,131]. Articles dealing with dihydroxylation [378] and microbial metabolism of aromatic compounds under anaerobic conditions [303] have also been published. Hydroxylation is often the initial step in the degradation of aromatic compounds by micro-organisms in the environment and by the liver. However, there are fundamental differences in the mechanism of the bio-hydroxylation reactions in these two situations.

Thus, prokaryotic organisms (bacteria) hydroxylate aromatic compounds (illustrated in Figure 4.4 by benzene) by dioxygenase enzymes which catalyse a cycloaddition reaction with molecular oxygen to yield a dioxetan (129). This is reduced to a *cis* diol (130)(which can be isolated in some cases, see later) whereafter oxidation [304] to catechols or products involving ring fission are formed. In contrast, eukaryotic organisms (fungi, yeasts, and higher organisms) utilize mono-oxygenases such as Cyt-P450 to give a dihydroarene epoxide (131) intermediate which is rapidly hydrolysed to a *trans* diol (132). Oxidation as before yields catechol. Alternatively, the epoxide (131) may rearrange as shown in Figure 4.5 with an NIH* shift to give a phenol. The latter reaction is facilitated by electron-donating substituents [305] which increase the rate of hydroxylation [306] and stabilize the intermediate carbonium ion. This results in the regiochemistry of hydroxylation being *ortho/para* to the electron-rich substituent. When deuterium is used as a "marker" for the NIH shift, the retention of the migrating group in

---

* NIH denotes the National Institutes of Health where the prototropic rearrangement was discovered.

the substrate is higher for *para* than for *ortho* hydroxylation. Although this phenomenon might indicate the presence of a specific isomerase, the effect can also be explained by the formation of 1,2-epoxides for which deuterium loss is obligatory in the aromatization step [307]. Semi-empirical molecular orbital calculations reproduce the observed degree of retention assuming the results are largely determined by the rate constant for the transformation of the dienone (133) into the phenol (134) [308][†]. The question of the involvement of 1,2-epoxides has not been resolved. Experiments designed to demonstrate the intermediacy of 1,2-epoxides in the hydroxylation of chlorobenzene showed the exclusive formation of 2,3-epoxides [309]. In a few cases, arene epoxides have been isolated [309,310].

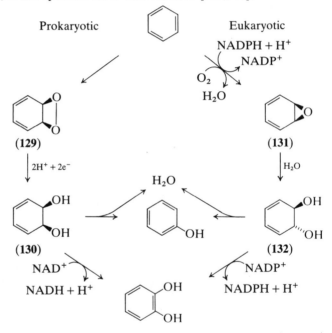

FIG. 4.4.   Oxidation of benzene by prokaryotic and eukaryotic organisms.

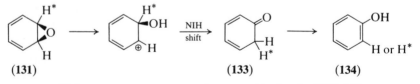

FIG. 4.5.   Rearrangement of arene oxides by the NIH shift.

† In the absence of any stereochemical preference for loss of hydrogen/deuterium a deuterium atom should be retained preferentially to a hydrogen atom because of a primary kinetic-isotope effect.

Although the dioxetan and arene epoxide mechanisms of aromatic hydroxylation have been characterized in many species, they are not the only possible pathways and other mechanisms have been demonstrated. For example, *meta*-hydroxylation of substituted aromatic compounds usually occurs by direct insertion of oxygen into the C—H bond [311–313].

The micro-organism *Beauveria sulfurescens* selectively hydroxylates propham (135) to give the phenol (136) and the corresponding 4-O-methylglucopyranosyl derivative. (4-O-Methyl-glycoside formation has also been observed with hydroxyquinolines [314] and the carbolines [294].) When 4-[D₁]-propham (137) was hydroxylated, 70% of the deuterium was retained in the product indicating an arene oxide intermediate which rearranges by an NIH shift. When incubated under the same conditions, *N*-methylpropham gave a product mixture similar to that obtained from propham suggesting facile enzyme-catalysed *N*-dealkylation. No ring hydroxylation was observed with 4-methylpropham (138), but the methyl group attached to the aromatic ring was hydroxylated to afford the compound (139) in 12% yield [299].

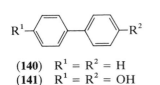

(135) R = H
(136) R = OH
(137) R = D
(138) R = Me
(139) R = CH₂OH

Aspergillus parasiticus converts biphenyl (140) into 4,4′-dihydroxybiphenyl (141) and biphenol-O-sulphonic acid, but problems typical in biotransformations were encountered. Biphenyl and monohydroxybiphenyls are toxic to fungal species and inhibit growth and oxygen uptake. Strains of *A. parasiticus* adapted to monohydroxybiphenyl were isolated and these performed the bioconversion at a two- to three-fold higher rate than the unadapted cells. Pulsed feeding of biphenyl and recycling of the mycelia increased the rate of hydroxylation and removed the lag period during which induction of activity occurs [315].

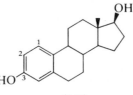

(140) R¹ = R² = H
(141) R¹ = R² = OH

(142)

Hydroxylation reactions involving the aromatic ring of certain steroids do not have synthetic utility as yet, but such reactions are obviously important in metabolism. In the 2-hydroxylation of labelled estradiol (142), 20% of the tritium is lost from C-1 and virtually all the tritium is lost from C-2. It appears that although a dihydro arene epoxide is the intermediate it does not rearrange via an NIH shift [316]. These hydroxylations are carried out in the presence of ascorbate to prevent over-oxidation to σ-quinones [317].

Phenol-, salicylate-, melilotate- and p-hydroxybenzoate hydroxylase are flavoproteins which carry out single hydroxylations of aromatic hydroxy compounds utilizing dioxygen and reduced FAD [318]. Recent evidence suggests that the reactive form of oxygen is a hydroxyl radical transferred from a flavin hydroperoxide [312]. Anthranilate hydroxylase (E.C. 1.14.12.2) is a dimeric flavoprotein (having one molecule of FAD per subunit). It catalyses the conversion of anthranilic acid (143) into 2,3-dihydroxybenzoic acid (144) as illustrated in Figure 4.6. Although the mechanism shown in Figure 4.6 indicates an electrophilic substitution, the same overall result would be obtained by transfer of a hydroxyl radical and single electron transfer (SET) back to the FAD-oxy radical [313].

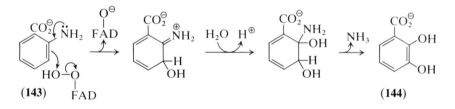

FIG. 4.6.   Conversion of anthranilic acid into 2,3-dihydroxybenzoic acid using anthranilate hydroxylase.

Rabbit liver aldehyde oxidase and bovine milk xanthine oxidase are molybdenum iron–sulphur flavin hydroxylases that oxidize nicotinium salts. Thus the alkyl nicotinium chlorides [(145) R = Me, Et, Pr] were oxidized by rat-liver aldehyde oxidase [319] to give the corresponding 6-oxo-3-carbox-amidopyridines (146). The isopropyl derivative [(145) R = Pr^i] gave a mixture of 6-oxo [(146) R = Pr^i] and 4-oxo product [(147) R = Pr^i] while the t-butyl derivative [(145) R = Bu^t] gave exclusively the corresponding 4-oxo-product [320]. Oxidation of arylnicotinium chlorides [(145) R = Ar] gave predominantly the 6-oxo products (146) plus some 4-oxo products (147). The 6-oxo products (146) are favoured by electron-withdrawing groups on the aryl ring ($\rho = +3.6$). Oxidations of the substrates [(146) R = Ar] with bovine milk xanthine oxidase were selective for formation of the 4-oxo products [(147) R = Ar] and were only slightly affected by substituents on the aryl ring [321].

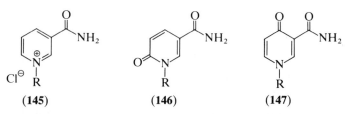

(145)   (146)   (147)

Xanthine oxidase catalyses the hydroxylation of purines and pyrimidines and the oxidation of benzaldehydes [322]. A yield of 98.6% has been claimed for the oxidation of nicotinic acid to 6-hydroxynicotinic acid by *Pseudomonas*, *Bacillus* and *Achromobacter* sp. [323].

The hydroxylation of alkaloids [118,119] often modifies their biological activity and makes available compounds that are otherwise difficult to synthesize. The β-carboline [(**148**) R = H] is hydroxylated by *B. sulfurescens* at the 6 position and to a lesser extent at the 8 position [294] with concomitant 4'-*O*-methylglucoside formation [294,299,314]. As the size of a substituent in the 4 position of the pyridine ring gets larger [(**148**) R = Me, Et, Pr] hydroxylation at the 8 position is increasingly preferred [294].

The heteroyohimbine alkaloids (**149**) are selectively hydroxylated by *Cunninghamella elegans* at the 10 position. On the other hand, *MRSS 10-IBI*, a mould isolated from the alkaloid bearing plant *Rauvolfia vomitoria*, hydroxylates tetrahydroalstonine at the 11 position, but fails to hydroxylate isoajmalicine or akuammigrine [324]. Acronycine (**150**), an anti-tumour alkaloid, is hydroxylated in 30% yield in the 9 position by *Cunninghamella echinulata* [325].

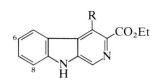

(148)

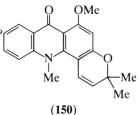

(150)

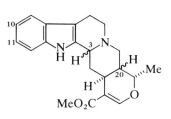

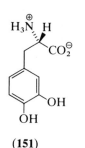

(149) 3α,20β . . . ajmalicine
3α,20α . . . tetrahydroalstonine
3β,20β . . . isoajmalicine
3β,20α . . . akuammigrine

(151)

There are aromatic hydroxylases available for the oxidation of virtually all aromatic amino acids. For example, L-phenylalanine hydroxylase (E.C. 1.14.16.1) [326] converts phenylalanine into tyrosine which, in turn, can be hydroxylated by tyrosine hydroxylase (E.C. 1.14.16.2) [327] to give the anti-parkinsonian drug L-dopa (151). These two enzymes, as well as tryptophan hydroxylase (E.C. 1.14.16.4)*, utilize tetrahydrobiopterin as a reducing cofactor for the hydroxylation that uses dioxygen as the oxidant. The second oxygen atom from O_2 is reduced to water [328]. L-Dopa (151) can also be prepared by hydroxylation of tyrosine using horseradish peroxidase, dioxygen and dihydroxyfumaric acid (as the hydrogen donor) at 0°C [301]. At room temperature, tyrosine is hydroxylated by dihydroxyfumaric acid and dioxygen in the absence of horseradish peroxidase [301]. Moreover, the reaction is not catalysed by hydrogen peroxide, the "normal" oxidant. The usual reactions of HRP such as the oxidative polymerization of diamines [171], the oxidation of polyhydric phenols to quinones [329] and oxidative coupling [330] do not seem to occur under these conditions.

Bouvardin, an anti-tumour cyclic hexapeptide from the plant *Bouvardia ternifola*, is *O*-demethylated by *Streptomyces rutgersensis* to give *O*-desmethylbouvardin and this phenol was in turn monohydroxylated by *Aspergillus ochraceous* to give bouvardin catechol in 34.5% yield [331].

Figure 4.4 shows the mechanism of the conversion of an aromatic compound to hydroxylated products. In all the transformations described so far, aromatization by loss of H_2O follows the dihydroxylation step. However, in some cases it is possible to isolate the appropriate *cis* (130) or *trans* diol (132) intermediates. The lower resonance energy of polycyclic aromatic hydrocarbons decreases the tendency to aromatize and diol intermediates may be isolated more readily. For example, the dimethylbenz[a]anthracene (152) is metabolized by *Cunninghamella elegans* and *Syncephalastrum racemosum* to produce the triols (153) and (154) plus several other triols and tetrols [332].

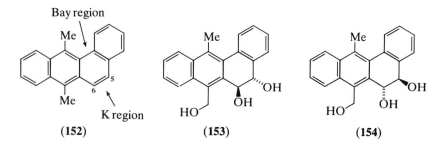

(152)   (153)   (154)

* Tryptophan hydroxylase converts tryptophan into 5-hydroxytryptophan.

Fifty-two micro-organisms were screened for their ability to transform precocene II (155). *Streptomyces griseus* produced the highest yields of three metabolites (156)–(158). The diol (156) clearly arises from *trans* opening of an epoxide, whereas the diol (157) is formed by an unusual *cis* ring opening. Only one oxygen atom was incorporated from $^{18}O_2$. The alcohol (158) arises from rearrangement and subsequent reduction of the ketone [333].

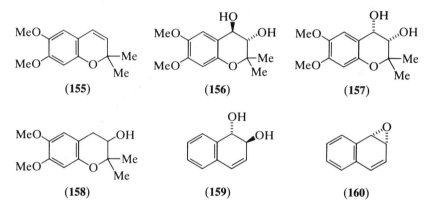

When naphthalene is incubated with *Cunninghamella elegans* the (1*S*,2*S*) diol (159) is produced; the enantiomer is formed by the action of rat, mouse, rabbit and pig liver microsomes on naphthalene. In the former case, labelling experiments indicated that the oxygen at C-1 was derived from dioxygen and that at C-2 from water. Therefore, the intermediate epoxide must have the (1*S*,2*R*) absolute stereochemistry shown in (160). 4-Hydroxy-1-tetralone, 1-naphthol and 2-naphthol were also isolated from the incubation. When 1-[D₁]-naphthalene was transformed in the same manner, the 1-naphthol isolated contained 78% of the initial deuterium indicating the operation of an NIH shift [334].

Benzene is bis-hydroxylated by a strain of *Pseudomonas putida* to give the cyclohexadienediol [(161) $R^1 = R^2 = H$] [335]. A range of substituted benzenes is transformed to give a single enantiomer of the corresponding diol [e.g. [(161) $R^1 = Me$, CF_3, Cl or Ph; $R^2 = H$] [(161) $R^1 = Me$; $R^2 = F$] and [(161) $R^1 = CO_2H$, $R^2 = Br$]]. However, the chloromethyl [(161) $R^3 = Me$; $R^2 = Cl$] and bromomethyl [(161) $R^1 = Me$; $R^1 = Br$] products were racemates [336,337]. *p*-Cymene (*p*-Pr^i-toluene) is oxidized at the methyl group before dihydroxylation occurs to give the diol [(161) $R^1 = CO_2H$; $R^2 = Pr^i$] [337].

There appears to be a reluctance for hydroxylation to take place at a centre possessing a fluoro substituent; for example, 3,5-difluorobenzoic acid gave the unexpected regioisomer (162) [338].

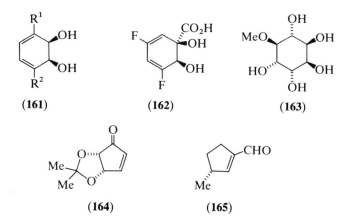

(161)   (162)   (163)

(164)   (165)

The diol [(161) $R^1 = R^2 = H$] has been used in a synthesis of ($\pm$)-pinitol (163) by selective epoxidation, epoxide ring opening and osmylation [339]. The methyl diol [(161) $R^1 = Me$, $R^2 = H$] has been employed in the preparation of a prostaglandin intermediate (164) and the useful synthon (165) [340].

Polyhydroxyaromatic compounds are labile species, susceptible to aerial oxidation and to bacterial degradation [107]. Orsellinic acid (166) is meta-bolized by *Penicillium cyclopium* to furnish penicillic acid (167) which may prove to be a useful synthetic intermediate. The metabolic pathway can be blocked so that 6-methyl-1,2-4-benzenetriol (168) accumulates by mutation of the organism with *N*-methyl-*N'*-nitro-*N*-nitrosoguanidine [341].

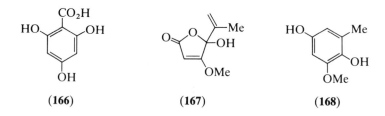

(166)   (167)   (168)

4.4. OXIDATION AT SULPHUR

Oxidation of sulphur by micro-organisms or isolated enzymes has been studied by several workers, but at the present state of knowledge, it cannot be recommended as a general synthetic method [342]. Oxidation of a sulphide to the corresponding sulphoxide is only stereospecific in certain cases and it is often difficult to predict the stereochemistry of the product.

Practical problems also arise from the high dilution of reactant(s) necessary in many of the oxidation reactions because large-scale apparatus is then needed to prepare suitable quantities of material. In some instances, however, it is possible that the ability to prepare sulphoxides of high enantiomeric purity will outweigh the disadvantages.

Early reports on the enantioselective oxidation of steroidal sulphides with *Calonectria decora* [343] and *Rhizopus stolonifer* [344] prompted a study of the oxidation of simpler unsymmetrical sulphides with *Aspergillus niger* [345]. Both methyl-2-naphthyl sulphide and phenylbenzyl sulphide produced optically active sulphoxides along with smaller amounts of the corresponding sulphones. The optical purity of the phenylbenzyl sulphoxide was later shown to be 18%. Further studies on simple sulphides with *Aspergillus niger* and also with *Mortierella isabellina* and *Helminthosporum* sp. NRRL 4671 showed that high enantiomeric excesses were possible, although the yields frequently left much to be desired (Table 4.2) [346,347]. Within this very limited series of compounds a trend is discernible for oxidations with *Aspergillus niger* in that stereoselectivity increases as the degree of substitution on the α-carbon atom increases.

TABLE 4.2

Oxidation of some sulphides with micro-organisms

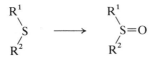

R^1	R^2	Species of micro-organism*	Yield (%)	e.e. (%)	Sign of rotation
p-CH$_3$Ph	Bu^t	A	25	99	+
	Pr^i	A	12	70	+
	Bu^n	A	14	32	+
	Me	A	48	35	+
PhCH$_2$	Bu^t	A	65	91	−
	Ph	A	14	26	−
p-CH$_3$–Ph	Me	M	60	100†	+
	Me	H	50	100	−

* A, *Aspergillus niger*; M, *Mortierella isabellina*; H, *Helminthosporum* NRRL 4671.
† See text and Table 4.3.

Perhaps the most notable feature of this series of reactions is the ability to prepare a sulphoxide with (R) or (S) configuration by a suitable choice of organism (last two entries of Table 4.2). A caveat is necessary to this work, however. The oxidation of methyl p-tolyl sulphide with *Mortierella isabellina* was repeated by Holland *et al.* [348], who estimated enantiomeric excesses by NMR studies with a chiral shift reagent rather than by comparison of rotation measurements. This showed the e.e. of the product sulphoxide to be 46% rather than the 100% reported by Abushanab *et al.* [347], an error caused by there being several different rotation values reported for the sulphoxide in the literature.

Preparation of both enantiomers of ethyl p-tolyl sulphoxide by the use of two different systems has been reported by Walsh and co-workers [349]. Thus purified mono-oxygenase from pig-liver microsomes gave the (R) sulphoxide with e.e. = 95%, while cyclohexanone oxygenase from *Acinetobacter* gave the (S) enantiomer (e.e. 82%). The use of (rat) liver microsomal fractions, rather than the purified enzyme, gave sulphoxide of very low enantiomeric excess. This was assumed to reflect oxidative contributions in an opposite enantiomeric sense from the FAD-containing mono-oxygenase and cyto-chrome P450 isozymes; the latter enzymes were shown to produce the (S) enantiomer preferentially [350].

TABLE 4.3

Oxidation of some aromatic sulphides with *Mortierella isabellina*

R	Yield (%)	e.e. (%)
H	65	58
Me	52	46*
Et	20	90
Pr^i	30	84
Bu^t	27	76
OCH$_3$	50	72
NO$_2$	6	20
F	45	70
Cl	69	90
Br	66	100
CN	35	80

* Note different result to Table 4.2 (see text).

Holland *et al.* [348] studied the oxidation of a range of methyl phenyl sulphides with four different substituents on the aromatic ring using *Mortierella isabellina* to gain an insight into the mechanism of the oxidation (Table 4.3). The results indicate that the enantioselectivity of the reaction increases as the electron-withdrawing nature of the 4-substituent increases. This was taken as evidence of a stepwise mechanism involving an electron-deficient sulphur intermediate. The oxidation of a homologous series of alkylphenyl sulphides was also studied (Table 4.4). No clear correlation of enantiospecificity and degree of substitution was possible in this case.

TABLE 4.4

Oxidation of a series of alkylphenyl sulphides using *Mortierella isabellina*

R	Yield (%)	e.e. (%)
Et	70	86
Pr^i	55	83
Bu^t	15	58
Pr^n	52	100

Oxidation of alkylaryl sulphides and allylaryl sulphides by *Corynebacterium equi* (IFO 3730) gave the respective (*R*) sulphoxides of high enantiomeric excess, although the yields are very variable and depend, at least partly, on the initial concentration of the substrate (Table 4.5) [351]. In the case of n-decylphenyl sulphide, the major product was the corresponding sulphone, a product not observed with the methyl sulphides, although observable to some extent in all the other reactions. Allylphenyl sulphides gave the sulphones as the major product, except for *p*-methoxyphenyl-cinnamyl sulphide which gave virtually no sulphone [352]. This is a further indication that the degree of electron density on the sulphur atom has a marked effect on the reaction path.

The synthetic utility of the above oxidation products (if the reactions could be performed on a suitable scale) lies in their use as chiral synthons, as exemplified by Solladie. Compounds of similar synthetic utility could conceivably be produced by the enantiospecific mono-oxidation of 1,3-dithiacetal systems to give chiral acyl anion equivalents [353]. Disappointingly, mono-oxidation of 1,3-dithiane with a variety of micro-organisms (*inter alia M.*

TABLE 4.5

Oxidation of alkylaryl sulphides and allylaryl sulphides using *Corynebacterium equi*

$$R^1 \diagdown \diagup S \diagdown R^2 \longrightarrow R^1 \diagdown \diagup \overset{\overset{O}{\|}}{S} \diagdown R^2$$

R^1	R^2	Yield (%)	e.e. (%)
H	n-$C_{10}H_{21}$	25	99
H	n-C_4H_9	29	100
H	CH_3	18	75
H	$CH \cdot CH = CH_2$	38	100
CH_3	n-$C_{10}H_{21}$	55	89
CH_3	n-C_4H_9	79	87
CH_3	CH_3	33	82
CH_3	$CH_2CH = CH_2$	67	92

isabellina, Helminthosporum NRRL 4671 and *Aspergillus foetida*) gave low yields of product with little, or no, enantiomeric enrichment [354]. The use of *Helminthosporum* NRRL 4671 to oxidize 2-substituted 1,3-dithianes gave products of higher e.e., but such substitution at the 2 position limits the synthetic utility of these compounds [355].

Several dithioacetals have been oxidized with *Corynebacterium equi* (IFO 3730) and in two cases chiral sulphoxides (**169**) and (**170**) of high e.e. were produced [356]. In the other examples studied, the products obtained were monosulphones. While the chiral sulphoxides could be valuable synthons, the method still suffers from a problem of scale, as the oxidation of 0.1 ml of substrate requires 100 ml of medium. For a final step in a synthetic sequence this may be acceptable, but for the larger quantities required at an early synthetic stage this would present difficulties in the average laboratory.

(R)-(**169**)   (R)-(**170**)

In an interesting "quasi-enzymic" process, mono-*S*-oxides of formaldehyde dithioacetals have been prepared by periodate oxidation in the presence of bovine serum albumin [357]. Optimization of the reaction conditions gave products with e.e. = 30–50% in moderate yields. Unfortunately the reaction conditions are such as to render large-scale preparations very difficult.

4.5. EPOXIDATION

Stereoselective epoxidation of double bonds by normal chemical methods is extremely unusual in the absence of an adjacent directing group (e.g. an adjacent hydroxyl group). In contrast, stereospecific enzyme-catalysed epoxidations of unactivated double bonds are often very facile reactions.

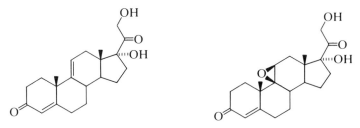

FIG. 4.7.   Epoxidation of a steroid.

Early work on the incubation of unsaturated steroids with micro-organisms capable of hydroxylation occasionally produced epoxides as products; an example is shown in Figure 4.7 [358]. Such epoxidizing activity was displayed by *Cunninghamella blakesleena*, *Curvularia lunata*, *Helicostylum pyriforme*, *Mucor griseocyanus*, *M. parasiticus*, *Nocardia* sp. ATCC 13934 and *Corynebacterium simplex* [359]. Despite the postulate of Bloom and Shull [358] that "a micro-organism capable of introducing an axial hydroxyl function at C-9 of a saturated steroid also effected the introduction of an epoxide group axial at C-9 in the corresponding unsaturated substrate", predictions as to whether an epoxide would be formed were not possible in the majority of cases. This unpredictability would render such a reaction useless as a general synthetic method.

The great strength of enzymic epoxidation lies in the production of small epoxides of high enantiomeric excess where there is little ambiguity in the types of product that can be formed. An example is the production of fosfomycin (**171**), the synthesis of which was studied by workers at Merck, Sharp and Dohme [360]. After surveying many organisms, they found that *Penicillium spinulosum* epoxidized *cis*-propenylphosphonate enantiospecifically in 90% yield. Although the highest concentration of alkene allowable was only $0.5 \, \text{g} \, \text{l}^{-1}$, the method allowed production of multigram quantities of a product whose enantiospecific synthesis by classical methods would have been extremely difficult.

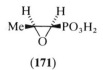

(171)

The most intensively studied microbiological epoxidizing agent is the ω-hydroxylase system of *Pseudomonas oleovorans*, first isolated by Peterson *et al.* [361]. In an extensive series of papers, May *et al.* have demonstrated that *P. oleovorans* epoxidizes terminal, acyclic alkenes to produce (*R*)-(+)-1,2-epoxides admixed with ω-ene-1-ols. Cyclic and non-terminal alkenes are not epoxidized, neither are aromatic compounds or styrenes [362]. There is thus a potential for selective epoxidation of only one double bond in a multiple unsaturated system. The ratio of hydroxylation to epoxidation products depends markedly upon the chain length of the alkene [362]. Epoxidation products are not observed with propene and butene, but reactivity for epoxidation reaches a maximum at octene and drops off slowly thereafter as the chain length increases. In contrast, with this micro-organism methyl group hydroxylation is little effected by chain length for short chains but decreases rapidly as the chain length increases. With 1-decene as substrate, epoxidation predominates over methyl group hydroxylation.

1,7-Octadiene, which does not possess a terminal methyl group, is converted exclusively into (*R*)-(+)-7,8-epoxy-1-octene (e.e. > 80%); further oxidation gave the 1,2-7,8-diepoxyoctane with a similar degree of optical purity [363]. *P. oleovorans* epoxidation of racemic 7,8-epoxy-1-octene, however, gave a product with poor optical purity (e.e. = 20%). This indicates that the configuration of the remote, preformed, epoxide group has a marked effect on the stereochemistry of the insertion of the second oxygen atom [364]. Further studies on the effect of remote asymmetric centres on the epoxidation reaction have yet to be performed, but are clearly desirable. Other α,ω-dienes with six to 12 carbons have also been epoxidized [365] with similar results, and a detailed study of the optimal reaction conditions has been reported [363]. This demonstrated that the product epoxide was rapidly degraded by viable cells in the absence of alkene. When alkene was present, however, there was a net increase in the amount of epoxide. A quantitative epoxidation is thus impossible, a point to be considered if the starting material is valuable.

The mechanism of the epoxidation reaction has been carefully studied by May *et al.* [366]. Analysis of the epoxide from *trans,trans*-1,8-dideuterio-1,7-octadiene by partially relaxed proton FT NMR spectroscopy demonstrated that the epoxidation reaction proceeded largely (70%) with inversion of the geometry of the alkene, thus discounting a simple *syn* addition of oxygen to the double bond [366]. Further studies on the epoxidation of *cis*- and *trans*-

1-deuterio-1-octene and 1,1-dideuterio-1-octene confirmed the inversion of the alkene geometry and pointed to a reaction mechanism which did not involve complex hydrogen abstraction steps [367]. In conjunction with other work on the aldehyde produced as a by-product in these reactions, the mechanism shown in Figure 4.8 was postulated. The nature of the iron–oxygen species is not known, and the reaction could proceed through homolytic or heterolytic processes (although only the heterolytic process is shown), but the essentials of the mechanism are consistent with the results obtained. The formation of (R) epoxides is accounted for by preliminary hydrophobic bonding of the alkene to the active site in such a way as to ensure that the epoxide closure must be from the *si*-face of C-2. Initial attack of the iron–oxygen species is on either face of the terminal carbon atom (this explains the specificity for terminal alkenes), and the resulting cationic (or radical) intermediate can then, if necessary, rotate before cyclization to form the epoxide.

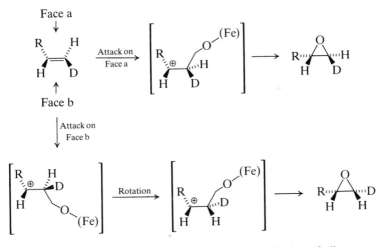

FIG. 4.8.   Postulated mechanism of the epoxidation of alkenes.

The above studies were performed using partially purified or crude cell-free systems, or whole-cell systems. Thus, an organic chemist unwilling to take on the isolation of the necessary enzymes need not be inhibited from trying this reaction. Indeed, this reaction is so easy to perform that it has been adapted for use in instructional organic chemistry practical courses [368].

Epoxidation in the presence of cyclohexane (hexadecane has also been used) as a water-immiscible cosolvent increases the efficiency of the conversion to 90 mol%, because the organic solvent contains the large majority of the epoxide and thus the cells are never exposed to inhibitory concentrations of the product [369]. However, the amount of epoxide isolable is very low, because the culture never contains more than 1% v/v alkene. This problem has been overcome for the epoxidation of 1-octene by utilizing the alkene itself as a second phase (20–50% v/v) [370]. This two-phase mixture has a deleterious effect on cell membranes, which means that the cells must be replaced after about 70 h incubation but the procedure will produce 5.6 g of (R)-1,2-epoxyoctane from 2.5 l of culture. At this time this is the sort of reaction where this method will be of greatest utility, i.e. producing bulk quantities of simple chirally enriched synthons.

Preparation 4.3   Preparation of 1,2-epoxyoctane

Two 5-l flasks each containing 1000 ml of P-1 medium [375] and 250 ml of oct-1-ene were inoculated with octane-grown cells of *P. oleovorans* (ATCC 29347) and were incubated at 30°C on a rotary shaker at 250 rpm. After 70 h the cultures were centrifuged (10 000 g, 30 min) at 0°C and the organic layer separated. The organic fraction was divided and treated with fresh cultures. After 70 h at 30°C, the mixture was worked-up as above and the organic layer was centrifuged (18 000 g, 30 min) to remove water. The organic layer was dried. Distillation gave oct-1-ene and 1,2-epoxyoctane: b.p. 60–61°C, 15 mmHg (5.6 g), $[\alpha]_D^{21} = 9.3°$ (c = 3.1, EtOH).

Other enzymic methods exist for epoxidizing alkenes but, in general, they have not been examined to the extent to allow them to be recommended as preparative procedures in the synthetic laboratory. Systems which fall into this category include *Corynebacterium equi* [371], *Streptomyces albus* NRRL 1865 [372], and several strains of *Mycobacteria* [373]. The NADPH-dependent oxygenase isolated from liver microsomes of phenobarbital-pretreated rats is a very effective system for the epoxidation of alkenes [374]. This gives products with almost complete specificity and has all the advantages associated with isolated enzyme reactions (viz. lower reaction volumes, easier work-up and greater reaction specificity). Access is needed, however, to animals, expertise and facilities to isolate microsomes.

REFERENCES

1. For general reviews, see Chapter 3 [1–6].
2. G. L. Lemière, J. A. Lepoivre and F. C. Alderweireldt, *Tetrahedron Lett.* **26**, 4527 (1985). G. L. Lemière, in *Enzymes as Catalysts in Organic Synthesis* (ed. M. P. Schneider), p. 19. Reidel, Dordrecht, 1986.
3. O. Jenkins and D. Jones, *J. Gen. Microbiol.* **133**, 453 (1978).
4. E. Bellion and G. T.-S. Wu, *J. Bacteriol.* **135**, 251 (1978); c.f. refs 121, 122.
5. K. Yamanaka and R. Minoshima, *Agric. Biol. Chem.* **48**, 171 (1984).
6. C. T. Hou, R. Patel, N. Barnabe and I. Marczak, *Eur. J. Biochem.* **199**, 359 (1981).

7. C. T. Hou, R. Patel, A. I. Laskin, N. Barnabe and I. Marczak, *Appl. Environ. Microbiol.* **38**, 135 (1979). S. F. Hiu, C.-X. Zhu, R.-T. Yan and J.-S. Chen, *Appl. Environ. Microbiol.* **53**, 697 (1987).

8. A. Steinbüchel and H. G. Schlegel, *Eur. J. Biochem.* **141**, 555 (1984).

9. C. H. Barrett, K. S. Dodgson and G. F. White, *Biochem. Biophys. Acta* **661**, 74 (1981).

10. D. J. Rigby, K. S. Dodgson and G. F. White, *J. Gen. Microbiol.* **132**, 35 (1986).

11. D. B. Janssen, S. Keuning and B. Witholt, *J. Gen, Microbiol.* **133**, 85 (1987).

12. H. Ohta, Y. Kato and G. I. Tsuchihashi, *J. Org. Chem.* **52**, 2735 (1987).

13. C. Shimizu, S. Hattori, H. Hata and Y. Yamada, *Appl. Environ. Microbiol.* **53**, 519 (1987).

14. A. J. Ganzhorn, D. W. Green, A. D. Hershey, R. M. Gould and B. V. Plapp, *J. Biol. Chem.* **262**, 3754 (1987).

15. J. Van Eys and N. O. Kaplan, *J. Am. Chem. Soc.* **79**, 2782 (1957).

16. Several enzyme systems have been proposed for the commercial production of acetaldehyde: M. S. A. Wecker and R. R. Zall, *Process Biochem.* **Oct.**, 135 (1987).

17. H. H. Günther, F. Biller, M. Kellner and H. Simon, *Angew. Chem. Int. Ed. Engl.* **12**, 146 (1973). H. R. Levy, F. A. Loewus and B. Vennesland, *J. Am. Chem. Soc.* **79**, 2949 (1957). D. Arigoni and E. L. Eliel, *Top. Stereochem.* **4**, 127 (1969). H. G. Floss and M.-D. Tsai, *Adv. Enzymol.* **50**, 243 (1979).

18. L. G. Lee and G. M. Whitesides, *J. Org. Chem.* **51**, 25 (1986).

19. C. Bally and F. Leuthardt, *Helv. Chim. Acta* **53**, 732 (1970).

20. E. Hochuli, K. E. Taylor and H. Dutler, *Eur. J. Biochem.* **75**, 433 (1977).

21. C. H. Blomquist, *Archiv. Biochem. Biophys.* **159**, 590 (1973). J. Kawamura, T. Tanimoto, H. Fukuda and T. Hayakawa, *Chem. Pharm. Bull.* **29**, 476 (1981). I. H. White and J. Jeffrey, *Biochem. Biophys. Acta* **296**, 604 (1973).

22. R. L. Root, J. P. Durrwachter and C.-H. Wong, *J. Am. Chem. Soc.* **107**, 2997 (1985).

23. A. M. Klibanov, B. N. Alberti and M. A. Marletta, *Biochem. Biophys. Res. Commun.* **108**, 804 (1982).

24. N. K. Richtmyer, L. C. Stewart and C. S. Hudson, *J. Am. Chem. Soc.* **72**, 4934 (1950). O. Touster and D. R. D. Shaw, *Physiol. Rev.* **42**, 181 (1962). R. Bentley, *Molecular Asymmetry in Biology*, Vol. II, p. 77. Academic Press, New York, 1970.

25. S. W. May and A. G. Katopodis, *Enzyme Microbiol. Technol.* **8**, 17 (1986).

26. J. Geigert, D. J. Dalietos, S. L. Neidlemann, T. D. Lee and J. Wandsworth, *Biochem. Biophys. Res. Commun.* **114**, 1104 (1983).

27. T. Suga, S. Izumi, T. Hirata and H. Hamada, *Chem. Lett.*, 425, 471, 903, 2053 (1987). *Nicotiana tabacum* is dealt with more fully in the section on carbon hydroxylation.

28. M. D. Legoy, H. S. Kim and D. Thomas, *Process Biochem.*, 145 (1985).

29. S. W. May, M. S. Steltenkamp, K. R. Borah, A. G. Katopodis and J. R. Thowsen, *J. Chem. Soc. Chem. Commun.*, 845 (1979).

30. J. R. Matos, M. B. Smith and C.-H. Wong, *Bioorg. Chem.* **13**, 121 (1985).

31. C.-H. Wong and J. R. Matos, *J. Org. Chem.* **52**, 1992 (1985).

32. J. B. Jones and K. E. Taylor, *Can. J. Chem.* **54**, 2969, 2974 (1976); *J. Chem. Soc. Chem. Commun.*, 205 (1973).

33. L. G. Lee and G. M. Whitesides, *J. Am. Chem. Soc.* **107**, 6999 (1985). G. M. Whitesides and C.-H. Wong, *Angew. Chem. Int. Ed. Engl.* **24**, 617 (1985).

34. S. W. Wang and C.-K. King, *Adv. Biochem. Eng.* **12**, 119 (1979).

35. D. G. Drueckhammer, V. W. Riddle and C.-H. Wong, *J. Org. Chem.* **50**, 5387 (1985).

36. J. M. Laval, C. Bourdillon and J. Moiroux, *Biotech. Bioeng. XXX*, 157 (1987).

37. H. Gunther and H. Simon, *Appl. Microbiol. Biotechnol.* **26**, 9 (1987).

38. A. J. Irwin, K. P. Lok, K. W.-C. Huang and J. B. Jones, *J. Chem. Soc. Perkin Trans. I*, 1636 (1978).

39. A. J. Irwin and J. B. Jones, *J. Am. Chem. Soc.* **98**, 8476 (1976); **99**, 556, 1625 (1977).
40. J. B. Jones and H. B. Goodbrand, *Can. J. Chem.* **55**, 2685 (1977).
41. M. A. Payton, *Trends. Biotechnol.* **2**, 153 (1984).
42. E. Keinan, K. K. Seth and R. Lamed, *J. Am. Chem. Soc.* **108**, 3474 (1986).
43. R. J. Lamed and J. G. Zeikus, *Biochem. J.* **195**, 183 (1981).
44. E. Keinan, E. K. Hafeli, K. K. Seth and R. Lamed, *J. Am. Chem. Soc.* **108**, 162 (1986).
45. F. Bryant and L. G. Ljungdahl, *Biochem. Biophys. Res. Commun.* **100**, 793 (1981).
46. S. Riva, G. Carrea, F. M. Veronese and A. F. Buckman, *Enzyme. Microbiol. Technol.* **9**, 556 (1986).
47. S. Riva, R. Bovara, P. Pasta and G. Carrea, *J. Org. Chem.* **51**, 2902 (1986).
48. J. D. Berg, H. I. Pandov and H. G. Sammons, *Clin. Chem.* **30**, 155 (1984); *Chem. Abs.* **100**, 82190g.
49. G. Carrea, R. Bovara, P. Cremonesi and R. Lodi, *Biotechnol. Bioeng. XXVI*, 560 (1984).
50. P. Pasta, G. Mazzola and G. Carrea, *Biochemistry* **26**, 1247 (1987). J. Kawamura, T. Tanimoto, H. Fukuda and T. Hayakawa, *Chem. Pharm. Bull.* **29**, 476 (1981).
51. I. Belič, R. Komel and H. Sočič, *Steroids* **29**, 271 (1977).
52. See Chapter 3, [82–85] and, in particular, the reviews in [82].
53. A. J. Irwin and J. B. Jones, *J. Am. Chem. Soc.* **99**, 556 (1977).
54. J. B. Jones and K. P. Lok, *Can. J. Chem.* **57**, 1025 (1979).
55. H. Ohta, H. Tetsukawa and N. Noto, *J. Org. Chem.* **47**, 2400 (1982).
56. F. C. Huang, L. F. H. Lee, R. S. D. Mittal, P. R. Ravikumar, J. A. Chan, C. J. Sih, E. Caspi and C. R. Eck, *J. Am. Chem. Soc.* **97**, 4144 (1975).
57. G. S. Y. Ng, L.-C. Yuan, I. J. Jokovac and J. B. Jones, *Tetrahedron* **40**, 1235 (1984).
58. J. B. Jones, M. A. W. Finch and I. J. Jakovac, *Can. J. Chem.* **60**, 2007 (1982).
59. W. Boland, K. Mertes, L. Jaenicke, D. G. Muller and E. Folster, *Helv. Chim. Acta* **66**, 1905 (1983).
60. H. B. Goodbrand and J. B. Jones, *J. Chem. Soc. Chem. Commun.*, 469 (1977); I. J. Jakovac, H. B. Goodbrand, K. P. Lok and J. B. Jones, *J. Am. Chem. Soc.* **104**, 4659 (1982).
61. J. B. Jones and I. J. Jakovac, *Organic Syntheses, 63* (ed. G. Saucy), p. 10. Wiley, New York, 1985.
62. J. R. Matos and C.-H. Wong, *J. Org. Chem.* **51**, 2388 (1986).
63. A. J. Bridges, P. S. Raman, G. S. Y. Ng and J. B. Jones, *J. Am. Chem. Soc.* **106**, 1461 (1984).
64. J. B. Jones and C. J. Francis, *Can. J. Chem.* **62**, 2578 (1984). K. P. Lok, I. J. Jakovac and J. B. Jones, *J. Am. Chem. Soc.* **107**, 2521 (1985).
65. Conversely, cyclohexanol is a substrate for HLAD but not for glycerol dehydrogenase. See also [18].
66. J. S. Wiseman and R. H. Abeles, *Biochemistry* **18**, 427 (1979).
67. N. Tamaki, M. Nakamura, K. Kimura and T. Hama, *J. Biochem. (Tokyo)* **82**, 73 (1977). N. Tamaki and T. Hama, *Methods Enzymol.* **89**, 649 (1982).
68. G. F. Betts, P. L. Poole, M. G. Springham and K. A. Bostian, *Biochem. J.* **183**, 633 (1979).
69. N. Takahashi, N. Kitabatake, R. Sasaki and H. Chiba, *Agric. Biol. Chem.* **43**, 1873, 1883, 1891 (1979).
70. C. E. Rietveld, M. De Zwart and P. G. F. Cox, *Biochem. Biophys. Acta* **911**, 162 (1987).
71. P. Klossek, M. Kirchner and J. Kurth, *J. Basic Microbiol.* **25**, 429 (1985).
72. C.-H. Wong and G. M. Whitesides, *J. Org. Chem.* **47**, 2816 (1982).
73. A. D. Mackerell Jr and R. Pietruszko, *Biochem. Biophys. Acta* **911**, 306 (1987).
74. H. Schutte, J. Flossdorf, H. Sahm and M.-R. Kula, *Eur. J. Biochem.* **62**, 151 (1976). N. Kato, M. Kano, Y. Tani and K. Ogata, *Agric. Biol. Chem.* **38**, 111 (1974).
75. C.-L. Liu and L. E. Mortenson, *J. Bacteriol.* **159**, 375 (1984).
76. J. B. Jones and T. C. Stadtman, *J. Biol. Chem.* **256**, 656 (1981).

77. C. Godfrey, A. Coddington, C. Greenwood, A. J. Thomson and P. M. A. Gadsbury, *Biochem. J.* **243**, 225 (1987).

78. T. C. Stadtman, *Ann. Rev. Biochem.* **49**, 93 (1980).

79. L. F. Wu and M. A. Mandrand-Berthelot, *J. Gen. Microbiol.* **133**, 2421 (1987).

80. K. H. Kroner, H. Schutte, W. Stach and M.-R. Kula, *J. Chem. Technol. Biotechnol.* **32**, 130 (1982). M.-R. Kula, K. H. Kroner, H. Hustedt and H. Shutte, in *Enzyme Engineering* (eds I. Chibata, S. Fukui and L. B. Wingard Jr), Vol. 6, p. 61. Plenum Press, New York, 1982.

81. Z. Shaked and G. M. Whitesides, *J. Am. Chem. Soc.* **102**, 7104 (1980). H. K. Chenault, M.-J. Kim, A. Akiyama, T. Miyazawa, E. S. Simon and G. M. Whitesides, *J. Org. Chem.* **52**, 2608 (1987). Y. Yamazaki and H. Maeda, *Agric. Biol. Chem.* **46**, 1571 (1982).

82. C. Wandrey, R. Wichmann, W. Lauchtenberger, M.-R. Kula and A. Buckmann, Ger. Offen. DE 2930087, 26 Feb. 1981; *Chem. Abs.* **94**, 172901b.

83. R. Wichmann, C. Wandrey, A. F. Buckmann and M.-R. Kula, *Biotechnol. Bioeng.* **23**, 2789 (1981).

84. S. Chao and M. S. Wrighton, *J. Am. Chem. Soc.* **109**, 5886 (1987).

85. W. Tischer, W. Tiemeyer and H. Simon, *Biochimie (Paris)* **62**, 331 (1980).

86. H. Guenther, S. Neumann and H. Simon, *J. Biotechnol.* **5**, 53 (1987).

87. A. Vischer and A. Wettstein, *Experientia* **9**, 371 (1953). D. H. Peterson, S. H. Eppstein, P. D. Meister, H. C. Murray, H. M. Leigh, A. Weintraub and L. M. Reineke, *J. Am. Chem. Soc.* **75**, 5768 (1953). J. Fried, R. W. Thoma and A. Klingsberg, *J. Am. Chem. Soc.* **75**, 5764 (1953).

88. R. L. Prairie and P. Talalay, *Biochemistry* **2**, 203 (1963).

89. M. A. Rahim and C. J. Sih, *J. Biol. Chem.* **241**, 3615 (1966).

90. A. I. Laskin, P. Grabowich, C. de L. Meyers and J. Fried, *J. Med. Chem.* **7**, 406 (1964).

91. F. W. Forney and A. J. Markovetz, *J. Bacteriol.* **96**, 1055 (1968). F. W. Forney and A. J. Markovetz, *Biochem. Biophys. Res. Commun.* **37**, 31 (1969). F. W. Forney, A. J. Markovetz and R. E. Kallio, *J. Bacteriol.* **93**, 649 (1967).

92. R. Shaw, *Nature* **209**, 1369 (1966). N. A. Donoghue, D. B. Morris and P. W. Trudgill, *Eur. J. Biochem.* **63**, 175 (1976). See also [12].

93. P. J. Chapman, G. Meerman and I. C. Gunsalus, *Biochem. Biophys. Res. Commun.* **20**, 104 (1965).

94. J. Meinwald and E. Frauenglass, *J. Am. Chem. Soc.* **82**, 5235 (1960).

95. H. E. Conrad, R. De Bus, M. J. Namtvedt and I. C. Gunsalus, *J. Biol. Chem.* **240**, 495 (1965) and references therein.

96. A. Siewinski, J. Dmochowska-Gladysz, T. Kolek, A. Zabza, K. Derdzinski and A. Nespiak, *Tetrahedron* **39**, 2265 (1983).

97. J. M. Schwab, W. Li and L. P. Thomas, *J. Am. Chem. Soc.* **105**, 4800 (1983). For a recent review on cyclohexanone oxygenase see: J. Latham and C. Walsh, *Ann. N.Y. Acad. Sci.* **471**, 208 (1986).

98. J. d'Angelo, G. Revial, R. Azerad and D. Buisson, *J. Org. Chem.* **51**, 40 (1986).

99. J. Ouazzani-Chahdi, D. Buisson and R. Azerad, *Tetrahedron Lett.* **28**, 1109 (1987).

100. C. C. Ryerson, D. P. Ballou and C. Walsh, *Biochemistry* **21**, 2644 (1982).

101. D. H. R. Barton, R. H. Hesse, M. M. Pechet, L. C. Smith, *J. Chem. Soc. Perkin Trans. I*, 1159 (1979).

102. R. Breslow, *Acc. Chem. Res.* **13**, 170 (1980).

103. J. Fossey, D. Lefort, M. Massoudi, J. -Y. Nedelec and J. Sorba, *Can. J. Chem.* **63**, 678 (1985). E. Trifilieff, L. Bang, A. S. Narula and G. Ourisson, *J. Chem. Res. (S)*, 64 (1978).

104. A. L. J. Beckwith and T. Duong, *J. Chem. Soc. Chem. Commun.*, 413 (1978).

105. R. A. Johnson, Oxygenations with Micro-organisms, in *Oxidation in Organic Synthesis*, Part C. (ed. W. S. Trahanovsky), pp. 131–210. Academic Press, New York, 1978.

106. G. Fonken and R. A. Johnson, *Chemical Oxidations with Microorganisms*. M. Dekker, New York, 1972.
107. K. Kieslich, Biotransformations, in *Biotechnology: A Comprehensive Treatise* (eds H.-J. Rehm and G. Reed), Vol. 6a, pp. 1–473. VCH Verlagsgesellschaft, Weinham, 1984.
108. K. Kieslich, *Microbial Transformations of Non-Steroid Cyclic Compounds (1976)*. Thieme, Stuttgart, 1976.
109. G. K. Skryabin and L. A. Golovleva, *Utilization of Microorganisms in Organic Synthesis*. Izdatel'stvo Nuaka, Moscow, 1976 (in Russian).
110. O. K. Sebek, *Mycologia* **75**, 383 (1983). K. Kieslich, *Bull. Soc. Chim. France* **11**, 9 (1980).
111. J. Schindler and R. D. Schmid, *Process Biochem.* **Sep./Oct.**, 2 (1982).
112. F. B. Kolot, *Process Biochem.* **Jan./Feb.**, 19 (1983).
113. K. Kieslich, Steroid conversions, in *Economic Microbiology, 5, Microbial Enzymes and Bioconversions* (ed. A. H. Rose), pp. 369–465. Academic Press, London, 1980.
114. W. Charney and H. L. Herzog, *Microbial Transformations of Steroids*. Academic Press, New York, 1968.
115. A. Capek, O. Hanc and M. Tadra, *Microbial transformations of steroids*. Academia, Prague, 1966. C. Vezina and S. Rakhit, Microbial transformations of steroids, in *CRC Handbook of Microbiology (4)*, pp. 117–441, CRC Press, Boca Raton, Florida, 1974. W. J. Marsheck Jr, *Prog. Ind. Microbiol.* **10**, 49 (1971). K. Kieslich, *Synthesis*, 120 (1969).
116. H. Iizuka and A. Naito, *Microbial Transformation of Steroids and Alkaloids*, pp. 1–294. University Park Press, State College, Pennsylvania and University of Tokyo Press, 1967.
117. H. Iizuka and A. Naito, *Microbial Conversion of Steroids and Alkaloids*, pp. 1–396. University of Tokyo Press and Springer-Verlag, Berlin, 1981.
118. H. L. Holland, *The Alkaloids* (ed. R. G. A. Rodrigo), Vol. 18. Academic Press, London, 1981.
119. L. C. Vining, Microbial transformations of alkaloids and related nitrogenous compounds, in *CRC Handbook of Microbiology (4)*, pp. 443–448. CRC Press, Boca Raton, Florida, 1974.
120. A. Ciegler, Microbial transformations of terpenes, in *CRC Handbook of Microbiology (4)*, pp. 449–458. CRC Press, Boca Raton, Florida, 1974. For the biohydroxylation of terpenes by mammals see: T. S. Santhanakrishnan, *Tetrahedron* **40**, 3597 (1984).
121. H. Dalton, Oxidation of hydrocarbons by methane monooxygenases from a variety of microbes, in *Advances in Applied Microbiology*, Vol. 26, pp. 71–87. Academic Press, London, 1980. A Ericson, B. Hedman, K. O. Hodgson, J. Green, H. Dalton, J. G. Bentsen, R. H. Beer and S. J. Lippard, *J. Am. Chem. Soc.* **110**, 2330 (1988).
122. C. Anthony, *The Biochemistry of Methylotrophs*. Academic Press, London, 1982.
123. A. Wiseman and D. J. King, *Topic Enzymol. Ferment. Biotechnol.* **6**, 151 (1982).
124. F. S. Sariaslani and J. P. N. Rosazza, *Enzyme Microbiol. Technol.* **6**, 242 (1984).
125. The following standard procedures have been described in J. B. Jones, C. J. Sih and D. Perlman (eds.), *Applications of Biochemical Systems in Organic Chemistry*, Part 1. Wiley, New York, 1976: (a) Conversion of progesterone to 11α-hydroxyprogesterone, p. 58; (b) 7-hydroxylation of nalidixic acid, p. 60; (c) hydroxylation of L-(+)-β-hydroxyisobutyric acid, p. 60; (d) 11β-hydroxylation of Reichstein's compound S, p. 61.
126. J. W. Gorrod and L. A. Damani (eds), *Biological Oxidation of Nitrogen in Organic Molecules, Chemistry, Toxicology and Pharmacology*. VCH, Vettagsgesellschaft, Weinham, 1985.
127. P. R. Ortiz de Montellano (ed.), *Cytochrome P-450: Structure, Mechanism and Biochemistry*, pp. 1–556. Plenum, New York, 1986.

128. S. Takemori, *TIBS* **12**, 118 (1987).
129. O. Kappeli, D. Sanglard and H. O. Laurila, Cytochrome P-450 of Yeasts, in *Cytochrome P-450 Biochemistry, Biophysics and Induction* (eds R. L. Vereczkey and K. Magyar), pp. 443–446. Akademiai Kaido, Budapest, 1985. F. P. Guengerich, T. L. MacDonald, L. T. Burka, R. E. Miller, D. C. Liebler, K. Zirvi, C. B. Frederick, F. F. Kadlubar and R. A. Prough, in *Cytochrome P-450: Biochemistry, Biophysics and Environmental Implications* (eds E. Heitanen, M. Laitinen, O. Hanninen), Elsevier/North Holland, New York, 1982. P. J. O'Brien and A. D. Rhimtula, The peroxidase function of cytochrome P-450 with possible implications for carcinogenesis, in *Biochemistry, Biophysics and Regulation of Cytochrome P-450* (eds J. A. Gustafsson, J. Carlstedt-Duke, A. Mode and J. Rafter), pp. 273–282. Elsevier/North Holland Biomedical Press, Amsterdam, 1980. J. Ahokas, C. Davies, P. Ravenscroft and B. Emmerson, Inhibition of cytochrome P-450 catalysed reactions by high molecular weight antibiotic, erythromycin, in *Biochemistry, Biophysics and Regulation of Cytochrome P-450* (eds J. A. Gustafsson, J. Carlstead-Duke, A. Mode and J. Rafter). Elsevier/North Holland Biomedical Press, 1980.
130. S. Sligar, M. Besman, M. Gelb, P. Gould, D. Heimbrook and D. Pearson, Occurrence and role of an acyl-peroxide intermediate in oxygen dependent P-450 hydroxylations, in *Biochemistry, Biophysics and Regulation of Cytochrome P-450* (eds J. A. Gustafsson *et al.*), pp. 379–382. Elsevier/North Holland Biomedical Press, 1980. R. Sato and T. Omura, *Cytochrome P-450*. Academic Press, New York, 1978 (Kodansha Ltd., Tokyo, 1978). S. Orrenius and L. Ernster, Microsomal Cytochrome P-450-Linked Systems in Mammalian Tissue, in *Molecular Mechanisms of Oxygen Activation* (ed. O. Hayashi), pp. 215–244. Academic Press, New York, 1974.
131. I. C. Gunsalus, J. R. Meeks, J. D. Lipscomb, P. Debrunner and E. Munck, Bacterial monooxygenases—The P-450 cytochrome system, in *Molecular Mechanisms of Oxygen Activation* (ed. O. Hayashi), pp. 561–614. Academic Press, New York, 1974.
132. C. Larroque and J. E. van Lier, *J. Biol. Chem.* **261**, 1083 (1986). M. B. McCarthy and R. E. White, *J. Biol. Chem.* **258**, 9153 (1983).
133. O. Kappeli, *Microbiol. Rev.* **50**, 244 (1986).
134. F. P. Guengerich and T. L. Macdonald, *Acc. Chem. Res.* **17**, 9 (1984).
135. R. E. White and M. J. Coon, *Ann. Rev. Biochem.* **49**, 315 (1980).
136. V. Ullrich, *Topics Curr. Chem.* **83**, 68 (1979).
137. S. W. May, *Enzyme Microbiol. Technol.* **1**, 15 (1979). A. J. Castellino and T. C. Bruice, *J. Am. Chem. Soc.* **110**, 158, 1313 (1988). T. G. Traylor and F. Xu, *J. Am. Chem. Soc.* **110**, 1953 (1988).
138. J. P. Rosazza and R. V. Smith, in *Advances in Applied Microbiology*, Vol. 25, pp. 169–208. Academic Press, London, 1979.
139. I. C. Gunsalus and S. G. Sligar, *Adv. Enzymol.* **47**, 1 (1978).
140. J. H. Dawson, L.-S. Kau, J. E. Penner-Hahn, M. Sono, K. Smith Eble, G. S. Bruce, L. P. Hager and K. O. Hodgson, *J. Am. Chem. Soc.* **108**, 8114 (1986).
141. E. L. Wheeler, *Biochem. Biophys. Res. Commun.* **110**, 646 (1983).
142. C. W. Jones and R. K. Poole, The Analysis of Cytochromes, in *Methods in Microbiology*, Vol. 18 (ed. G. Gottschalk), pp. 285–328. Academic Press, London, 1985. H. G. Muller, W.-H. Schunck, P. Riege and H. Honeck, Cytochrome P-450 of Microorganisms, in *Cytochrome P-450* (eds K. Ruckpaul and H. Rein), pp. 337–369. Akademie-Verlag, Berlin, 1984.
143. J. E. Tomaszewski, D. M. Jerina and J. W. Daly, Cytochrome P-450 monooxygenases and drug metabolism, in *Annual Reports of Medicinal Chemistry* (ed. R. A. Wiley), Vol. 9, pp. 290–299. Academic Press, New York, 1974.

144. D. A. Haugen and M. J. Coon, *J. Biol. Chem.* **251**, 7929 (1976). T. A. van der Hoeven, D. A. Haugen and M. J. Coon, *Biochem. Biophys., Res. Commun.* **60**, 569 (1974). F. S. Heinemann and J. Ozols, *J. Biol. Chem.* **258**, 4195 (1983).

145. J. Ozols, *J. Biol. Chem.* **261**, 3965 (1986).

146. S. Byung-Joon, H. V. Gelboin, S.-S. Park, C. S. Yang and F. J. Gonzalez, *J. Biol. Chem.* **261**, 16689 (1986).

147. M. Haniu, L. G. Armes, M. Tanaka and K. T. Yasunobu, *Biochem. Biophys. Res. Commun.* **105**, 889 (1982).

148. T. L. Poulos, B. C. Finzel, I. C. Gunsalus, G. C. Wagner and J. Kraut, *J. Biol. Chem.* **260**, 16122 (1985).

149. T. L. Poulos, B. C. Finzel and A. J. Howard, *Biochemistry* **25**, 5314 (1986).

150. B. P. Unger, I. C. Gunsalus and S. G. Sligar, *J. Biol. Chem.* **261**, 1158 (1986).

151. P. P. Tamburini, G. G. Gibson, W. L. Backes, S. G. Sligar and J. B. Schenkman, *Biochemistry* **23**, 4526 (1984).

152. A.-W. Wahlefeld, L. Jaenicke and G. Hein, *Lieb. Ann. Chem.* **715**, 52 (1968).

153. W. M. Atkins and S. G. Sligar, *J. Am. Chem. Soc.* **109**, 3754 (1987). See also [223].

154. H. Rein, S. Maricic, G.-R. Janig, S. Vuk-Pavlovic, B. Benko, O. Ristau and K. Ruckpaul, *Biochem. Biophys. Acta* **446**, 325 (1976).

155. O. Bangcharoenpaurpong, A. K. Rizos, P. M. Champion, D. Jollie and S. G. Sligar, *J. Biol. Chem.* **261**, 8089 (1986). B. G. Malmstrom, *Ann. Rev. Biochem.* **51**, 21 (1982).

156. P. R. Ortiz de Montellano and R. A. Stearns, *J. Am. Chem. Soc.* **106**, 3415 (1987).

157. M. Hayano, M. Gut, R. I. Dorfman, O. K. Sebek, D. H. Peterson, *J. Am. Chem. Soc.* **80**, 2336 (1958). S. Bergstrom, S. Lindstredt, B. Samuelson, E. J. Corey and G. A. Gregoriou, *J. Am. Chem. Soc.* **80**, 2337 (1958). E. J. Corey, G. A. Gregoriou and D. H. Peterson, *J. Am. Chem. Soc.* **80**, 2338 (1958).

158. H. L. Holland and E. Riemland, *Can. J. Chem.* **63**, 1121 (1985). R. H. McClanahan, A. C. Huitric, P. G. Pearson, J. C. Desper and S. D. Nelson, *J. Am. Chem. Soc.* **110**, 1979 (1988).

159. F. P. Guengerich, *J. Biol. Chem.* **262**, 8459 (1987).

160. E. F. Johnson, G. E. Schwab, J. Singh and L. E. Viskery, *J. Biol. Chem.* **261**, 10204 (1986).

161. K. Sugiyama and W. F. Trager, *Biochemistry* **25**, 7336 (1986).

162. A. Parkinson, D. E. Ryan, P. E. Thomas, D. M. Jerina, J. M. Sayer, P. J. van Bladeren, M. Haniu, J. E. Shively and W. Levin, *J. Biol. Chem.* **261**, 11478 (1986). A. Parkinson, P. E. Thomas, D. E. Ryan, L. D. Gorsky, J. E. Shively, J. M. Sayer, D. M. Jerina and W. Levin, *J. Biol. Chem.* **261**, 11487 (1986).

163. H. B. Dunford and J. S. Stillman, *Coord. Chem. Rev.* **19**, 187 (1976). H. B. Dunford, *Adv. Inorg. Biochem.* **4**, 41 (1982). H. B. Dunford and A. D. Nadezhdin, in *The Past Eight Years of Peroxidase Research in Oxidases and Related Redox Systems* (eds T. E. King, H. S. Mason and M. Morrison), pp. 653–670. Pergamon Press, Elmsford, New York, 1982.

164. J. S. Dordick, A. M. Klibanov and M. A. Marletta, *Biochemistry* **25**, 2946 (1986).

165. T. L. Poulos, S. T. Freer, R. A. Alden, S. L. Edwards, U. Skogland, K. Takio, B. Eriksson, N.-H. Xuong, T. Yonetani and J. Kraut, *J. Biol. Chem.* **255**, 575 (1980). B. C. Finzel, T. L. Poulos and J. Kraut, *J. Biol. Chem.* **259**, 13027 (1984). V. Thanabal, J. S. de Ropp and G. N. La Mer, *J. Am. Chem. Soc.* **109**, 7516 (1987).

166. M. A. Ator, S. K. David and P. R. Ortiz de Montellano, *J. Biol. Chem.* **262**, 14954 (1987). T. G. Traylor and R. Popovitz-Biro, *J. Am. Chem. Soc.* **110**, 239 (1988).

167. S. L. Edwards, N.-H. Xuong, R. C. Hamlin and J. Kraut, *Biochemistry* **26**, 1503 (1987). B. Chance, L. Powers, Y. Ching, T. Poulos, G. R. Schonbaum, I. Yamazaki and K. G. Paul, *Archiv. Biochem. Biophys.* **235**, 596 (1984).

168. J. E. Penner-Hahn, K. Smith Eble, T. J. McMurray, M. Renner, A. L. Balch, J. T. Groves, J. H. Dawson and K. O. Hodgson, *J. Am. Chem. Soc.* **108**, 7819 (1986).

169. I. Morishima, M. Kurono and Y. Shiro, *J. Biol. Chem.* **261**, 9391 (1986).

170. P. Douzo and F. Leterrier, *Biochim. Biophys. Acta* **220**, 338 (1970).

171. R. Z. Kazandjian, J. S. Dordick and A. M. Klibanov *Biotechnol. Bioeng.* **28**, 417 (1986).

172. C. Vezina, S. N. Sehgal and K. Singh, *Adv. Appl. Microbiol.* **10**, 221 (1968). C. Vezina and K. Singh, Transformations of organic compounds by fungal spores, in *The Filamentous Fungi, Vol. 1, Industrial Mycology* (eds J. E. Smith and D. R. Berry), pp. 158–192. Edward Arnold, London, 1975. A. Jaworski, L. Sedlaczek, A. Sasiak and J. Dlugonski, *Eur. J. Appl. Microbiol. Biotechnol.* **16**, 63 (1982). A. Jaworski, L. Sedlaczek, D. Wilmanska, A. Sasiak and A. Strycharski, *Z. Allge. Mikrobiol.* **22**, 327 (1982). A. Jaworski, L. Sedlaczek, J. Dlugonski and A. Zajaczkowska, *J. Basic Microbiol.* **25**, 423 (1985).

173. R. A. Johnson, M. E. Herr, H. C. Murray and G. S. Fonken, *J. Org. Chem.* **33**, 3217 (1968).

174. R. Furstoss, A. Archelas, J. D. Fourneron and B. Vigne, Biohydroxylation of non-activated carbon atoms. A model of the hydroxylation site of the fungus *Beauveria sulfurescens*, in *Enzymes as Catalysts in Organic Synthesis* (ed. M. P. Schneider), pp. 361–370. D. Reidel, New York, 1986. R. Furstoss, A. Archelas, J. D. Fourneron and B. Vigne, Biohydroxylation of non-activated carbon atoms by the fungus *Beauveria sulfurescens*, in *Organic Synthesis Interdisciplinary Challenge, Proceedings of the 5th IUPAC Symposium on Organic Synthesis* (eds J. Streith, H. Prinzback and G. Schill), pp. 215–226. Blackwell, Oxford, 1984. R. Furstoss, A. Archelas, B. Waegell, J. Le Petit and L. Deveze, *Tetrahedron Lett.* **21**, 451 (1980).

175. A. Kramli and J. Horvath, *Nature* **163**, 219 (1949), and references therein.

176. O. Hechter, R. P. Jacobsen, R. Jeanloz, H. Levy, C. W. Marshall, G. Pincus and V. Schenker, *J. Am. Chem. Soc.* **71**, 3261 (1949).

177. D. H. Peterson and H. C. Murray, *J. Am. Chem. Soc.* **74**, 1871 (1952). D. H. Peterson, H. C. Murray, S. H. Eppstein, L. M. Reineke, A. Weintraub, P. D. Meister and H. M. Leigh, *J. Am. Chem. Soc.* **74**, 5933 (1952).

178. J. Fried, R. W. Thoma, J. R. Gerke, J. E. Herz, M. N. Donin and D. Perlman, *J. Am. Chem. Soc.* **74**, 3692 (1952).

179. G. M. Schull and D. A. Kita, *J. Am. Chem. Soc.* **77**, 763 (1955). L. Velluz, J. Valls and G. Nomine, *Angew. Chem. Int. Ed. Engl.* **4**, 181 (1965).

180. V. Bihari, P. P. Goswami, S. H. M. Rizvi, A. W. Kahn, S. K. Basu and V. C. Vora, *Biotechnol. Bioeng.* **26**, 1403 (1984). D. F. Broad, J. Foulkes and P. Dunnill, *Biotechnol. Lett.* **6**, 357 (1984). T. A. Clark, I. S. Maddox and R. Chong, *Eur. J. Appl. Microbiol. Biotechnol.* **17**, 211 (1983). K. Sonomoto, K. Nomura, A. Tanaka and S. Fukui, *Eur. J. Appl. Microbiol. Biotechnol.* **16**, 57 (1982).

181. A. Tanaka, K. Sonomoto, M. M. Hoq, N. Usui, K. Nomura and S. Fukui, *Enzyme Eng.* **6**, 131 (1982).

182. K. Sonomoto, N. Usui, A. Tanaka and S. Fukui, *Eur. J. Appl. Microbiol. Biotechnol.* **17**, 203 (1983). E. W. Weiler, M. Droste, J. Eberle, H. J. Halfman and A. Weber, *Appl. Microbiol. Biotechnol.* **27**, 252 (1987).

183. D. Perlman, E. Titius and J. Fried, *J. Am. Chem. Soc.* **74**, 2126 (1952).

184. S. Hayakawa, *Z. Allge. Microbiol.* **22**, 309 (1982).

185. S. Sawada, S. Kulprecha, N. Nilubol, T. Yoshida, S. Kinoshita and H. Taguchi, *Appl. Env. Microbiol.* **44**, 1249 (1982).

186. Y. Jodoi, T. Nihira, H. Naoki, Y. Yamada and H. Taguchi, *Tetrahedron* **43**, 487 (1987).

187. K. Kieslich, G.-A. Hoyer, A. Seeger, R. Wiechert and U. Kerb, *Chem. Ber.* **113**, 203
      (1980). K. K. Kieslich, H. Weiglepp and G.-A. Hoyer, *Chem. Ber.* **112**, 979 (1979). H.-J.
      Vidic, D. Rosenberg and K. Kieslich, *Chem. Ber.* **111**, 2143 (1978). K. Kieslich and G.
      Schulz, *Annaler.* **726**, 152 (1969). K Kieslich, H. Weiglepp, K. Petzoldt and P. Hill,
      Tetrahedron **27**, 445 (1971).
188. K. Kieslich, H.-D. Berndt, R. Wiechert, U. Kerb, G. Schulz and H.-J. Koch, *Annaler.*
      726, 161 (1969).
189. K. Kieslich, K. Petzoldt, H. Kosmol and W. Koch, *Annaler.* **726**, 168 (1969).
190. P. C. Cherry, E. R. H. Jones and G. D. Meakins, *J. Chem. Soc. Chem. Commun.*, 587
      (1966). J. E. Bridgeman, P. C. Cherry, W. R. T. Cottrell, E. R. H. Jones, P. W. Le Quesne
      and G. D. Meakins, *J. Chem. Soc. Chem. Commun.*, 560 (1966). J. E. Bridgeman, P. C.
      Cherry, E. R. H. Jones and G. C. Meakins, *J. Chem. Soc. Chem. Commun.*, 482 (1967). J.
      E. Bridgeman, J. W. Browne, P. C. Cherry, M. G. Combe, J. M. Evans, E. R. H. Jones,
      A. Kasal, G. D. Meakins, Y. Morisawa and P. D. Woodgate, *J. Chem. Soc. Chem.*
      Commun., 463 (1969). J. E. Bridgeman, C. E. Butchers, E. R. H. Jones, A. Kasal, G. D.
      Meakins and P. D. Woodgate, *J. Chem. Soc. (C)*, 244 (1970). J. W. Blunt, I. M. Clark,
      J. M. Evans, E. R. H. Jones, G. D. Meakins and J. T. Pinhey, *J. Chem. Soc. (C)*, 1136
      (1971). A. M. Bell, P. C. Cherry, I. M. Clark, W. A. Denny, E. R. H. Jones, G. D.
      Meakins and P. D. Woodgate, *J. Chem. Soc. Perkin Trans. I.*, 2081 (1972). A. M. Bell,
      W. A. Denny, E. R. H. Jones, G. D. Meakins and W. E. Muller, *J. Chem. Soc. Perkin*
      Trans. I., 2759 (1972). A. M. Bell, J. W. Browne, W. A. Denny, E. R. H. Jones, A. Kasal
      and G. D. Meakins, *J. Chem. Soc. Perkin Trans. I.*, 2930 (1972). J. W. Browne, W. A.
      Denny, E. R. H. Jones, G. D. Meakins, Y. Morisawa, A. Pendlebury and J. Pragnell, *J.*
      Chem. Soc. Perkin Trans. I., 1493 (1973). V. E. M. Chambers, W. A. Denny, J. M. Evans,
      E. R. H. Jones, A. Kasal, G. D. Meakins and J. Pragnell, *J. Chem. Soc. Perkin Trans. I.*,
      1500 (1973). A. M. Bell, I. M. Clark, W. A. Denny, E. R. H. Jones, G. D. Meakins, W. E.
      Muller and E. E. Richards, *J. Chem. Soc. Perkin Trans. I.*, 2131 (1973). A. S. Clegg, W. A.
      Denny, E. R. H. Jones, G. D. Meakins and J. T. Pinhey, *J. Chem. Soc. Perkin Trans. I.*,
      2137 (1973). A. M. Bell, V. E. M. Chambers, E. R. H. Jones, G. D. Meakins, W. E.
      Muller and J. Pragnell, *J. Chem. Soc. Perkin Trans. I.*, 312 (1974). M. J. Ashton, A. S.
      Bailey and E. R. H. Jones, *J. Chem. Soc. Perkin Trans. I.*, 1658 (1974). V. E. M.
      Chambers, E. R. H. Jones, G. D. Meakins, J. O. Miners and A. L. Wilkins, *J. Chem. Soc.*
      Perkin Trans. I., 55 (1975). A. M. Bell, E. R. H. Jones, G. D. Meakins, J. O. Miners and
      A. Pendlebury, *J. Chem. Soc. Perkin Trans. I.*, 357 (1975). J. M. Evans, E. R. H. Jones,
      G. D. Meakins, J. O. Miners, A. Pendlebury and A. L. Wilkins, *J. Chem. Soc. Perkin*
      Trans. I., 1356 (1975). V. E. M. Chambers, W. A. Denny, E. R. H. Jones, G. D. Meakins,
      J. O. Miners, J. T. Pinhey and A. L. Wilkins, *J. Chem. Soc. Perkin Trans. I.*, 1359 (1975).
      A. M. Bell, A. D. Boul, E. R. H. Jones, G. D. Meakins, J. O. Miners and A. L. Wilkins,
      J. Chem. Soc. Perkin Trans. I., 1364 (1975). E. R. H. Jones, G. D. Meakins, J. O. Miners,
      J. H. Pragnell and A. L. Wilkins, *J. Chem. Soc. Perkin Trans. I.*, 1552 (1975). A. M. Bell,
      E. R. H. Jones, G. D. Meakins, J. O. Miners and A. L. Wilkins, *J. Chem. Soc. Perkin*
      Trans. I., 2040 (1975). E. R. H. Jones, G. D. Meakins, J. O. Miners and A. L. Wilkins, *J.*
      Chem. Soc. Perkin Trans. I., 2308 (1975). E. R. H. Jones, G. D. Meakins, J. O. Miners,
      R. N. Mirrington and A. L. Wilkins, *J. Chem. Soc. Perkin Trans. I.*, 1842 (1976).
191. J. E. Bridgeman, P. C. Cherry, A. S. Clegg, J. M. Evans, E. R. H. Jones, A. Kasal, V.
      Kumar, G. D. Meakins, Y. Morisawa, E. E. Richards and P. D. Woodgate, *J. Chem.*
      Soc., Ser. C., 250 (1970).
192. A. D. Boul, J. W. Blunt, J. W. Browne, V. Kumar, G. D. Meakins, J. T. Pinhey and
      V. E. M. Thomas, *J. Chem. Soc., Ser. C*, 1130 (1971).

193. M. J. Ashton, A. S. Bailey and E. R. H. Jones, *J. Chem. Soc. Perkin Trans. I.*, 1665 (1974).
194. A. S. Bailey, M. L. Gilpin and E. R. H. Jones, *J. Chem. Soc. Perkin Trans. I.*, 259 (1977).
195. A. S. Bailey, M. L. Gilpin and E. R. H. Jones, *J. Chem. Soc. Perkin Trans. I.*, 265 (1977).
196. E. R. H. Jones, *Pure Appl. Chem.* **33**, 39 (1973).
197. H. L. Holland and B. J. Auret, *Can. J. Chem.* **53**, 2041 (1975). H. L. Holland, P. C. Chenchaiah, E. M. Thomas, B. Mader and M. J. Dennis, *Can. J. Chem.* **62**, 2740 (1984). H. L. Holland and P. C. Chenchaiah, *Can. J. Chem.* **63**, 1127 (1985).
198. H. L. Holland, *Chem. Soc. Rev.* **11**, 371 (1982).
199. H. L. Holland, *Acc. Chem. Res.* **17**, 398 (1984).
200. H. L. Holland and E. Riemland, *Can. J. Chem.* **63**, 981 (1985).
201. H. L. Holland, P. R. P. Diakow and G. J. Taylor, *Can. J. Chem.* **56**, 3121 (1978).
202. T. A. Crabb and N. M. Ratcliffe, *J. Chem. Res. (S)*, 48 (1986). S. A. Campbell, T. A. Crabb and R. O. Williams, *J. Chem. Res. (S)*, 208 (1986). T. A. Crabb, J. A. Saul and R. O. Williams, *J. Chem. Soc. Perkin Trans. I.*, 1041 (1981). T. A. Crabb, P. J. Dawson and R. O. Williams, *J. Chem. Soc. Perkin Trans. I.*, 2535 (1980). T. A. Crabb, J. A. Saul and R. O. Williams, *J. Chem. Soc. Perkin Trans. I.*, 2599 (1977). T. A. Crabb, P. J. Dawson and R. O. Williams, *J. Chem. Soc. Perkin Trans. I.*, 571 (1982). T. A. Crabb, P. J. Dawson and R. O. Williams, *Tetrahedron Lett.*, 3623 (1975).
203. N. R. Orme-Johnson, D. R. Light, R. W. White-Stevens and W. H. Orme-Johnson, *J. Biol. Chem.* **254**, 2103 (1979).
204. S. Nakajin, M. Shinoda, M. Haniu, J. E. Shively and P. F. Hall, *J. Biol. Chem.* **259**, 3971 (1984).
205. J. M. Trzaskos, R. T. Fischer and M. F. Favata, *J. Biol. Chem.* **261**, 16937 (1986). Y. Aoyama, Y. Yoshida, Y. Sonoda and Y. Sato, *J. Biol. Chem.* **262**, 1239 (1987). L. L. Fyre and C. H. Robinson, *J. Chem. Soc. Chem. Commun.*, 129 (1988).
206. E. Caspi, J. Wicha, T. Arunachalam, P. Nelson and G. Spiteller, *J. Am. Chem. Soc.* **106**, 7282 (1984). J. T. Kellis Jr and L. E. Vickery, *J. Biol. Chem.* **262**, 8840 (1987). P. A. Cole and C. H. Robinson, *J. Am. Chem. Soc.* **110**, 1284 (1988).
207. B. Szukalski, *Post. Biochem.* **23**, 95 (1977).
208. J. W. Blunt and J. B. Strothers, *Org. Magn. Res.* **9**, 439 (1977). J. R. Hanson and M. Siverns, *J. Chem. Soc. Perkin Trans. I.*, 1956 (1975).
209. J. Williamson, D. Van Orden and J. P. Rosazza, *Appl. Env. Microbiol.* **49**, 563 (1985).
210. D. de Marcano, J. F. del Giorgio, J. M. Evans, E. Garcia, L. Kohout, I. Ludovic and M. Narvaez, *Steroids* **41**, 1 (1983).
211. M. Zakelj-Mavric, I. Belic and H. E. Gottlieb, *FEMS Microbiol. Lett.* **33**, 117 (1986).
212. D. R. Brannon, F. W. Parrish, B. J. Wiley and L. Long Jr, *J. Org. Chem.* **32**, 1521 (1967).
213. F. R. Hanson, K. M. Mann, E. D. Nielson, H. V. Anderson, M. P. Brunner, J. N. Karnemaat, D. R. Colingsworth and W. J. Haines, *J. Am. Chem. Soc.* **75**, 5369 (1953).
214. M. E. John, T. Okamura, A. Dee, B. Adler, M. C. John, P. C. White, E. R. Simpson and M. R. Waterman, *Biochemistry* **25**, 2846 (1986).
215. J. Fuska, J. Prousek, J. Rosazza and M. Budesinsky, *Steroids* **40**, 157 (1982).
216. J. Fried, R. W. Thoma, A Klingsberg, *J. Am. Chem. Soc.* **75**, 5764 (1953).
217. S. Kulprecha, T. Nihira, C. Shimomura, K. Yamada, N. Nilubol, T. Yoshida and H. Taguchi, *Tetrahedron* **40**, 2843 (1984). S. B. Mahato and A. Mukherjee, *J. Steroid Biochem.* **21**, 341 (1984).
218. T. A. Clark, R. Chong and I. S. Maddox, *Appl. Microbiol. Biotechnol.* **21**, 132 (1985).
219. M. Iida, T. Murohisa, A. Yoneyama and H. Iizuka, *J. Ferment. Technol.* **63**, 559 (1985).
220. R. L. Horst, T. A. Reinhardt, C. F. Ramberg, N. J. Koszewski and J. L. Napoli, *J. Biol. Chem.* **261**, 9250 (1986).

221. S. Andersson and H. Jornvall, *J. Biol. Chem.* **261**, 16932 (1986).

222. J. Trzaskos, S. Kawata and J. L. Gaylor, *J. Biol. Chem.* **261**, 14651 (1986).

223. M. H. Gelb, D. C. Heimbrook, P. Malkonen and S. G. Sligar, *Biochemistry* **21**, 370 (1982). J. T. Groves, G. A. McClusky, R. E. White and M. J. Coon, *Biochem. Biophys. Res. Commun.* **81**, 154 (1978).

224. R. E. White, M.-B. McCarthy, K. D. Egeberg and S. G. Sligar, *Arch. Biochem. Biophys.* **228**, 493 (1984).

225. J. T. Groves and D. V. Subramanian, *J. Am. Chem. Soc.* **106**, 2177 (1984).

226. H. Ziffer and D. T. Gibson, *Tetrahedron Lett.*, 2137 (1975).

227. W. R. Abraham, H. M. R. Hoffman, K. Kieslich, K. G. Reng and B. Stumpf, Microbial transformations of some monoterpenoids and sesquiterpenes, in *Enzymes in Organic Synthesis* (eds. R. Porter and S. Clark), pp. 146–160. *Ciba Foundation Symposium* 111, Pitman, London (1985).

228. R. E. White and M.-B. McCarthy, *J. Am. Chem. Soc.* **106**, 4922 (1984).

229. S. Kurozumi, T. Toru and S. Ishimoto, *Tetrahedron Lett.*, 4959 (1973).

230. B. Tabenkin, R. A. Le Mahieu, J. Berger and R. W. Kierstead, *Appl. Microbiol.* **17**, 714 (1969).

231. H. G. Davies, M. J. Dawson, G. C. Lawrence, J. Mayall, D. Noble, S. M. Roberts, M. K. Turner and W. F. Wall, *Tetrahedron Lett.* **27**, 1089 (1986).

232. G. S. Fonken, M. E. Herr, H. C. Murray and L. M. Reineke, *J. Am. Chem. Soc.* **89**, 672 (1967).

233. G. S. Fonken, M. E. Herr, H. C. Murray and L. M. Reineke, *J. Org. Chem.* **33**, 3182 (1968).

234. R. A. Johnson, M. E. Herr, H. C. Murray and G. S. Fonken, *J. Org. Chem.* **33**, 3187 (1968).

235. R. A. Johnson, M. E. Herr, H. C. Murray, L. M. Reineke and G. S. Fonken, *J. Org. Chem.* **33**, 3195 (1968). M. E. Herr, R. A. Johnson, H. C. Murray, L. M. Reineke and G. S. Fonken, *J. Org. Chem.* **33**, 3201 (1968).

236. R. A. Johnson, M. E. Herr, H. C. Murray and G. S. Fonken, *J. Org. Chem.* **35**, 622 (1970).

237. R. A. Johnson, H. C. Murray, L. M. Reineke and G. S. Fonken, *J. Org. Chem.* **33**, 3207 (1968).

238. A. Archelas, R. Furstoss, B. Waegell, J. Le Petit and L. Deveze, *Tetrahedron* **40**, 355 (1984).

239. A. Archelas, J. D. Fourneron, B. Vigne and R. Furstoss, *Tetrahedron* **42**, 3863 (1986).

240. A. Archelas and C. Morin, *Tetrahedron Lett.* **25**, 1277 (1984). A. Archelas, J.-D. Fourneron and R. Furstoss, *J. Org. Chem.* **53**, 1797 (1988).

241. J. P. N. Rosazza, J. J. Steffens, S. Sariaslani, A. Goswami, J. M. Beale, S. Reeg and R. Chapman, *Appl. Env. Microbiol.* **53**, 2482 (1987).

242. Y. Yamazaki and H. Maeda, *Tetrahedron Lett.* **26**, 4775 (1985).

243. T. Suga, T. Hirata and Y. S. Lee, *Chem. Lett.*, 1595 (1982).

244. T. Hirata, T. Aoki, Y. Hirano, T. Ito and T. Suga, *Bull. Chem. Soc. Jpn.* **54**, 3527 (1981).

245. Y. S. Lee, T. Hirata and T. Suga, *J. Chem. Soc. Perkin Trans. I.*, 2475 (1983). See also T. Suga, T. Hirata and H. Hamada, *Bull. Chem. Soc. Jpn.* **59**, 2865 (1986). T. Suga, H. Hamada and T. Hirata, *Chem. Lett.*, 471 (1987). T. Suga, S. Izumi and T. Hirata, *Chem. Lett.*, 2053 (1986). T. Suga, S. Izumi, T. Hirata and H. Hamada, *Chem. Lett.*, 425, 471, 903 (1987).

246. N. Serizawa, S. Serizawa, K. Nakagawa, K. Furuya, T. Okazaki and A. Terahara, *J. Antibiotics* **36**, 887 (1983). H. Oikawa, A. Ichihara and S. Sakamura, *J. Chem. Soc. Chem. Commun.*, 600 (1988).

247. N. Serizawa, K. Nakagawa, K. Hamano, Y. Tsujita, A. Terahara and H. Kuwano, *J. Antibiotics* **36**, 604, 608 (1983). N. Serizawa, K. Nakagawa, Y. Tsujita, A. Terahara, H. Kuwano and M. Tanaka, *J. Antibiotics* **36**, 918 (1983).

248. T. Okazaki, N. Serizawa, R. Enokita, A. Torikata and A. Terahara, *J. Antibiotics* **36**, 1176 (1983).

249. J. A. Nelson, M. R. Czarny and T. A. Spencer, *Bioorg. Chem.* **11**, 371 (1982).

250. D. M. Hollinshead, S. C. Howell, S. V. Ley, M. Mahon, N. M. Ratcliffe and P. A. Worthington, *J. Chem. Soc. Perkin Trans. I.*, 1579 (1983).

251. A. Schubert, A. Riecke, G. Hilgetag, R. Siebert and S. Schwarz, *Naturwissenschaften* **45**, 623 (1958).

252. S. R. Nadkarni, P. M. Akut, B. N. Ganguli, Y. Khandelwal, N. J. de Souza and R. H. Rupp, *Tetrahedron Lett.* **27**, 5265 (1986).

253. J. Ipsen, J. Fuska, A. Foskova and J. P. Rosazza, *J. Org. Chem.* **47**, 3278 (1982). J. F. Gordon, J. R. Hanson and A. H. Ratcliffe, *J. Chem. Soc. Chem. Commun.*, 6 (1988).

254. J. MacMillan, Gibberellin metabolism beyond Gibberellin A12-aldehyde: 2-oxoglutarate-coupled oxidases. In *Biochemical Aspects of Synthetic and Naturally Occurring Plant Growth Regulators* (eds R. Menhenett and D. K. Lawrence), pp. 13–31. Society of Chemical Industry, British Plant Growth Regulator Group Monograph No. 11, London, 1984. See also J. MacMillan and C. L. Willis, *J. Chem. Soc. Perkin Trans. I.*, 2177 (1985). M. Takahashi, Y. Kamiya, N. Takahashi and J. E. Graebe, *Planta* **168**, 190 (1986).

255. S. Englard, J. S. Blanchard and C. F. Midelfort, *Biochemistry* **24**, 1110 (1985).

256. M. H. Beale, J. MacMillan, C. R. Spray, D. A. Taylor and B. O. Phinney, *J. Chem. Soc. Perkin Trans. I.*, 541 (1984).

257. Y. Okumura, M. Ohnishi, R. Okamoto and T. Ishikura, *Agric. Biol. Chem.* **46**, 3063 (1982).

258. K. Majamaa, V. Gunzler, H. N. Hanauske-Abel, R. Myllya and K. I. Kivirikko, *J. Biol. Chem.* **261**, 7819 (1986); J. Koivu, R. Myllyla, T. Helaakaski, T. D. Wajaniemi, K. Tasaneu and K. I. Kivirikko, *J. Biol. Chem.* **262**, 6447 (1987).

259. E. Eich and R. Sieben, *Planta Med.*, 282 (1985).

· 260. F. M. Eckenrode, *J. Nat. Prod.* **47**, 882 (1984).

261. R. M. Adlington, J. E. Baldwin, B. Chakravarti, M. Jung, S. Moroney, J. A. Murphy, P. D. Singh, J. J. Usher and C. Vallejo, *J. Chem. Soc. Chem. Commun.*, 153 (1983). J. E. Baldwin, R. M. Adlington, R. T. Aplin, N. P. Crouch, G. Knight, C. J. Schofield and H.-T. Ting. *J. Chem. Soc. Chem. Commun.*, 1651, 1654 (1987).

262. A. Garcia-Granados, A. Martinez, M. E. Onorato and J. M. Arias, *J. Nat. Prod.* **48**, 371 (1985).

263. J. P. Beilby, E. L. Ghisalberti, P. R. Jefferies, M. A. Sefton and P. N. Sheppard, *Tetrahedron Lett.*, 2589 (1973).

264. J. M. Arias, A. Garcia-Granados, A. Martinez and E. Onorato, *J. Nat. Prod.* **47**, 59 (1984).

265. S. W. May and B. J. Abbott, *J. Biol. Chem.* **245**, 1725 (1973). M.-J. de Smet, B. Witholt and H. Wynberg, *J. Org. Chem.* **46**, 3128 (1981).

266. Z.-H. Yi and H.-J. Rehm, *Eur. J. Appl. Microbiol. Biotechnol.* **16**, 1 (1982). A. G. Katopodis, H. A. Smith Jr and S. W. May, *J. Am. Chem. Soc.* **110**, 897 (1988).

267. G. Eggink, P. H. van Lelyveld, A. Arnberg, N. Arfman, C. Witteveen and B. Witholt, *J. Biol. Chem.* **262**, 6400 (1987).

268. J. Green, S. D. Prior and H. Dalton, *Eur. J. Biochem.* **153**, 137 (1985). H. Matsuyama, T. Nakahara and Y. Minoda, *Agric. Biol. Chem.* **45**, 9 (1981).

269. S. Shapiro, J. U. Piper and E. Caspi, *J. Am. Chem. Soc.* **104**, 2301 (1982).

270. J. P. Jones and W. F. Trager, *J. Am. Chem. Soc.* **109**, 2171 (1987); **110**, 2018 (1988); H. Matsuyama, T. Nakahara and Y. Minoda, *Agric. Biol. Chem.* **45**, 9 (1981).

271. C. T. Goodhue and J. R. Schaeffer, *Biotechnol. Bioeng.* **13**, 203 (1971).
272. R. S. Robinson and L. J. Szarko, *Eur. Pat. Appl.*, EP 151,419 (14 August 1985).
273. Q. Branca and A. Fischli, *Helv. Chim. Acta* **60**, 925 (1977).
274. N. Cohen, W. F. Eichel, R. J. Lopresti, C. Neukom and G. Saucy, *J. Org. Chem.* **41**, 3505 (1976).
275. J. Hasegawa, S. Hamaguchi, M. Ogura and K. Watanabe, *J. Ferment. Technol.* **59**, 257 (1981). J. Hasegawa, M. Ogura, H. Kanema, H. Kawaharada and K. Watanabe, *J. Ferment. Technol.* **61**, 37 (1983).
276. K. Mori and S. Kuwahara, *Tetrahedron* **42**, 5545 (1986).
277. S. Maemoto and K. Mori, *Chem. Lett.*, 109 (1987).
278. T. Iimori and M. Shibasaki, *Tetrahedron Lett.* **27**, 2149 (1986). D. M. Tschaen, L. M. Fuentes, J. E. Lynch, W. L. Laswell, R. P. Volante and I. Shinkai, *Tetrahedron Lett.* **29**, 2779 (1988). See also Chapter 3, [17].
279. C. Bohne, I. D. MacDonald and H. B. Dunford, *J. Am. Chem. Soc.* **108**, 7867 (1986).
280. R. E. McMahon, H. R. Sullivan, J. C. Craig and W. E. Pereira Jr, *Arch. Biochim. Biophys.* **132**, 575 (1969).
281. R. E. White, J. P. Miller, L. V. Favreau and A. Bhattacharyya, *J. Am. Chem. Soc.* **108**, 6024 (1986).
282. R. P. Hanzlik, K. Hogberg, J. B. Moon and C. M. Judson, *J. Am. Chem. Soc.* **107**, 7164 (1985).
283. H. L. Holland, I. M. Carter, C. Chenchaiah, S. H. Khan, B. Munoz, R. W. Ninniss and D. Richards, *Tetrahedron Lett.* **26**, 6409 (1985).
284. H. L. Holland, E. J. Bergen, P. C. Chenchaiah, S. H. Khan, B. Munoz, R. W. Ninniss and D. Richards, *Can. J. Chem.* **65**, 502 (1987).
285. K. Wimalasena and S. W. May, *J. Am. Chem. Soc.* **109**, 4036 (1987). T. M. Bargar, R. J. Broersma, L. C. Creemer, J. R. McCarthy, J.-M. Hornsperger, P. V. Attwood and M. J. Jung, *J. Am. Chem. Soc.* **110**, 2975 (1988).
286. A. R. Battersby, P. W. Sheldrake, J. Staunton and D. C. Williams, *J. Chem. Soc. Perkin Trans. I.*, 1056 (1976).
287. A. R. Battersby, J. E. Kelsey, J. Staunton and K. E. Suckling, *J. Chem. Soc. Perkin Trans. I.*, 1609 (1973).
288. M. G. Peter and W. Vaupel, *J. Chem. Soc. Chem. Commun.*, 848 (1985).
289. T. A. Crabb, S. L. Soilleux, *J. Chem. Soc. Perkin Trans. I.*, 1381 (1985).
290. D. Rosi, G. Peruzzotti, E. W. Dennis, D. A. Berberian, H. Freele, B. F. Tullar and S. Archer, *J. Med. Chem.* **10**, 867 (1967). See also H. C. Richards, in *Medicinal Chemistry: the Role of Organic Chemistry in Drug Research* (eds. S. M. Roberts and B. J. Price). Academic Press, London (1985).
291. N. S. Egorev, L. I. Vorobeva, E. V. Dovgilevich and L. V. Modyanova, *Prikl. Biokhim. Mikrobiol.* **21**, 349 (1985).
292. G. T. Phillips, G. W. J. Matcham, M. A. Bertola, A. F. Marx and H. S. Koger, *Eur. Pat. Appl.*, EP 205,21517 (December 1986).
293. E. A. Komives and P. R. Ortiz de Montellano, *J. Biol. Chem.* **262**, 9793 (1987).
294. G. Neef, U. Eder, K. Petzoldt, A. Seeger and H. Weiglepp, *J. Chem. Soc. Chem. Commun.*, 366 (1982).
295. A. Goswami, J. P. Schaumberg, M. W. Duffel and J. P. Rosazza, *J. Org. Chem.* **52**, 1500 (1987).
296. F. Eckenrode, W. Peczynska-Czoch and J. P. Rosazza, *J. Pharm. Sci.* **71**, 1246 (1982).
297. K. M. Kerr and P. J. Davis, *J. Org. Chem.* **51**, 1741 (1986).
298. D. C. Heimbrook, R. I. Murray, K. D. Egeberg and S. G. Sligar, *J. Am. Chem. Soc.* **106**, 1514 (1984).

299. B. Vigne, A. Archeles, J. D. Fourneron and R. Furstoss, *Tetrahedron* **42**, 2451 (1986).

300. J. S. Walsh and G. T. Miwa, *Biochem. Biophys. Res. Commun.* **121**, 960 (1984).

301. A. M. Klibanov, Z. Berman and B. N. Alberti, *J. Am. Chem. Soc.* **103**, 6263 (1981).

302. O. K. Sebek and L. A. Dolak, *J. Antibiotics* **37**, 136 (1984).

303. D. F. Berry, A. J. Francis and J.-M. Bollag, *Microbiol. Rev.* **51**, 43 (1987).

304. T. E. Smithgall, R. G. Harvey and T. M. Penning, *J. Biol. Chem.* **261**, 6184 (1986).

305. R. V. Smith and J. P Rosazza, *Archiv. Biochem. Biophys.* **161**, 551 (1974).

306. S. Kojo and K. Fukunishi, *Chem. Lett.*, 1707 (1983).

307. B. J. Auret, S. K. Balani, D. R. Boyd and R. M. E. Greene, *J. Chem. Soc., Chem. Commun.*, 1585 (1971).

308. M. Tsuda, S. Oikawa, Y. Okumara, K. Kimura, T. Urabe and M. Nakajima, *Chem. Pharm. Bull.* **34**, 4457 (1986).

309. B. J. Auret, S. K. Balani, D. R. Boyd, R. M. E. Greene and G. Berchtold, *J. Chem. Soc. Perkin Trans. I.*, 2659 (1984).

310. S. G. Jezequel and I. J. Higgins, *J. Chem. Technol. Biotechol.* **33B**, 139 (1983). W. Adam and M. Balci, *Tetrahedron* **36**, 833 (1980).

311. J. E. Tomaszewski, D. M. Jerina and J. W. Daly, *Biochemistry* **14**, 2024 (1975).

312. R. F. Anderson, K. B. Patel and M. R. L. Stratford, *J. Biol. Chem.* **262**, 17475 (1987). B. Entsch, V. Massey and A. Claiborn, *J. Biol. Chem.* **262**, 6060 (1987).

313. J. B. Powlowski, S. Dagley, V. Massey and D. P. Ballou, *J. Biol. Chem.* **262**, 69 (1987).

314. K. Kieslich, H.-J. Vidic, K. Petzoldt and G.-A. Hoyer, *Chem. Ber.* **109**, 2259 (1976).

315. J. C. Cox and J. H. Golbeck, *Biotechnol. Bioeng.* **27**, 1395 (1985).

316. P. H. Jellinck, E. F. Hahn and J. Fishman, *J. Biol. Chem.* **261**, 7729 (1986).

317. J. G. Liehr, A. A. Ulubelen and H. W. Strobel, *J. Biol. Chem.* **261**, 16865 (1986).

318. F. Muller, *Biochem. Soc. Trans.* **13**, 443 (1985).

319. S. A. G. F. Angelino, F. Muller and H. C. van der Plas, *Biotechnol. Bioeng.* **XXVII**, 447 (1985).

320. S. A. G. F. Angelino, D. J. Buurman, H. C. van der Plas and F. Muller, *Recueil.* **101**, 342 (1982).

321. S. A. G. F. Angelino, D. J. Buurman, H. C. van der Plas and F. Muller, *Recueil.* **102**, 331 (1983).

322. G. Pelsy and A. M. Klibanov, *Biochim. Biophys. Acta* **742**, 352 (1983).

323. P. Lehky, H. Kulla and S. Mischer, *Eur. Pat. Appl.*, EP 152948 (28th August 1985).

324. G. Lesma, G. Palmisano, S. Tollari, P. Ame, P. Martelli and U. Valcavi, *J. Org. Chem.* **48**, 3825 (1983).

325. R. E. Betts, D. E. Walters and J. P. Rosazza, *J. Med. Chem.* **17**, 599 (1974).

326. S. Benkovic, D. Wallick, L. Bloom, B. J. Gaffney, P. Domanico, T. Dix and S. Pamber, *Biochem. Soc. Trans.* **13**, 436 (1985). J. McCraken, S. Pember, S. J. Benkovic, J. J. Villafranca, R. J. Miller and J. Peisach, *J. Am. Chem. Soc.* **110**, 1069 (1988).

327. K. Oka, T. Kato, T. Sugimoto, S. Matsuura and T. Nagatsu, *Biochim. Biophys. Acta* **661**, 45 (1981).

328. S. Kaufman, *Biochem. Soc. Trans* **13**, 433 (1985).

329. D. W. Potter and J. A. Hinson, *J. Biol. Chem.* **262**, 966 (1987).

330. D. M. X. Donnelly, F. G. Murphy, J. Polonski and T. Prange, *J. Chem. Soc. Perkin Trans. I.*, 2719 (1987). M. J. Begley, L. Crombie, M. London, J. Savin and D. A. Whiting, *J. Chem. Soc. Chem. Commun.*, 1319 (1982).

331. R. J. Petroski, R. B. Bates, G. S. Linz and J. P. Rosazza, *J. Pharm. Sci.* **72**, 1291 (1983).

332. D. C. McMillan, P. P. Fu and C. E. Cerniglia, *Appl. Environ. Microbiol.* **53**, 2560 (1987).

333. F. S. Sariaslani, L. R. McGee and D. W. Ovenall, *Appl. Environ. Microbiol.* **53**, 1780 (1987).

334. C. E. Cerniglia, J. R. Althaus, F. E. Evans, J. P. Freeman, R. K. Mitchum and S. K. Yang, *Chem. Biol. Int.* **44**, 119 (1983).

335. D. G. H. Ballard, A. Courtis, I. M. Shirley and S. C. Taylor, *J. Chem. Soc. Chem. Commun.*, 954 (1983). For aza-arenes see D. R. Boyd, R. A. S. McMordie, H. P. Porter, H. Dalton, R. O. Jenkins and O. W. Howarth, *J. Chem. Soc. Chem. Commun.*, 1722 (1987).

336. H. Ziffer, K. Kabuto, D. T. Gibson, V. M. Kobal and D. M. Jerina, *Tetrahedron* **33**, 2491 (1977).

337. S. J. C. Taylor, D. W. Ribbons, A. M. Z. Slawin, D. A. Widdowson and D. J. Williams, *Tetrahedron Lett.* **28**, 6391 (1987).

338. J. T. Rossiter, S. R. Williams, A. E. G. Cass and D. W. Ribbons, *Tetrahedron Lett.* **28**, 5173 (1987). See also D. W. Ribbons, A. E. G. Cass, J. T. Rossiter, S. J. C. Taylor, M. P. Woodland, D. A. Widdowson, S. R. Williams, R. B. Baker and R. E. Martin, *J. Fluorine Chem.* **37**, 299 (1987).

339. S. V. Ley, F. Sternfeld and S. Taylor, *Tetrahedron Lett.* **28**, 225 (1987).

340. T. Hudlicky, G. Barbieri, H. Luna and L. D. Kwart, *Proc. 193rd ACS Nat. Meeting*, Denver, Colorado (1987). T. Hudlicky, H. Luna, G. Barbieri and L. D. Kwart, *J. Am. Chem. Soc.* **110**, 4735 (1988).

341. J. Sekiguchi, S. Katayama and Y. Yamada, *Appl. Environ. Microbiol.* **53**, 1531 (1987).

342. For a biologically orientated review see R. S. Phillips and S. W. May, *Enzyme Microbiol. Technol.* **3**, 9 (1981).

343. C. E. Holmlund, K. J. Sax, B. E. Nielsen, R. E. Hartman, R. H. Evans and R. H. Blank, *J. Org. Chem.* **27**, 1468 (1962).

344. R. M. Dobson and P. B. Sollman, *US Patent* 2,999,101 (1961).

345. R. M. Dobson, N. Newman and H. M. Tsuchiya, *J. Org. Chem.* **27**, 2707 (1962).

346. B. J. Auret, D. R. Boyd, H. B. Henbest and S. Ross, *J. Chem. Soc.*, 2371 (1968). S. W. May and R. S. Phillips, *J. Am. Chem. Soc.* **102**, 5981 (1980).

347. E. Abushanab, D. Reed, F. Suzuki and C. J. Sih, *Tetrahedron Lett.*, 3415 (1978).

348. H. L. Holland, H. Popperl, R. W. Ninniss and P. C. Chenchaiah, *Can. J. Chem.* **63**, 1118 (1985).

349. D. R. Light, D. J. Waxman and C. Walsh, *Biochemistry* **21**, 2490 (1982). See also B. P. Branchaud and C. T. Walsh, *J. Am. Chem. Soc.* **107**, 2153 (1985).

350. D. J. Waxman, D. R. Light and C. Walsh, *Biochemistry* **21**, 2499 (1982). See also Y. Watanabe, T. Numata, T. Iyanagi and S. Oae, *Bull. Chem. Soc. Jpn.* **54**, 1163 (1981).

351. H. Ohta, Y. Okamoto and G.-I. Tsuchihashi, *Chem. Lett.*, 205 (1984). H. Ohta, Y. Okamoto and G.-I. Tsuchihashi, *Agric. Biol. Chem.* **49**, 671 (1985).

352. H. Ohta, Y. Okamoto and G.-I. Tsuchihashi, *Agric. Biol. Chem.* **49**, 2229 (1985).

353. K. Ogura, M. Fujita, T. Inaba, K. Takahashi and H. Iida, *Tetrahedron Lett.* **24**, 503 (1983).

354. B. J. Auret, D. R. Boyd, F. Breen, R. M. E. Greene and P. M. Robinson, *J. Chem. Soc. Perkin Trans. I.*, 930 (1981).

355. B. J. Auret, D. R. Boyd, E. S. Cassidy, F. Turley, A. F. Drake and S. F. Mason, *J. Chem. Soc. Chem. Commun.*, 282 (1983).

356. Y. Okamoto, H. Ohta and G.-I. Tsuchihashi, *Chem. Lett.*, 2049 (1986).

357. K. Ogura, M. Fujita and H. Iida, *Tetrahedron Lett.* **21**, 2233 (1980). See also T. Sugimoto, T. Kokubo, J. Miyazaki, S. Tanimoto and M. Okano, *J. Chem. Soc. Chem. Commun.*, 402 (1979); 1052 (1979).

358. B. M. Bloom and G. M. Shull, *J. Am. Chem. Soc.* **77**, 5767 (1955).

359. C. Coronelli, D. Kluepfel and P. Sensi, *Experientia* **20**, 208 (1964). C. J. Sih, *J. Bacteriol.* **84**, 382 (1962) and references therein. H. Hikino, C. Konno, T. Nagashima, T. Kohama and T. Takemoto, *Tetrahedron Lett.*, 337 (1971).
360. R. F. White, J. Birnbaum, R. T. Meyer, J. ten Broeke, J. M. Chemerda and A. L. Demain, *Appl. Microbiol.* **22**, 55 (1971).
361. J. A. Peterson, D. Basu and M. J. Coon, *J. Biol. Chem.* **241**, 5162 (1966).
362. S. W. May, R. D. Schwartz, B. J. Abbott and O. R. Zaborsky, *Biochem. Biophys. Acta* **403**, 245 (1975).
363. S. W. May and R. D. Schwartz, *J. Am. Chem. Soc.* **96**, 4031 (1974).
364. S. W. May, M. S. Steltenkamp, R. D. Schwartz and C. J. McCoy, *J. Am. Chem. Soc.* **98**, 7856 (1976).
365. B. J. Abbott and C. T. Hou, *Appl. Microbiol.* **26**, 86 (1973).
366. S. W. May, S. L. Gordon and M. S. Steltenkamp, *J. Am. Chem. Soc.* **99**, 2017 (1977).
367. A. G. Katopodis, K. Wimalasena, J. Lee and S. W. May, *J. Am. Chem. Soc.* **106**, 7928 (1984).
368. P. L. Kumler and P. J. De Jong, *J. Chem. Educ.* **52**, 475 (1975).
369. R. D. Schwartz and C. J. McCoy, *Appl. Microbiol.* **34**, 47 (1977).
370. M.-J. de Smet, B. Witholt and H. Wynberg, *J. Org. Chem.* **46**, 3128 (1981). M.-J. de Smet, J. Kingma, H. Wynberg and B. Witholt, *Enzyme Microbiol. Technol.* **5**, 352 (1983).
371. H. Ohta and H. Tetsukawa, *J. Chem. Soc. Chem. Commun.*, 849 (1978).
372. L. David and H. Veschambre, *Tetrahedron Lett.* **25**, 543 (1984).
373. A. Q. H. Habets-Crutzen, S. J. N. Carlier, J. A. M. de Bont, D. Wistuba, V. Schurig, S. Hartmans and J. Tramper, *Enzyme Microbiol. Technol.* **7**, 17 (1985).
374. V. Schurig and D. Wistuba, *Angew. Chem. Int. Ed. Engl.* **23**, 796 (1984).
375. R. D. Schwartz, *Appl. Microbiol.* **25**, 574 (1973).
376. K. Nagata, T. Matsunaga, J. Gillette, H. V. Gelboin and F. J. Gonzalez, *J. Biol. Chem.* **262**, 2787 (1987); T. Ogishima, S. Deguichi and K. Okuda, *ibid.* **262**, 7646 (1987).
377. E. W. Weiler, M. Droste, T. Eberle, H. J. Halfman and A. Weber, *Appl. Microbiol. Biotechnol.* **27**, 252 (1987).
378. V. Subramanian, M. Sugumaran and C. S. Vaidyanathan, *J. Ind. Inst. Sci.* **60**, 143 (1978).

—5—

Other Biotransformations

5.1. REACTIONS INVOLVING THE FORMATION OF A CARBON–CARBON BOND

As enzymes are condensation catalysts *sine qua non*, it is, perhaps, surprising that the deliberate use of enzymes to catalyse specific carbon–carbon bond forming reactions in the laboratory is not widespread. This situation is rapidly changing, however, as the necessary enzymes are becoming commercially available. For reactions involving isolated enzymes, the problems of cofactor regeneration have been solved for some processes on a scale of several moles of substrate; for whole-cell reactions, the necessary cofactors are already present.

5.1.1. Yeast Catalysed Acyloin Condensations

Neuberg and his co-workers were the first to report the yeast catalysed reaction of benzaldehyde with fermentatively generated acetaldehyde to give an optically active acyloin [1]. This method was later patented by Groger *et al.* [2] who claimed yields of up to 76% of (*R*)-1-hydroxy-1-phenylpropanone using added acetaldehyde (rather than relying solely on that produced fermentatively), a refinement suggested by Becvarova *et al.* [3]. Acyloin reactions with substituted benzaldehydes have also been reported (Scheme 5.1) [1,4] and 1-naphthaldehyde and 2-furaldehyde have also been shown to undergo this reaction [5]. Indeed, the yeast–acyloin reaction represents one

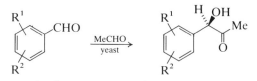

$R^1 = R^2 = H; R^1, R^2 = 2,4\text{-di-Me}; R^1 = 2\text{-Cl}, R^2 = 4\text{-OMe};$
$R^1, R^2 = OCH_2O; R^1, R^2 = 3,4\text{-di-OMe}; R^1 = 3\text{-OMe}, R^2 = 4\text{-OH}$

SCHEME 5.1

221

of the first industrial uses of a microbiological process, because the acyloin produced from benzaldehyde is the first intermediate in a two-step process for producing ephedrine [6]. Japanese workers have reported that the formation of the acyloin reduction product, 1-phenylpropane-1,2-diol, can be suppressed by the addition of acetone to the fermentation medium. The added acetone is reduced to isopropanol and competitively inhibits the reduction of the acyloin [7].

In recent years, the yeast catalysed acyloin reaction as applied to cinnamaldehydes has been used extensively by Fuganti and his co-workers as a means of preparing chiral fragments for natural product syntheses. They first described this reaction in their studies of yeast reduction of cinnamaldehyde (see Section 3.2 on carbon–carbon double bond reduction) when they isolated the reduced acyloin (**1**, R = H) [8] in 8% yield. Reduced acyloins were also isolated from α-bromocinnamaldehyde (**1**, R = Br) and α-methylcinnamaldehyde (**1**, R = Me) in 50 and 30% yield, respectively. Also isolated were the expected cinnamyl alcohols and the corresponding saturated alcohols. Cinnamyl alcohols also gave reduced acyloins upon fermentation (presumably by preliminary oxidation to the aldehyde with a yeast alcohol dehydrogenase) but in lower yield. α-Substituents larger than methyl on the double bond sufficed to inhibit the acyloin reaction [9]; unsaturated ketones such as (**2**) were similarly unreactive, but β-methylcinnamaldehyde did give a 10% yield of reduced acyloin. This paper includes a comment that added acetaldehyde approximately doubles the yield of the reduced acyloins obtained from cinnamaldehyde and α-methylcinnamaldehyde. The homologous compounds, 2- and 3-methyl-5-phenylpenta-2,4-dien-1-al, gave the reduced acyloins (**3**) and (**4**) in 15 and 10% yield, respectively. The configurations of the diol (**3**) and the α-methylcinnamaldehyde condensation product (**1**, R = Me) were proved to be (2S,3R) by conversion into the natural product L-olivomycose (**5**) [9,10].

Preparation 5.1.   Preparation of (2S,3R)-5-phenyl-4-pentene-2,3-diol (**1**)

To a stirred suspension of bakers' yeast (1 kg) in tap water (2000 ml) containing D-glucose (200 g), a solution of cinnamaldehyde (30 g) in ethanol (50 ml) was added over 2 h at 20–30°C. After a further 1 h, additional portions of yeast (500 g) and D-glucose (100 g) were added and stirring was continued overnight. The mixture was filtered through Celite and the filtrate extracted with ethyl acetate/ethanol (9:1; 3 × 500 ml). The extract was dried with sodium sulphate, the solvent evaporated *in vacuo*, and the residue column chromatographed on silica gel (150 g). Elution with hexane/ethyl acetate (9/1) gave cinnamyl alcohol and 3-phenylpropanol; subsequent elution with hexane/ethyl acetate (1/9) afforded (**1**) as a thick oil which solidifies on standing: yield, 6–10 g (16–25%); m.p. 42–44°C; $[\alpha]_D^{20}$ + 16.4° (c = 1.05, ethanol).

Many other conversions into natural products [11] have shown that the yeast catalysed acyloin reaction always gives this configuration, and also

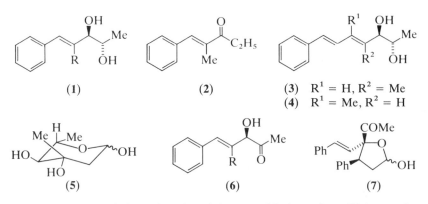

(1)   (2)   (3) R^1 = H, R^2 = Me
                                                (4) R^1 = Me, R^2 = H

(5)   (6)   (7)

proved the utility of the reduced acyloin as a chiral synthon. This reaction can be rationalized as the addition of an acetaldehyde equivalent to the *si*-face of the carbonyl carbon atom and reduction of this intermediate on the *re*-face of the resulting α-ketol [12]. Deuterium-labelling studies have shown that the formyl proton is retained in the condensation step.

When the α-methylcinnamaldehyde reaction is conducted under the conditions described by Groger *et al.* [2], with acetaldehyde added to the mixture fermenting on sugar beet molasses at pH 5, it is possible to isolate the ketol (**6**, R = Me) in about 20% yield before it is reduced to the diol [13]. Under these conditions, cinnamaldehyde gave the abnormal product (**7**), which is thought to arise from a Michael reaction of the ketol (**6**, R = H) on a second mole of cinnamaldehyde. Under the "normal" conditions, however, as stated above, the usual condensation product is obtained in 25–30% yield [11].

Preparation 5.2.   Preparation of (3*R*)-3-hydroxy-4-methyl-5-phenyl-4-penten-2-one (**6**, R = Me)

In a 30-l glass jar a mixture was made up composed of 2.5 kg of commercial bakers' yeast, 3.2 kg of sugar beet molasses, 20 g of KH_2PO_4, 10 g of $MgSO_4 \cdot 7H_2O$ and 64 g $(NH_4)_2HPO_4$ in 20 l of tap water at 32°C. As the fermentation started, the pH was adjusted with 10% H_3PO_4 to 5 and, under stirring, 115 g of α-methylcinnamaldehyde and 200 ml of acetaldehyde were added from two dropping funnels. After 4 h at 28–32°C, 1 kg of Celite was added, the reaction mixture was filtered on a large Büchner funnel, the solid pad was washed with 2 l of ethyl acetate, and the filtrate was extracted twice with 4-l portions of ethyl acetate. The organic phase, once dried (Na_2SO_4), was evaporated, leaving a residue of ca. 100–110 g. A 40-g sample of the above crude extract was chromatographed on 350 g of silica with hexane as the eluent to give ca. 30 g of unreacted aldehyde and with hexane/ethyl acetate (98 : 2 to 90 : 10) to give (**6**, R = Me) in 10–20% yield as a yellowish oil; $[\alpha]_D^{20}$ −412° (c = 1, $CHCl_3$). An analytical sample was obtained by bulb-to-bulb distillation [oven temperature 80°C (0.1 mmHg)]. ^{1}H-NMR ($CDCl_3$; δ: 4.06 (OH), 6.70 (H-5), 4.65 (H-3), 2.19 (CH_3CO), 2.71 (C-3 CH_3).

Cinnamaldehydes are not the only substrates for yeast catalysed acyloin reactions. α-Methyl-β-(2-furyl)acrolein, when fermented with yeast, gave the reduced acyloin (8) in 15–20% yield and about 10% yield of the reduced furyl alcohol (9) [14]. These two products were then used in a very neat synthesis of α-tocopherol (10) (Scheme 5.2).

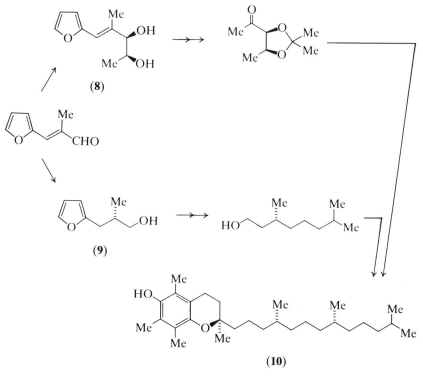

SCHEME 5.2

A yeast catalysed reaction which is formally related to the above acyloin reaction has been used as a means of preparing chiral trifluoromethyl compounds [15]. Fermentation of trifluoroethanol and an α,β-unsaturated ketone (11) with yeast at pH 5.9 gave trifluoromethyl ketols (12) in yields of 26–41% and with an e.e. of >90% (Scheme 5.3). Cyclohex-2-enone gave a similar product. α,β-Unsaturated esters (13), when fermented with yeast in the presence of trifluoroethanol, give the lactones (14) with high optical purity (Scheme 5.4).

Very few of these acyloin-type reactions proceed in high yield, but this is offset by the ease of the reaction and the cheapness of the reagents used. The resulting chiral fragments are valuable chirons for synthesis, as they do not necessarily correspond to those easily obtained from the "chiral pool" of natural products.

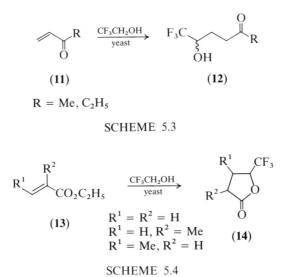

(11)   **(12)**

R = Me, C₂H₅

SCHEME 5.3

(13) $R^1 = R^2 = H$
            $R^1 = H, R^2 = Me$ **(14)**
            $R^1 = Me, R^2 = H$

SCHEME 5.4

5.1.2. Aldol Condensations

Rabbit muscle aldolase is a cheap, commercially available enzyme which has been much studied in recent years. When immobilized it is stable and often shows no inhibition even in the presence of ca. 0.1 M concentration of reactants [16]. While being quite catholic in the aldehydes it will accept, it is quite specific in the ketone required. The enzyme's natural substrate is dihydroxyacetone phosphate (DHAP) **(15)**, and the only known effective replacement for DHAP is the phosphonate **(16)** [16].

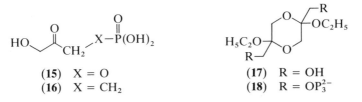

(15) X = O   **(17)** R = OH
(16) X = CH₂   **(18)** R = OP₃²⁻

Large-scale syntheses using aldolase have been problematic until recently, as DHAP is prohibitively expensive to buy, and biochemical preparations, based on the phosphorylation of dihydroxyacetone using glycerol kinase, have only recently been worked out on a suitably large scale [17]. This is, however, a very good method giving yields of over 80% on a 0.3 mol scale. *In situ* generation of DHAP from fructose-1,6-diphosphate and triose phosphate isomerase (coimmobilized with aldolase) is possible, but tends to result in lower yields in the subsequent condensation reaction [18]. It has been reported that DHA can be used directly in aldolase reactions if inorganic arsenate is added to the reaction mixture [19]. This has been noted with other enzymes, for which the natural substrates bear a phosphate group, and is presumed to be a consequence of the enzyme accepting arsenate esters, formed *in situ*, instead of phosphates. The chemical preparation of DHAP by direct phosphorylation of dihydroxyacetone with phosphoryl chloride has been used, with success, by Wong and Whitesides [16], but Effenberger and Straub [20] have reported difficulties in reproducing this work. They prepared DHAP by phosphorylating the protected DHA (**17**) and converting the product into the barium salt of (**18**), from which DHAP could be generated easily by treatment with acid. With the advent of facile and reliable methods of preparing DHAP in quantity, aldolase catalysed reactions on a very large scale are entirely feasible [21].

The stereochemistry of aldolase catalysed condensations can be predicted with great accuracy. The DHAP forms an enamine with a lysine residue in the active site of the enzyme and the aldehyde is constrained to approach from only one side of the activated complex (Scheme 5.5) [22]. The consequence of this is that the two new chiral centres are formed stereospecifically and that the stereochemistry at position C-5 of the product is directly related to that of the starting aldehyde.

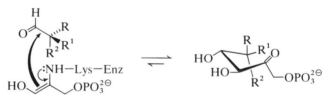

SCHEME 5.5

Numerous examples now exist in the literature of preparative-scale aldolase reactions (Figure 5.1). Lower primary and secondary aldehydes (**19**, R = n-C$_3$H$_7$ or i-C$_3$H$_7$) give high yields of condensation products but, predictably, a tertiary carbon atom α to the aldehyde stops the reaction completely [20]. Increasing chain length or branching of aliphatic aldehydes leads to lower yields of adducts. The reaction is tolerant of a wide range of functionality (Figure 5.1) [20,23]; particularly notable is the mono-functionalization of the dials (**19**, R = CHO, –(CH$_2$)$_3$CHO). The high yields obtained in these reactions (94 and 89%, respectively) would be very difficult to reproduce by classical chemistry unless resort was made to the use of protecting groups. Many aromatic aldehydes and α,β-unsaturated aldehydes are not substrates for aldolase, but 2- and 3-pyridyl aldehydes give good yields (95 and 84%, respectively).

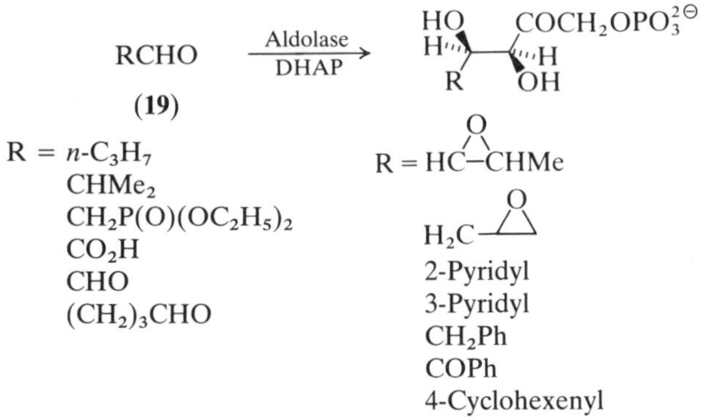

$$RCHO \xrightarrow[\text{DHAP}]{\text{Aldolase}} \quad \overset{HO}{\underset{R}{\text{H}}} \begin{matrix} COCH_2OPO_3^{2\ominus} \\ \\ OH \end{matrix} H$$

(19)

R = n-C₃H₇
CHMe₂
CH₂P(O)(OC₂H₅)₂
CO₂H
CHO
(CH₂)₃CHO

R = HC—CHMe (epoxide)

H₂C (epoxide)

2-Pyridyl
3-Pyridyl
CH₂Ph
COPh
4-Cyclohexenyl

FIG. 5.1. A selection of the aldehydes that react with DHAP under aldolase catalysis.

The utility of aldolase in sugar chemistry has been amply demonstrated by the extensive, elegant and pioneering researches of Wong and Whitesides. Isomerization of DHAP with triosephosphate isomerase (TPI) produces an equilibrium mixture containing 4% D-glyceraldehyde-3-phosphate and 96% DHAP (Scheme 5.6). Aldolase then condenses these two units to give fructose-1,6-bisphosphate (20) in 80% yield after one day. By a combination of chemical and enzymic procedures, this compound was converted into fructose-6-phosphate, glucose-6-phosphate and the corresponding dephosphorylated sugars [16]. Repetition of this condensation with D,L-glyceraldehyde gave an equimolar mixture of the fructose (20) and sorbose (21), two sugars isomeric at C-5. It should be noted that, while phosphate derivatives are good substrates for aldolase, the enzyme will equally well accept unphosphorylated aldehydes; for example 6-deoxy-D-fructose (22) and 6-deoxy-L-sorbose (23) can be prepared enantiospecifically in high yield from D- or L-lactaldehyde, respectively [24]. These procedures are, of course, ideally suited for the preparation of ¹³C-labelled sugars [16,25]. The hydroxy aldehydes (24) have been used in condensations with DHAP to generate unnatural ketose-1-phosphates which were later dephosphorylated and isomerized to aldose derivatives with the commercially available glucose isomerase [19]. Elaboration of sugar monophosphates (25) and (26) with aldolase and DHAP is a high yielding reaction, but, in contrast to the usual situation, the use of the phosphate precursors is essential as the non-phosphorylated sugars are not good substrates for aldolase [26]. By this means, C_8 and C_9 sugars, such as (27), were produced in high yield, illustrating the general applicability of such a reaction.

Preparation 5.3.   Preparation of D-fructose-1,6-bisphosphate (20)

To a DHAP (15) solution prepared from dihydroxyacetone and $POCl_3$ (1 l of solution containing 0.1 mol of DHAP) were added magnesium(II) chloride (2 mmol) and mercaptoethanol (0.1 ml). The mixture was neutralized with 5 N sodium hydroxide to pH 7.0. Coimmobilized TPI (104 U) and aldolase (60 U) in 5 ml of PAN gel was added and the mixture was kept at room temperature under argon with stirring for one day. Enzymic analysis indicated that the reaction was complete and that 45 mmol of compound (20) was present in solution (90% yield based on DHAP). Barium chloride (26.3 g) was added with stirring and the precipitated material was filtered and discarded. More barium chloride (17.4 g) was added to the filtrate followed by ethanol (600 ml). The solid product (28.3 g) containing 40 mmol of the dibarium salt of D-fructose-1,6-bisphosphate (20) was added (86% purity, 80% yield based on DHAP).

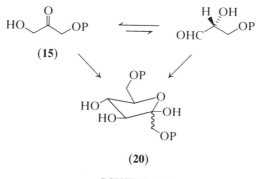

(20)

SCHEME 5.6

As Whitesides and Wong have pointed out in a review [27], there are a large number of readily available enzymes which can catalyse aldol reactions. As many of these enzymes are entities other than DHAP (for example, pyruvic acid, phosphoenolpyruvate or even acetaldehyde), they should extend the scope of the enzyme catalysed aldol reaction. Further work is needed, however, to define the synthetic utility of these enzymes on a preparative scale. An example of the results which may be anticipated is provided by the work of Auge *et al.* [28]. By the use of commercially available *N*-acetylneuraminate pyruvate lyase, they have prepared the *N*-acetylneuraminic acids (28) by condensation of pyruvic acid with the appropriate mannosamine. Although these reactions were conducted only on a millimolar scale, their potential is clear.

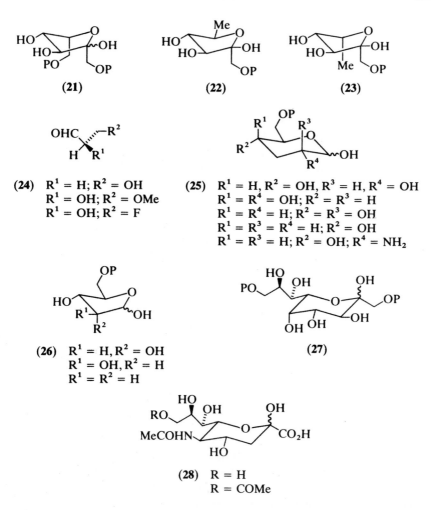

(21)

(22)

(23)

(24) R¹ = H; R² = OH
R¹ = OH; R² = OMe
R¹ = OH; R² = F

(25) R¹ = H, R² = OH, R³ = H, R⁴ = OH
R¹ = R⁴ = OH; R² = R³ = H
R¹ = R⁴ = H; R² = R³ = OH
R¹ = R³ = R⁴ = H; R² = OH
R¹ = R³ = H; R² = OH; R⁴ = NH₂

(26) R¹ = H, R² = OH
R¹ = OH, R² = H
R¹ = R² = H

(27)

(28) R = H
R = COMe

In a recent paper, Whitesides and co-workers have described a new technique for conducting enzyme catalysed reactions such as the aldolase reactions described above [29]. This avoids the inconvenience of immobilizing the enzymes by using the enzymes in soluble form, stored in dialysis tubing. This eliminates any loss in enzyme activity on immobilization. In addition, as large quantities of immobilizing supports are not needed, a higher concentration of enzyme is found at the reaction interface. A further advantage is that the enzyme could be readily recovered after reaction, stored at 4°C, and re-used in later reactions. The only real disadvantage is that reactions may be slower in some cases because the rate of diffusion across the membrane becomes the major determinant of the overall reaction rate. This method is reminiscent of an earlier method, described by Rony, where the enzymes were adsorbed into semi-permeable fibres [30].

5.1.3. Cyanohydrin Synthesis

Mandelonitrile lyase, an enzyme isolated from bitter cherries, catalyses the enantiospecific addition of hydrogen cyanide to the *si*-face of aldehyde carbonyl carbon atoms. A wide range of aliphatic, aromatic and heterocyclic aldehydes will react to give chiral cyanohydrins, frequently in excellent yield (Figure 5.2) [31]. Notable among these results is that even pivaldehyde gives a good yield of cyanohydrin.

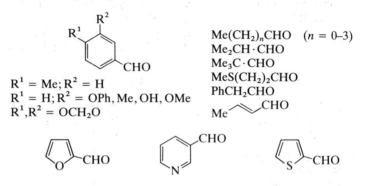

$R^1 = Me; R^2 = H$
$R^1 = H; R^2 = OPh, Me, OH, OMe$
$R^1, R^2 = OCH_2O$

$Me(CH_2)_n CHO$ $(n = 0-3)$
$Me_2 CH \cdot CHO$
$Me_3 C \cdot CHO$
$MeS(CH_2)_2 CHO$
$PhCH_2 CHO$

FIG. 5.2. Some aldehydes which undergo enzyme catalysed addition of HCN.

The enzyme is now commercially available and is highly efficient; milligram quantities of enzyme are sufficient to catalyse formation of kilograms of cyanohydrin. It is normally used in a bound form on a cellulose support to facilitate recovery of the enzyme for re-use and this supported enzyme can be used in a column for continuous production of cyanohydrins [32]. These reactions are very easy to perform on a large scale; formation of (R)-mandelonitrile on a 0.4-M scale can be performed in normal chemical apparatus in 20 min to give a 96% yield of 90% optically pure product. (Effenberger later reported an enantiomeric excess of 86% as measured by GLC of the (R)-(+)-MTPA derivatives [31,32].) It is important to note for this reaction that very pure benzaldehyde is necessary, as benzoic acid is a strong inhibitor of the enzyme [33]. Large-scale syntheses must be performed under a nitrogen atmosphere.

As pointed out by Effenberger and his co-workers, the chemical reaction to produce cyanohydrins occurs in competition with the enzyme catalysed reaction [32]. This frequently results in a diminution in chiral purity of the products. Variation in the pH, temperature or concentration gave no improvement, but the use of a water-immiscible solvent, particularly ethyl acetate, was advantageous. In this solvent the chemical reaction was much slower, whereas the enzyme catalysed reaction was only slightly affected. The reaction was thus effected in ethyl acetate, saturated with 0.01 M acetate buffer, with cellulose-bound enzyme. By this method, benzaldehyde gave (R)-mandelonitrile in 99% optical purity, and the optical purity of products from 3-phenoxybenzaldehyde and butyraldehyde increased from 10.5% and 69% to 98% and 96%, respectively.

As a method of preparing chiral cyanohydrins on a large scale, this method must now be the method of choice over the usual chemical procedures. The only drawback is, simultaneously, its greatest advantage in that only cyanohydrins of one configuration are available; i.e. that conforming to si-face addition to the carbonyl group.

5.2. CARBOHYDRATE, NUCLEOSIDE AND NUCLEOTIDE CHEMISTRY

5.2.1. Carbohydrate Chemistry

Carbohydrate chemistry has perhaps the longest history of useful enzyme applications both commercially and in the laboratory. The degradation of cornstarch to glucose with glycosidases [34] and the equilibration of glucose and fructose with glucose isomerase [35] are well-known, large-scale, commercial applications of enzymes. Elucidation of the structures of oligosaccharides in the laboratory, particularly before NMR methods became widely available, was commonly performed by the use of specific enzymes [36].

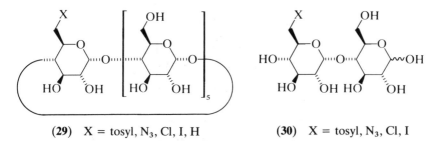

(29) X = tosyl, N₃, Cl, I, H   (30) X = tosyl, N₃, Cl, I

An amusing cycle is represented by the degradation of starch with cyclodextrin-glycosyltransferases (E.C. 2.4.1.19). This gives cyclic oligosaccharides containing mainly six, seven, or eight glucose units, commonly called α-, β- and γ-cyclodextrins. These, in turn, have been studied as synthetic enzymes as they possess a central, hydrophobic cavity somewhat similar to the active site of a natural enzyme [37]. Cyclodextrins are degraded by the commercially available amylase from *Aspergillus oryzae*, commonly called Taka-amylase, but it is known that the presence of a C-6 substituent on a glucopyranosyl residue prevents hydrolysis of the glucosidic bond of the substituted residue. Thus, amylase catalysed degradation of the monosubstituted α-cyclodextrins (**29**, X = Cl, I, N$_3$ or tosyl) gave, in addition to large quantities of glucose, the corresponding 6'-substituted maltoses (**30**) [38]. However, the monodeoxycyclodextrin (**29**, X = H) was completely degraded to 6-deoxyglucose and glucose. This agrees with previous studies

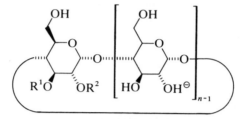

(**31**)   R^1 = α- or β-naphthalenesulphonyl or tosyl; R^2 = H; n = 6, 7
(**32**)   R^1 = H; R^2 = α- or β-naphthalenesulphonyl or tosyl; n = 6, 7

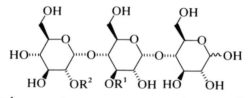

(**33**)   R^1 = α- or β-naphthalenesulphonyl or tosyl; R^2 = H
(**34**)   R^1 = H; R^2 = α- or β-naphthalenesulphonyl or tosyl

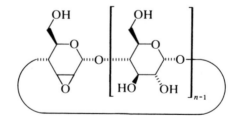

(**35**)   n = 6, 7; α- or β-epoxide

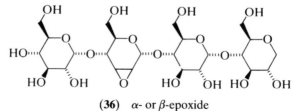

(**36**)   α- or β-epoxide

which showed that the amylase could not distinguish between natural and deoxy substrates. This method of preparing 6'-substituted maltoses represents a considerable improvement over other chemical procedures, as it requires no protection steps and yields isomerically pure product. An extension of this method to the production of 6',6''-disubstituted maltotrioses has also been reported [39]. Similarly, substitution at the 3 or 2 position of α- (31) or β-cyclodextrins (32) inhibited the action of the amylase so as to produce the trisaccharides (33) and (34), respectively, in good yield [40]. In contrast, the α- and β-epoxides of cyclodextrins (35) gave the tetrasaccharides (36) in excellent yield.

The preparation of individual saccharide units by condensation reactions using aldolases is discussed elsewhere. Enzyme catalysed coupling of monosaccharides to give oligosaccharides has been reported, such as the use of β-galactosidase (E.C. 3.2.1.23) from *Escherichia coli* to prepare a disaccharide by transfer of a β-D-galactopyranosyl residue from lactose [41]. In another instance 1-fluoroglucosides were used as donors of sugar moieties catalysed by α-amylase (E.C. 3.2.1.1) [42]. Unnatural sugars are also accessible by enzyme catalysed methods as has been demonstrated in two very pleasing papers by Card and co-workers. The coupling of the fluorofructose (37) with UDP-glucose, mediated by the commercially available sucrose synthetase, gave the easily isolated fluorosugar (38, R = F) in good yield [43]. The corresponding 1'-azido-1'-deoxysucrose (38, R = N$_3$) could also be prepared by this method, but only in 15% yield, whereas fructoses substituted at the 1 position with a hydrogen atom, a chlorine atom, methoxy or sulphydryl groups gave no coupled products at all [44]. The synthetase enzyme is thus reasonably specific in the groups it will tolerate at the 1 position of the fructose.

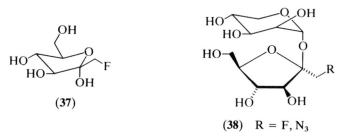

(37)

(38)   R = F, N$_3$

Preparation of 6'-substituted sucroses with sucrose synthetase requires access to suitable 6-substituted fructoses. Synthesis of these substrates by usual synthetic methods is tedious, so Card and his co-workers circumvented this problem by using glucose isomerase (E.C. 5.3.1.5) to convert the readily available 6-substituted glucoses into fructoses (Scheme 5.7) [44]. As this enzyme only affords a 10–15% equilibrium concentration of the fructose, this step was coupled with the sucrose synthetase reaction, which is irreversible under these conditions and drives the equilibrium reaction in the required direction. By this means, 6'-deoxy (**39**, R = H) and 6'-deoxy-6'-fluorosucrose (**39**, R = F) were prepared in 73% and 53% isolated yields, respectively.

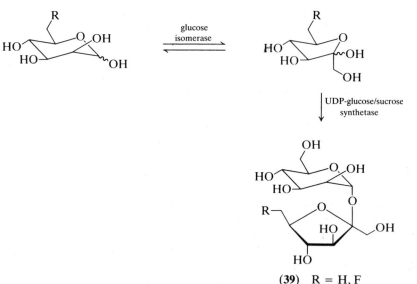

(**39**)   R = H, F

SCHEME 5.7

This coupled-enzyme procedure is not suitable for the preparation of 4-deoxy-sucroses/fructoses, as glucose isomerase requires the presence of a 4-hydroxy group on the glucose for it to be a substrate. This was overcome by phosphorylating the 4-fluorinated glucose (**40**) with hexokinase (E.C. 2.7.1.1) to provide a suitable substrate (**41**) for phosphoglucose isomerase (E.C. 5.3.1.9) (Scheme 5.8) [44]. This 6'-phosphate isomerizes to the fructose (**42**), which in turn is irreversibly phosphorylated by fructose-6-phosphate kinase to drive the isomerization equilibrium reaction in the desired direction. The fructose-bis-phosphate (**43**) was not isolated but converted into the required 4-deoxy-4-fluorofructose (**44**) in 30% overall yield by treatment with an alkaline phosphatase (E.C. 3.1.3.1). Sucrose synthetase catalysed coupling of compound (**44**) with UDP-glucose gave 4'-deoxy-4'-fluorosucrose (**45**), albeit in only 16% yield. Although this yield may appear low, it should be borne in mind that this simple two-flask procedure replaces the many steps required by other more traditional procedures.

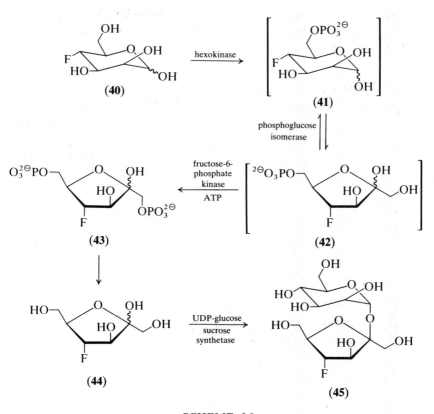

SCHEME 5.8

The yeast hexokinase/ATP/phosphoenolpyruvate/pyruvate kinase system has also been used by Drueckhammer and Wong to phosphorylate 3-deoxy-3-fluoro-D-glucose at the 6 position [45].

The preparation of oligosaccharides on a large scale by enzymic methods was not exploited until recently because neither the enzymes nor the required nucleoside diphosphate sugars were readily available in quantity. This area has been expertly developed by Whitesides, Wong and their co-workers, who used a multiple-enzyme system for the *in situ* regeneration and reaction of uridine-5'-diphosphate glucose and UDP-galactose under the conditions required for oligosaccharide synthesis. By this means, it was possible to prepare N-acetyl-lactosamine (46) (Scheme 5.9) in 70% overall yield on a scale of greater than 10 g [46]. All six enzymes used in this synthesis were commercially available. After use in this reaction in their PAN-immobilized state, they still retain >80% of their activity and could be re-used. This one

example, on its own, should be sufficient to exemplify the advantages of
enzyme mediated synthesis. A preparation by more normal chemical means
could never be so facile.

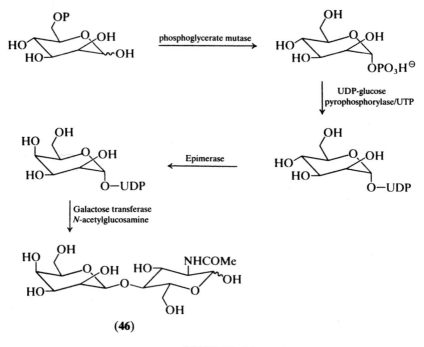

SCHEME 5.9

Thiem and Treder [47] synthesized N-acetyl-lactosamine (**46**) essentially by
the above method but with the enzymes immobilized on silica gel by means
of glutardialdehyde. This was then coupled with cytidine 5'-monophospho-
sialate (**47**) using β-D-galactoside-α-(2,6)-sialyl transferase (E.C. 2.4.99.1) to
give the trisaccharide (**48**) [48]. The trisaccharides (**49**) and (**50**) were also
prepared from chemically synthesized disaccharides using the multiple-
enzyme method of Wong *et al.* to form UDP-galactose *in situ* [49]. In these
cases, Ultrogel was used as an insoluble support for the enzymes. Ribulose-
1,5-diphosphate (**51**) has also been prepared using a multiple-enzyme
approach [50]. In this case the kinase required for the final step is oxygen
sensitive, which means the *in situ* regeneration of NAD from NADH cannot
be carried out by an aerobic system. This was circumvented by the use of the
anaerobic α-ketoglutarate/glutamate dehydrogenase system.

These multiple-enzyme preparations demonstrate an important philo-
sophical point about the use of enzymes in synthesis. A one-pot chemical

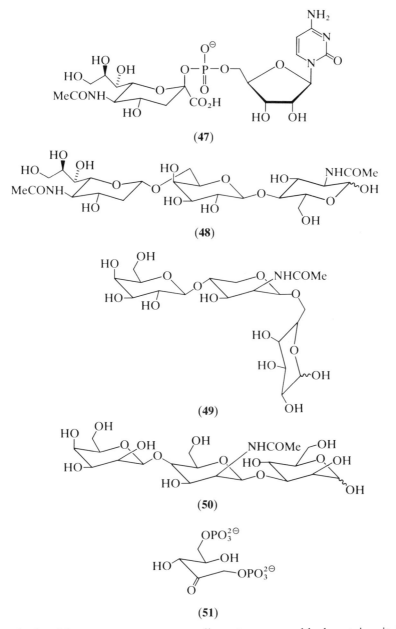

(47)

(48)

(49)

(50)

(51)

synthesis with so many steps proceeding at once would almost inevitably produce a very crude product, if it produced any product at all. By the use of specific enzymes, however, such a sequence is perfectly feasible, although it does require a considerable amount of background knowledge to plan the synthetic pathway.

The use of glycosidases to liberate glycosides from the associated aglycones is a very well known procedure [51]. The reverse reaction, to couple glycosides and alcohols, is also known [52]. This has proved useful in the synthesis of acid- and base-sensitive cardiac glycosides such as compound (52) [53]. β-Galactosidase catalysed glycosylation in aqueous acetonitrile is reported to give the required glycosides in a very rapid reaction (20 min at 20°C), but unfortunately no yields are given in the paper.

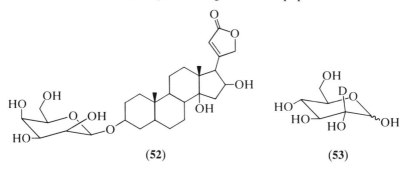

(52)   (53)

Some preparations of labelled saccharides are discussed in another section. Another example is provided by the preparation of [2-^{2}H]-glucose (53) [54]. Isomerization of fructose-6-phosphate with glucose isomerase in buffered deuterium oxide gives labelled glucose-6-phosphate, which was cleaved by alkaline phosphatase to give the sugar (53).

5.2.2. Nucleoside and Nucleotide Chemistry

Modern nucleotide chemistry is replete with enzyme catalysed synthetic methods. Indeed, gene synthesis would be almost impossible by any other technique [55]. As this is the province of a more specialized text, only a few general examples will be presented here [56].

Specific oxidation of the aromatic amino group in 3'-amino-3'-deoxyadenosine (54) by a commercially available (albeit crude) sample of Takadiastase gave the nucleoside (55) in 75% yield [57]. This undoubtedly utilizes an adenosine deaminase present in the crude enzyme preparation. Adenosine deaminase from calf intestinal mucosa is now commercially available and was used in an analogous transformation to produce 1-hydroxyinosine in 88% yield [58]. Another deaminase, cytidine deaminase, was used in the first step of a high-yielding synthesis of the D arabinonucleosides (56) (Scheme 5.10) [59].

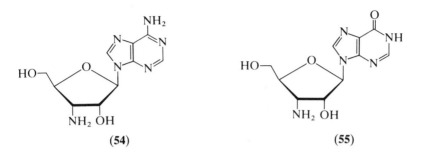

(54)   (55)

While nucleotides and nucleosides can be prepared by direct enzymic coupling of the base and 5-phosphoribose pyrophosphate (**57**) [60], or by reversal of an enzyme's usual hydrolytic pathway [61], work in this area has tended to concentrate upon enzyme catalysed exchange of the base. The preparation of 9-β-D-arabinofuranosyl adenine (**58**) provides an example of this type of reaction, whereby the bacterium *Enterobacter aerogenes* catalyses the transfer of the arabinofuranosyl moiety from uracil to adenine (Scheme 5.11) [62]. The commercially available NAD-ase from pig brain was used in a similar base-exchange reaction to give the β-NAD analogues (**59**) [63]. The free nucleoside could be prepared from these NAD analogues by hydrolytic cleavage of the diphosphate with phosphodiesterase followed by incubation with 5'-nucleotidase.

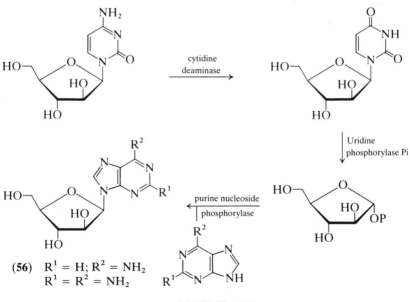

(**56**) R¹ = H; R² = NH₂
R¹ = R² = NH₂

SCHEME 5.10

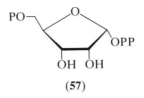

(57)

Polymeric nucleotides can readily be prepared on an impressively large scale, as in the preparation of polyriboinosinic and polyribocytidylic acids. These can be prepared on a 30-g scale by the intermediacy of the polyribonucleotide phosphorylase from *Micrococcus luteus* [64]. The trinucleotides UAA, UAG and UGA, which are protein synthesis termination codons, can be prepared by the enzyme ribonuclease immobilized on CM-cellulose [65].

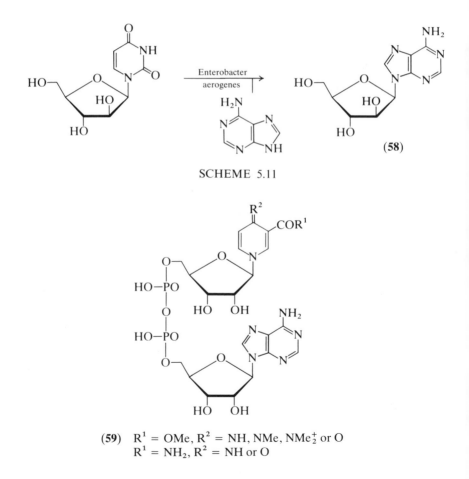

SCHEME 5.11

(59) R^1 = OMe, R^2 = NH, NMe, NMe_2^+ or O
         R^1 = NH_2, R^2 = NH or O

This is an interesting example because the enzyme is used in a flow system to minimize the phosphate hydrolysis which would normally prevail upon continued exposure of the products to the enzyme.

5.3. PREPARATION AND REACTIONS OF AMINO ACIDS

The ready availability of large amounts of amino acids at low prices is largely a result of the application of microbiological fermentation processes. L-Phenylalanine, L-methionine [66], L-alanine [67] and L-aspartic acid [68] are just some examples of the amino acids produced enzymically on scales of several tonnes a month. The willingness of the Japanese chemical industry to invest heavily in this relatively new methodology has paid dividends and ensured them a leading place in this technology.

A discussion of industrial amino acid production would be outside the scope of this book, but many reviews are available [69]. This section will concentrate upon some of the reactions which are considered to be useful in the synthetic laboratory, although it must be stressed that work up to the present time has barely scratched the surface of available transformations [70].

An early example of the use of enzymes to prepare enantiomerically pure amino acids by selective destruction of one enantiomer in a racemic mixture with L or D amino acid oxidase was provided by Parikh *et al.* [71]. Although this inevitably results in the destruction of half of the starting material, the ease of the reaction and the commercial availability of the enzymes (*Crotalus adamanteus* venom L amino acid oxidase and pig kidney D amino acid oxidase) make this a very attractive procedure in many cases. Hydrogen peroxide is produced as a by-product in these reactions and, if this is allowed to accumulate, undesirable side-reactions can occur. This can be prevented by the addition of catalase (E.C. 1.11.1.6). In the case of hog kidney D amino acid oxidase, the use of crude, rather than pure, enzyme preparations is to be recommended, as they possess residual catalase activity, whereas L amino acid oxidases from snake venoms have little or no catalase activity and the addition of catalase is necessary. The specificities of the D and L amino acid oxidases have been carefully defined. They require the presence of α-carboxyl and α-amino groups, while the presence of a bulky group at the α position precludes binding at the active site of the enzyme. The L amino acid oxidases are slightly more sensitive to structural variations of the substrate. Snake venom oxidases will not tolerate substitution on the amine nitrogen, while the D oxidases will accept secondary amines [72].

Further examples of the use of amino acid oxidases are provided by the preparation of the bleomycin precursor L-(60) from racemic material by the use of a D amino acid oxidase [73], and the resolution of the acetyllysine (**61**) with L amino acid oxidase [71].

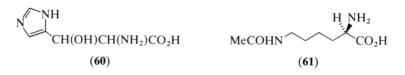

(**60**)   (**61**)

The reverse reaction, that of reductive amination, is a well-established procedure for the laboratory preparation of amino acids. This is illustrated by the preparation of a separable mixture of diastereomerically pure 4-hydroxyglutamic acids (**62**), with the natural (*S*) stereochemistry at the amine centre, using glutamate dehydrogenase (Scheme 5.12) [74]. The *erythro*-diastereoisomer was isolated in 30% yield from 4-hydroxy-2-keto-glutaric acid while the *threo*-diastereoisomer was isolated in 35% yield. This procedure has also been used to prepare diastereomerically pure samples of 4(*R*) and 4(*S*) deuterioglutamic acids [75]. It should be noted that, in such transamination reactions, the glutamate or aspartate need not be enantio-merically pure, as these enzymes only utilize the L enantiomer. To push the reaction in the desired direction NADH is needed. This has been provided by introducing a NADH-recycling procedure involving ethanol and yeast alcohol dehydrogenase [75].

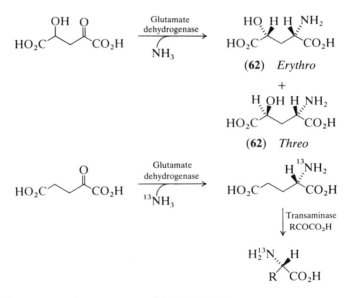

SCHEME 5.12

In a recent paper, the use of *Escherichia coli* aspartate transaminase from cloned cells has been described [76]. Either aspartate or glutamate could be used as the transaminating agent with this enzyme and a wide range of L amino acids of high enantiomeric excess (>90%) was produced, although alkyl-substituted keto-acid starting materials required the use of greater quantities of enzyme (Figure 5.3).

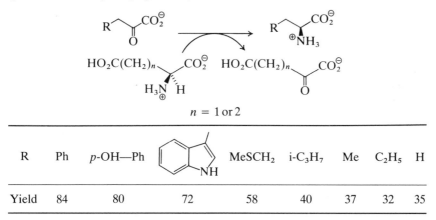

$n = 1$ or 2

R	Ph	*p*-OH—Ph	(indole)	MeSCH$_2$	i-C$_3$H$_7$	Me	C$_2$H$_5$	H
Yield	84	80	72	58	40	37	32	35

FIG. 5.3. Preparation of L amino acids using *E. coli* aspartate transaminase.

Enzyme catalysed reductive amination is an ideal procedure for the introduction of the short-lived ^{13}N isotope needed in modern nuclear medicine ($t^{1/2} = 10$ min). By this means, ^{13}N-glutamic acid can be prepared and used, with the aid of a transaminase, as an intermediate in the preparation of various L amino acids (Scheme 5.12) [77]. As an example, the use of immobilized enzymes packed in columns and eluted under pressure, allowed the preparation of ^{13}N-alanine from α-ketoglutarate in 4 min in a yield of 70% [78]. Introduction of other isotopes is also possible; the above reaction conditions were established by the use of ^{14}C-labelled substrates.

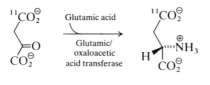

SCHEME 5.13

The exchange of amino acid side chains by the use of transaminases (amine transferases) represents an extremely useful synthetic method in the laboratory. [4-^{11}C]Oxaloacetic acid can be readily transaminated with

L-glutamic acid under catalysis by glutamic/oxaloacetic acid transferase to give L-[4-^{11}C]aspartic acid which was radiochemically pure (Scheme 5.13) [79]. Such reactions can be performed on a large scale as exemplified, once again, by the transaminase catalysed preparation of 4-hydroxyglutamic acid (Scheme 5.14) [80]. In this example, the equilibrium position favours the reverse reaction, but this was neatly circumvented by the use of cysteine sulphinic acid (63) as the amine transfer agent. The cysteine transamination product, being very unstable, decomposes into sulphur dioxide and pyruvic acid and thus drives the equilibrium in the desired direction. A useful adjunct of this method is that no separation of starting and product amino acids is necessary. By this means, multigram quantities of the product were readily attainable without the necessity of using NADH as in the example described previously. The use of cysteine sulphinic acid reflects the same strategy as the use of L-aspartic acid as an amine donor inasmuch as the α-keto acid produced in the latter case can spontaneously decarboxylate to give pyruvic acid thus driving the reaction to completion.

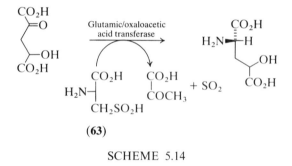

(63)

SCHEME 5.14

Pyridoxal phosphate-dependent transaminases function through the formation of conjugated imines as intermediates, a mechanism which has been thoroughly studied and recently reviewed [81]. These enzymes only accept a single enantiomer of an amino acid and only generate one enantiomer of the product amino acid by enantiospecific hydrolytic cleavage of the imine intermediate. The pyridoxal phosphate-dependent enzymes tyrosinase and tryptophanase have demonstrated considerable synthetic utility in recent years [82]. From pyruvic acid, ammonia, and the appropriate phenol or indole derivative, the tyrosines and tryptophans listed in Figure 5.4 could all be prepared in good yield under very mild conditions. L-Tyrosine labelled with ^{15}N and/or ^{13}C has also been prepared using aspartate transaminase from cloned E. coli with very high levels of incorporation (>95%) [83].

The lyase catalysed addition of ammonia to fumaric acid represents a valuable industrial method for the production of L-aspartic acid; the reac-

tion is performed by passage of the reactants through a column of immobilized enzyme [68,84]. The potential of lyases has not been defined (except for
the stereospecific preparation of labelled compounds [85]); however, it is
noteworthy that the products obtained from lyase catalysed reactions can
readily be obtained by other methods of synthesis; in addition, lyases
generally have a narrow substrate specificity and will only accept small
changes in the substrates offered to them. Thus some lyases will only be of
limited use in synthesis.

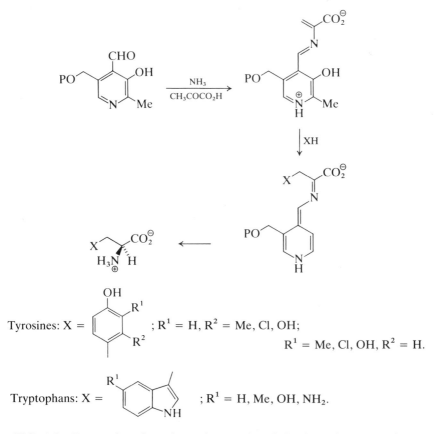

FIG. 5.4.   Preparation of tyrosine and tryptophan derivatives using transaminase.

However, the enzyme β-methylaspartase from *Clostridium tetanomorphum*
holds promise as a means of preparing small, multifunctional, chiral synthons. This enzyme catalyses the addition of ammonia to mesaconic acid by
stereospecific attack of nitrogen on the *si*-face of C-2 with proton addition to
the *re*-face of C-3 [86]. When applied to [methyl-^{13}C]mesaconic acid (**64**), this

gave (2S,3R)-3-methylaspartic acid (65) in high yield after recycling of the recovered starting material [87]. The real synthetic advantage, however, may be indicated by a study of the action of β-methylaspartase on 3-halogeno-fumaric acids and acetylene dicarboxylate [88]. While the acetylene and 2-iodofumarate gave no products, and 2-fluorofumarate only reacted poorly, 2-chlorofumarate gave a 60% yield of 3-chloroaspartic acid, believed to be the (2R,3S) diastereoisomer. 2-Bromofumarate was similarly converted into the bromoaspartate but the product cyclized *in situ* to 2,3-aziridinedicarboxylic acid. Further work is needed to assess the range of substrates that these types of enzymes will accept.

(64)   (65)

The use of enzymes to decarboxylate amino acids regio- and enantio-specifically is well established as a useful industrial method, as in the production of L-alanine from L-aspartic acid with immobilized aspartate decarboxylase [89]. This can be coupled with lyase catalysed addition of ammonia to fumarate to produce L-alanine in a single-stage process from fumarate [90]. These decarboxylases are, as yet, of limited use synthetically apart from the preparation of enantiospecifically labelled compounds. This latter use is illustrated by a stereospecific synthesis of (R)-[^{2}H]-tyramine (66) with tyrosine decarboxylase [91]. Glutamate decarboxylase has been used in a synthesis of the neurotransmitter γ-aminobutyric acid, labelled with ^{13}N (67) [92].

(66)   (67)

5.4. FORMATION OF HALOHYDRINS AND DIHALIDES

The addition of hypohalous acids to unsaturated systems catalysed by chloroperoxidase, lactoperoxidase, or horseradish peroxidase has not been studied in depth as a general synthetic method in the laboratory. On an industrial scale, however, the ability to produce large quantities of halogen-containing materials under mild conditions without the necessity of handling dangerously reactive halogens or halogen/halide equivalents has led to a great deal of interest in this type of reaction. This Section has been included in this Volume in the hope that it will stimulate studies of the peroxidases as synthetic reagents, facilitated by the commercial availability of the necessary enzymes.

The chloroperoxidase of *Caldariomyces fumago* catalyses the oxidation of all halide ions (except fluoride) [93]; lactoperoxidase (a bromoperoxidase) oxidizes bromide and iodide ions, while horseradish peroxidase only oxidizes iodide ion [94]. The primary oxidant required for this process, hydrogen peroxide, can be generated by chemical or enzymic methods, or added directly to the reaction mixture [95]. Chloroperoxidase will tolerate up to 10% of methanol or dimethyl sulphoxide while still retaining about a third of its activity at the optimum pH, which for reactions using chloroperoxidase is pH 3 [96]. At higher pH levels (pH 5–6) peroxidation becomes the major process [96,97].

In the presence of an unsaturated acceptor a reaction generally ensues (the enzymes have a broad substrate specificity), and the products formed by haloperoxidase catalysed reactions are consistent with an initially formed hypohalous acid intermediate [98]. For chloroperoxidase mediated reactions, differences in kinetics and product distribution between the enzymic reaction and normal hypochlorous acid addition were taken to indicate that the acid was not the actual reactive species. Further work implicated an enzyme bound halogenating intermediate which reacted with the substrate to produce an unstable enzyme–product complex; this then rapidly decomposed to release the product [99].

However, this hypothesis does not agree with the work of Geigert and his co-workers [100] on the halogenation of alkynes and cyclopropanes with chloro- and lacto-peroxidases. From their results they deduce that "(the reaction of haloperoxidase) with both alkenes and alkynes means either that this enzyme is a unique enzyme . . . . . or this enzyme generates a chemical intermediate that can react upon substrates" and they postulate free hypohalous acid as the halogenating agent. The lack of stereoselectivity of some chloroperoxidase reactions may also be taken to indicate the intermediacy of free hypohalous acid unless the activated enzyme/substrate complex is exceptionally loose.

As it is known that in the absence of a halogen acceptor the haloperoxidases will produce hypohalous acid, an intermediate explanation is possible. If the acceptor is a good substrate for haloperoxidase then an enzyme bound intermediate is quite feasible. Alternatively, if the acceptor is a poor substrate, the rate of production of free hypohalous acid may exceed the rate at which the substrate is complexed with the enzyme, leading to products compatible with a non-enzyme catalysed reaction.

Most of the reported work on enzyme catalysed halogenations has involved the use of chloroperoxides from *Caldariomyces fumago* in the presence of an alkali metal salt (normally a potassium salt). Thus, reaction of the β-diketo steroids (**68**) and (**69**) in the presence of potassium chloride or bromide gave the corresponding α-halo steroids in 50% yields (Scheme 5.15) [101]. 15-Keto-1-dehydrotestolactone (**70**) was dibrominated under these conditions (Scheme 5.16) [101]. Similar reactions have been observed for other β-dicarbonyl systems [102].

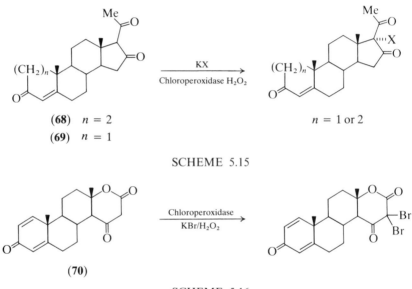

SCHEME 5.15

SCHEME 5.16

Reaction of steroids with double bonds remote from ring A, such as the $\Delta^{9,11}$-steroid (**71**), produced the bromoalcohol (**72**) by *anti* addition of hypobromous acid to the double bond. With an extended reaction time, this bromoalcohol is non-enzymically converted into the corresponding epoxide [103]. Under these reaction conditions, the Δ^5-steroid pregnenolone (**73**) and its acetate gave the corresponding epoxide (**74**) directly, although the intermediacy of the bromohydrin was indicated by TLC. Such stereospecificity is not the norm for such reactions, however, as is indicated by the action of chloroperoxidase on *cis*-prop-1-enylphosphonic acid (**75**) to give a racemic mixture of *threo*-1-chloro-2-hydroxypropylphosphonate (**76**) [104]. Similarly, reaction of the *trans* isomer of (**75**) gave the racemic *erythro* product.

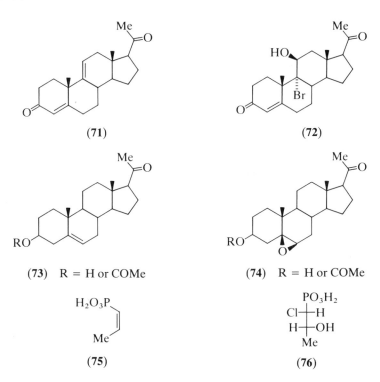

(71)

(72)

(73)   R = H or COMe

(74)   R = H or COMe

(75)

(76)

Halogenation of phenols [105] or heterocycles such as thiazolines [106], antipyrine [97] and barbituric acids [107] is possible under very mild conditions with haloperoxidases.

Recent work with haloperoxidases has demonstrated that halonium ions derived from allylic alcohols can be intercepted by high concentrations of a second nucleophile to produce, for example, mixed dihalides (Scheme 5.17) [108]. A notable feature of this work is the preparation of fluoroiodo-1-propanols by incubation of allyl alcohol with horseradish peroxidase, iodide, hydrogen peroxide and fluoride ion [109]. Although it is still in its infancy, this method shows great potential.

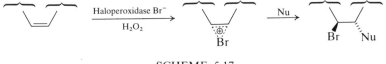

SCHEME 5.17

It could reasonably be argued from the results described above that the use of haloperoxidases in the laboratory has no advantage over the more usual chemical methods. However, in view of the mild conditions of these reactions and the high yields frequently obtained, an examination of their selectivity in polyfunctional systems could pay great dividends.

5.5. *O*- AND *N*-DEALKYLATION

Biosynthetic elaboration of metabolites by *O*- and *N*-methylation is well known and could be adapted for synthetic use. However, the range of conventional methylation procedures is so large that the use of enzymic procedures would, in most cases, be totally unnecessary. Simple demethylation reactions of aliphatic methyl ethers by mono-oxidase enzymes are known [110] but, again, usual chemical procedures are available to deal with these cases. However, some examples are known of selective enzymic or microbial demethylation of complex substrates under mild conditions which could be of use in synthesis. It should be borne in mind, however, that the procedures described here are very specific examples.

An early indication of the use of micro-organisms to accomplish such demethylations was provided by Boothroyd *et al.* [111]. Examination of a wide range of fungi led to the discovery of three fungi which could selectively demethylate griseofulvin (**77**) at each of three possible positions (Scheme 5.18). Although the low concentrations used render these specific transformations of limited use, it is a good example of the selectivity possible with this method.

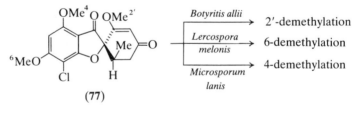

SCHEME 5.18

Many of the dealkylation procedures that have been reported have come from the work of Rosazza and his co-workers as a consequence of studies on microbial models of mammalian metabolism. These studies largely concern demethylations of various alkaloids with bacteria and fungi. However, the first example of microbial alkaloid demethylation was reported by Taylor and his co-workers, who found that aspidospermine could be *O*-demethylated in 32% yield by a subspecies of *Streptomyces griseus* [112]. This same culture failed to demethylate *N*-deacetylaspidospermine, *N*-ethyl-*N*-deacetylaspidospermine, 7-methoxyindole and 7-methoxytryptophan, leading to the

supposition that an acetyl group on the indole nitrogen is necessary for
O-demethylation to occur. This requirement appears to be specific to this
strain of S. griseus as fermentation of glaucine (78) with the UI1158 strain of
S. griseus gives 2-O-desmethylglaucine [113]. N-Demethylation of glaucine is
observed with S. griseus and is also seen when the anti-parkinsonian drug,
lergonitrile (79) is treated with S. platensis [114].

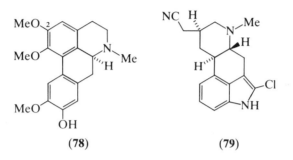

(78)   (79)

De-ethylation reactions are also possible with S. griseus, as in the dealkylation of 7-ethoxycoumarin [115], although none of the yields in these
latter examples is high enough to warrant consideration as a viable synthetic
procedure.

Vindoline can be O-demethylated in 33% yield by Sepedonium chrysosper-
mum (ATCC 13378) [116], whereas S. griseus does not catalyse this reaction.
In contrast, fermentation of dihydrovindoline with S. griseus does produce
an O-demethylated product, albeit in low yield [117]. Such diverse reaction
behaviour in response to a small change in the structure of the substrate is an
indication that further study is needed to define the structural parameters
necessary for a successful reaction.

10,11-Dimethoxyaporphine has been selectively O-demethylated at the 10
position in 59% yield with Cunninghamella elegans (ATCC 9245); this stands
in contrast to chemical monodemethylation which gives only the 11-O-
demethylated product [118]. S. griseus demethylates 10,11-dimethoxyapor-
phine non-selectively to give 20 and 24% yields of 10- and 11-demethylated
products [119]. These results led Rosazza and his co-workers to study the O-
demethylation of papaverine (80) with a range of 60 micro-organisms [120].
Notable results from this study were that S. griseus gave an unusable
mixture of 4'-, 6-, and 7-demethylated products whereas Aspergillus
alliaceus gave 6-desmethylpapaverine in reasonable yield. Cunninghamella
echinulata was also selective in giving 4'-desmethylpapaverine.

The quinone imine (82) has been prepared in 64% isolated yield by
horseradish peroxidase/hydrogen peroxide O-demethylation of 9-methoxy-
ellipticine (81) [121]. Studies with $H_2{}^{18}O$ as a reaction medium demonstrated

that the reaction was not a simple demethylation but rather a replacement of the methoxy group by OH from the solvent. As this enzyme is readily available and the reaction occurs under very mild conditions, this could well be of general synthetic use.

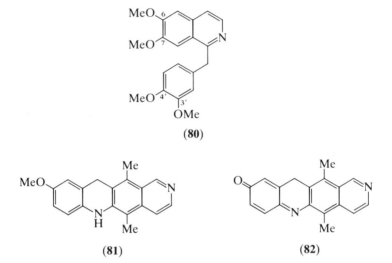

(80)

(81)   **(82)**

The limited range of preparative examples illustrates that enzyme catalysed dealkylation is a topic in its infancy and further work is necessary to delineate mechanisms and the scope of the reaction. Rosazza has made a start on this task and has proposed a topographical model of a binding site in an *S. griseus* enzyme in an excellent short review of the subject [122]. If such studies could be extended so that only a very limited range of enzymes/ microbes need be tested for a specific substrate, then the mild conditions of this reaction would commend such dealkylations as a useful synthetic method.

REFERENCES

1. C. Neuberg and J. Hirsch, *Biochem. Z.* **115**, 282 (1921).
2. D. Groger, H. P. Schmader and H. Frommel, *Ger. Offen.* 1, 543, 691 (1966).
3. H. Becvarova, O. Hanc and K. Mauk, *Folia Microbiol.* **8**, 165 (1963).
4. C. Neuberg and L. Liebermann, *Biochem. Z.* **121**, 311 (1921); **143**, 553 (1923). M. Behrens and N. N. Iwanoff, *Biochem. Z.* **196**, 478 (1926). Merck and Co., U.S. Patent, 3,339,796 (1967). J. W. Rothrock (Merck and Co.), U.S. Patent 3,338,796 (1964); *Chem. Abs.* **67**, 89797. See also [7].
5. D. Groger, H. P. Schmauder and H. Frommel, East German Patent, 62, 554 (1967); *Chem. Abs.* **70**, 86297.

6. A. H. Rose, *Industrial Microbiology*, p. 264. Butterworths, Washington, 1961.
7. Y. Shimazu, *J. Chem. Soc. Jpn.* **71**, 503 (1950); *Chem. Abs.* **45**, 9004.
8. C. Fuganti and P. Grasselli, *Chem. Ind.*, 983 (1977).
9. C. Fuganti, P. Grasselli and G. Marinoni, *Tetrahedron Lett.* **20**, 1161 (1979).
10. C. Fuganti and P. Grasselli, *J. Chem. Soc., Chem. Commun.*, 299 (1978).
11. R. Bernardi, C. Fuganti, P. Grasselli and G. Marinoni, *Synthesis*, 50 (1980). C. Fuganti, P. Grasselli, F. Spreafico and C. Zirotti, *J. Org. Chem.* **49**, 543 (1984). G. Fronza, C. Fuganti and P. Grasselli, *J. Chem. Soc., Chem. Commun.*, 442 (1980). C. Fuganti, S. Servi and C. Zirotti, *Tetrahedron Lett.* **24**, 5285 (1983). C. Fuganti, P. Grasselli and S. Servi, *J. Chem. Soc., Perkin Trans. 1*, 241 (1983).
12. C. Fuganti, P. Grasselli, S. Servi, F. Spreafico and C. Zirotti, *J. Org. Chem.* **49**, 4087 (1984).
13. G. Bertolli, G. Fronza, C. Fuganti, P. Grasselli, L. Majori and F. Spreafico, *Tetrahedron Lett.* **22**, 965 (1981).
14. C. Fuganti and P. Grasselli, *J. Chem. Soc., Chem. Commun.*, 205 (1982).
15. T. Kitazume and N. Ishikawa, *Chem. Lett.*, 1815 (1984).
16. C.-H. Wong and G. M. Whitesides, *J. Org. Chem.* **48**, 3199 (1983).
17. D. C. Crans and G. M. Whitesides, *J. Am. Chem. Soc.* **107**, 7019 (1985).
18. M. D. Bednarski, H. J. Waldmann and G. M. Whitesides, *Tetrahedron Lett.* **27**, 5807 (1986). See also N. Bischofberger, H. Waldmann, T. Saito, E. S. Simon, W. Lees, M. D. Bednarski and G. M. Whitesides, *J. Org. Chem.* **53**, 3457 (1988).
19. J. R. Durrwachter, D. G. Drueckhammer, K. Nozaki, H. M. Sweers and C.-H. Wong, *J. Am. Chem. Soc.* **108**, 7812 (1986).
20. F. Effenberger and A. Straub, *Tetrahedron Lett.* **28**, 1641 (1987).
21. For a good review see: C.-H. Wong, in *"Enzymes as Catalysts in Organic Synthesis"* (ed. M. P. Schneider), pp. 199–216. Reidel, Dordrecht (1986).
22. I. A. Rose, *J. Am. Chem. Soc.* **80**, 5835 (1958).
23. E. L. O'Connell and I. A. Rose, *J. Biol. Chem.* **248**, 2225 (1973).
24. C.-H. Wong, F. P. Mazenod and G. M. Whitesides, *J. Org. Chem.* **48**, 3493 (1983).
25. A. S. Serianni, E. Cadman, J. Pierce, M. L. Hayes and R. Barker, *Methods Enzymol.* **89**, 83 (1982).
26. J. K. N. Jones and H. H. Septon, *Can. J. Chem.* **38**, 753 (1960). M. Kapuscinski, F. P. Franke, J. Flanigan, J. K. McLeod and J. F. Williams, *Carbohydrate Res.* **140**, 69 (1985).
27. G. M. Whitesides and C.-H. Wong, *Angew. Chem. Int. Ed. Engl.* **24**, 617 (1985).
28. C. Auge, S. David and C. Gautheron, *Tetrahedron Lett.* **25**, 4663 (1984). C. Auge, S. David, C. Gautheron and A. Veyrieres, *Tetrahedron Lett.* **26**, 2439 (1985).
29. M. D. Bednarski, H. K. Chenault, E. S. Simon and G. M. Whitesides, *J. Am. Chem. Soc.* **109**, 1283 (1987).
30. P. R. Rony, *J. Am. Chem. Soc.* **94**, 8247 (1972).
31. W. Becker, H. Freund and E. Pfeil, *Angew. Chem. Int. Ed. Engl.* **4**, 1079 (1965). See also J. B. Jones and J. F. Beck, in *Techniques of Chemistry* (eds J. B. Jones, C. J. Sih and D. Perlman), Vol. 10, pp. 234–236. Wiley, New York (1976). W. Becker, U. Benthin, E. Eschenhof and E. Pfeil, *Angew. Chem.* **75**, 93 (1963). W. Becker, U. Benthin, E. Eschenhof and E. Pfeil, *Biochem. Z.* **337**, 156 (1963). W. Becker and E. Pfeil, *Naturwissenschaften* **51**, 193 (1964).
32. F. Effenberger, T. Ziegler and S. Forster, *Angew. Chem. Int. Ed. Engl.* **26**, 458 (1987).
33. W. Becker and E. Pfeil, *J. Am. Chem. Soc.* **88**, 4299 (1966).
34. R. J. H. Wilson and M. D. Lilly, *Biotechnol. Bioeng.* **11**, 349 (1969). J. J. Marshall and W. J. Whelan, *Chem. Ind. (London)* **25**, 701 (1971). C. Gruesbeck and H. F. Rase, *Ind. Eng. Chem. Proc. Res. Dev.* **11**, 74 (1972).

35. H. H. Weetall, *Process Biochem.* **10**, 3 (1975). H. H. Weetall, W. P. Vann, W. H. Pitcher,
     D. D. Lee, Y. Y. Lee and G. T. Tsao, *Methods Enzymol.* **44**, 776 (1976). G. W.
     Strandberg and K. L. Smiley, *Appl. Microbiol.* **21**, 588 (1971). N. B. Havewala and W. H.
     Pitcher, *Enzyme Eng.* **2**, 315 (1974). N. H. Mermelstein, *Food Technol. Chicago* **29**, 20
     (1975). P. Fulbrook and B. Vabo, *Ind. Chem. Eng. Symp. Ser.* **51**, 31 (1977). C. Bucke, in
     Topics in Enzyme and Fermentation Biotechnology, Vol. 1, ed. A. Wiseman, p. 147. Ellis
     Horwood, Chichester (1977).
36. For a review see R. W. Bailey and J. B. Pridham, *Adv. Carbohydr. Chem.* **17**, 140 (1962).
37. R. Breslow, *Science (Washington, D.C.)* **218**, 532 (1982). For a review see J. P. Guthrie,
     in *Techniques of Chemistry* (eds J. B. Jones, C. J. Sih and D. Perlman), Vol. 10, pp. 702–
     708. Wiley, New York (1976).
38. L. D. Melton and K. N. Slessor, *Can. J. Chem.* **51**, 327 (1973).
39. K. Fujita, A. Matsunaga and T. Imoto, *Tetrahedron Lett.* **25**, 5533 (1984). I. Tabushi, T.
     Nabeshima, K. Fujita, A. Matsunaga and T. Imoto, *J. Org. Chem.* **50**, 2638 (1985). K.
     Fujita, A. Matsunaga and T. Imoto, *J. Am. Chem. Soc.* **106**, 5740 (1984). K. Fujita, A.
     Matsunaga, Y. Ikeda and T. Imoto, *Tetrahedron Lett.* **26**, 6439 (1985).
40. K. Fujita, T. Tahara, S. Nagamura, T. Imoto and T. Koga, *J. Org. Chem.* **52**, 636 (1987).
41. L. Hedbys, P.-O. Larsson, K. Mosbach and S. Svensson, *Biochem. Biophys. Res.
     Commun.* **123**, 8 (1984).
42. E. J. Hehre, D. S. Ganghof and G. Okada, *Arch. Biochem. Biophys.* **142**, 382 (1971).
43. P. J. Card and W. D. Hitz, *J. Am. Chem. Soc.* **106**, 5348 (1984).
44. P. J. Card, W. D. Hitz and K. G. Ripp, *J. Am. Chem. Soc.* **108**, 158 (1986).
45. D. G. Drueckhammer and C.-H. Wong, *J. Org. Chem.* **50**, 5912 (1985).
46. C.-H. Wong, S. L. Haynie and G. M. Whitesides, *J. Org. Chem.* **47**, 5416 (1982).
47. J. Thiem and W. Treder, *Angew. Chem. Int. Ed. Engl.* **25**, 1096 (1986).
48. H. Paulsen and H. Tietz, *Carbohydrate Res.* **125**, 47 (1984). H. Paulsen and H. Tietz,
     Angew. Chem. Int. Ed. Engl. **24**, 128 (1985).
49. C. Auge, S. David, C. Mathieu and C. Gautheron, *Tetrahedron Lett.* **25**, 1467 (1984).
50. C.-H. Wong, A. Pollak, S. D. McCurry, J. M. Sue, J. R. Knowles and G. M. Whitesides,
     Methods Enzymol. **89**, 108 (1982).
51. For reviews see T. Reichstein and E. Weiss, *Adv. Carbohydr. Chem.* **17**, 65 (1962). K.
     Nisizawa and T. Hashimoto, in *The Carbohydrates* (eds W. Pigman and D. Horton),
     Vol. 2A, p. 241. Academic Press, New York (1970). J. B. Jones and J. F. Beck, in
     Techniques of Chemistry (eds J. B. Jones, C. J. Sih and D. Perlman), Vol. 10, pp. 225–231.
     Wiley, New York (1976).
52. G. J. Dutton, *Biochem. J.* **64**, 693 (1956). W. Eichenberger and D. W. Newman, *Biochem.
     Biophys. Res. Commun.* **32**, 366 (1968). S. C. Pan, *Biochemistry* **9**, 1833 (1970).
53. Y. Ooi, T. Hashimoto, N. Mitsuo and T. Satoh, *Tetrahedron Lett.* **25**, 2241 (1984).
54. C. E. Snipes, C.-J. Chang and H. G. Floss, *J. Am. Chem. Soc.* **101**, 701 (1979).
55. For example see: H. G. Khorana *et al.*, *J. Mol. Biol.* **72**, 209 (1972); *J. Biol. Chem.* **251**,
     565 (1976). A. G. Bruce and O. C. Uhlenbeck, *Biochemistry* **21**, 855 (1982). P. Carbon, E.
     Haumont, S. de Henau, G. Keith and H. Grosjean, *Nucleic Acids Res.* **10**, 3715 (1982).
     S. M. Zhenodarova, V. P. Klyagina, E. A. Sedelnikova, O. A. Smolyaninova, M. I.
     Khabarova, E. N. Belova and A. S. Mankin, *Bioorg. Khim.* **9**, 764 (1983). See also [64].
56. J. Engels, M. Leineweber and E. Uhlmann, in *Organic Synthesis, an Interdisciplinary
     Challenge*, ed. J. Streith. Blackwell, Oxford, 1984. For a good practical text on oligo-
     nucleotide synthesis see M. J. Gait (ed.), *Oligonucleotide Synthesis*. IRL Press, Oxford
     (1984).
57. N. N. Gerber, *J. Med. Chem.* **7**, 204 (1964).
58. M. H. Robins and B. Uznanski, *Can. J. Chem.* **59**, 2601 (1981).

59. T. A. Krenitsky, G. W. Koszalka, J. V. Tuttle, J. L. Rideout and G. B. Elion, *Carbohydrate Res.* **97**, 139 (1981).
60. A. Gross, O. Abril, J. M. Lewis, S. Geresh and G. M. Whitesides, *J. Am. Chem. Soc.* **105**, 7428 (1983).
61. S. Shimizu, S. Shiozaki, T. Oshiro and H. Yamada, *Agric. Biol. Chem.* **48**, 1383 (1984).
62. T. Utagawa, H. Morisawa, F. Yoshinaga, A. Yamazaki, K. Mitsugi and Y. Hirose, *Agric. Biol. Chem.* **49**, 1053 (1985).
63. S. Tono-oka, Y. Sasahara, A. Sasaki, H. Shirahama, T. Matsumoto and S. Kakimoto, *Bull. Chem. Soc. Jpn.* **54**, 212 (1981). S. Tono-oka, *Bull. Chem. Soc. Jpn.* **55**, 1531 (1982).
64. C. H. Hoffman, E. Harris, S. Chodroff, S. Michelson, J. W. Rothrock, E. Peterson and W. Reuter, *Biochem. Biophys. Res. Commun.* **41**, 710 (1970).
65. H. G. Gassen and R. Nolte, *Biochem. Biophys. Res. Commun.* **44**, 1410 (1971).
66. T. Tosa, T. Mori, N. Fuse and I. Chibata, *Agric. Biol. Chem.* **33**, 1047 (1969). M. D. Lilly and P. Dunnill, *Proc. Biochem.* **6**, 29 (1971).
67. K. Yamamoto, T. Tosa and I. Chibata, *Biotechnol. Bioeng.* **22**, 2045 (1980).
68. I. Chibata, T. Tosa and T. Sato, *Methods Enzymol.* **44**, 739 (1976).
69. I. Chibata, *Pure Appl. Chem.* **50**, 667 (1978). Y. Izumi, I. Chibata and T. Itoh, *Angew. Chem. Int. Ed. Engl.* **17**, 176 (1978). C. Wandrey, in *Enzymes as Catalysts in Organic Synthesis* (ed. M. P. Schneider), pp. 263–284. Reidel, Dordrecht (1986). For a review on the use of immobilized cells and enzymes see I. Chibata, T. Tosa and T. Sato, *J. Mol. Catal.* **37**, 1 (1986).
70. Preparation of enantiomerically pure amino acids by enzymatic resolution is described in the section on amidases (Section 2.2). See also W. H. J. Boesten *et al.*, in *Enzymes as Catalysts in Organic Synthesis* (ed. M. P. Schneider), pp. 355–360. Reidel, Dordrecht (1986).
71. J. R. Parikh, J. P. Greenstein, M. Winitz and S. M. Birnbaum, *J. Am. Chem. Soc.* **80**, 953 (1958). J. P. Greenstein and M. Winitz, *The Chemistry of the Amino Acids*, Vol. 1, p. 728. Wiley, New York (1961).
72. J. B. Jones and J. F. Beck, in *Techniques of Chemistry* (eds J. B. Jones, C. J. Sih and D. Perlman), Vol. 10, pp. 236–243. Wiley, New York (1976).
73. S. M. Hecht, K. M. Rupprecht and P. M. Jacobs, *J. Am. Chem. Soc.* **101**, 3982 (1979).
74. L. Benoiton, M. Winitz, S. M. Birnbaum and J. P. Greenstein, *J. Am. Chem. Soc.* **79**, 6192 (1957). E. Adams and A. Goldstone, *Biochem. Biophys. Acta* **77**, 133 (1963).
75. C. Du Crocq, P. Decottignies-Le Marechal and R. Azerad, *J. Labelled Compounds Radiopharmacol.* **22**, 61 (1985).
76. J. E. Baldwin, R. L. Dyer, S. C. Ng, A. J. Pratt and M. A. Russell, *Tetrahedron Lett.* **28**, 3745 (1987).
77. A. S. Gelbard, *J. Labelled Compounds Radiopharmacol.* **18**, 933 (1981). A. S. Gelbard and A. J. L. Cooper, *J. Labelled Compounds Radiopharmacol.* **16**, 92 (1979). A. J. L. Cooper and A. S. Gelbard, *Anal. Biochem.* **111**, 42 (1981).
78. M. B. Cohen, L. Spolter, C. C. Chang, N. S. MacDonald, J. Takahashi and D. D. Bobinet, *J. Nucl. Med.* **15**, 1192 (1974).
79. J. R. Barrio, J. E. Egbert, E. Henze, H. R. Schelbert and F. J. Baumgartner, *J. Med. Chem.* **25**, 93 (1982).
80. N. Passerat and J. Bolte, *Tetrahedron Lett.* **28**, 1277 (1987).
81. J. C. Vederas and H. G. Floss, *Acc. Chem. Res.* **13**, 455 (1980).
82. S. Fukui, S. Ikeda, M. Fujimara, H. Yamada and H. Kumagai, *Eur. J. Biochem.* **51**, 155 (1975); *Eur. J. Appl. Microbiol.* **1**, 25 (1975). H. Yamada and H. Kumagai, *Pure Appl. Chem.* **50**, 1117 (1978) and references therein. K. Ogata, H. Yamada, H. Enei and S. Okumara, GB Patent 1,268,721 (1972). H. Yamada and H. Kumagai, *Adv. Appl. Microbiol.* **19**, 249 (1975).

83. J. E. Baldwin, S. C. Ng, A. J. Pratt, M. A. Russell and R. L. Dyer, *Tetrahedron Lett.* **28**, 2303 (1987).

84. T. Tosa, T. Sato, T. Mori, Y. Matuo and I. Chibata, *Biotechnol. Bioeng.* **15**, 69 (1973). See also [68].

85. K. Bartl, C. Cavalar, T. Krebs, E. Ripp, J. Retey, W. E. Hull, H. Gunther and H. Simon, *Eur. J. Biochem.* **72**, 247 (1977). W. Stocklein, A. Eisgruber and H.-L. Schmidt, *Biotechnol. Lett.* **5**, 703 (1983).

86. I. A. Rose and K. R. Hanson, in *Techniques of Chemistry*, Vol. 10 (eds. J. B. Jones, C. J. Sih and D. Perlman), p. 507. Wiley, New York (1976). I. A. Rose, *CRC Crit. Rev. Biochem.* **1**, 33 (1972). K. R. Hanson and E. A. Havir, in *The Enzymes*, 3rd edn., Vol. 7 (ed. P. D. Boyer), p. 75. Academic Press, New York (1972).

87. H. Kluender, C. H. Bradley, C. J. Sih, P. Fawcett and E. P. Abraham, *J. Am. Chem. Soc.* **95**, 6149 (1973).

88. M. Akhtar, M. A. Cohen and D. Gani, *J. Chem. Soc., Chem. Commun.*, 1290 (1986).

89. K. Yamamoto, T. Tosa and I. Chibata, *Biotechnol. Bioeng.* **22**, 2045 (1980). I. Chibata, T. Tosa, T. Sato and K. Yamamoto, Japanese Patent 7,475,782 (1975).

90. S. Takamatsu, I. Umemura, K. Yamamoto, T. Sato, T. Tosa and I. Chibata, *Eur. J. Appl. Microbiol. Technol.* **15**, 147 (1982) and references therein.

91. R. Belleau and J. Burba, *J. Am. Chem. Soc.* **82**, 5751 (1960).

92. R. H. D. Lambrecht, G. Slegers, G. Mannens and A. Claeys, *Enzyme Microbiol. Technol.* **9**, 221 (1987).

93. J. A. Thomas, D. R. Morris and L. P. Hager, *J. Biol. Chem.* **245**, 3135 (1970). See also [105].

94. For a thorough review of earlier work see M. Morrison and G. R. Schonbaum, *Ann. Rev. Biochem.* **45**, 861 (1976).

95. S. L. Neidleman, W. F. Amon and J. Geigert, U.S. Patent 4,247,641 (1981).

96. C. L. Cooney and J. Hulter, *Biotechnol. Bioeng.* **16**, 1045 (1974).

97. P. L. Ashley and B. W. Griffin, *Arch. Biochem. Biophys.* **210**, 167 (1981).

98. S. L. Neidleman, W. F. Amon and J. Geigert, U.S. Patent 4,247,641 (1981).

99. R. D. Libby, J. A. Thomas, L. W. Kaiser and L. P. Hager, *J. Biol. Chem.* **257**, 5030 (1982).

100. J. Geigert, S. L. Neidleman and D. J. Dalietos, *J. Biol. Chem.* **258**, 2273 (1983).

101. S. L. Neidleman, P. A. Diassi, B. Junta, R. M. Palmere and S. C. Pan, *Tetrahedron Lett.* **44**, 5337 (1966).

102. J. R. Beckwith and L. P. Hager, *J. Biol. Chem.* **238**, 3091 (1963). S. D. Levine, S. L. Neidleman and M. Oberc, *Tetrahedron* **24**, 2979 (1968).

103. S. L. Neidleman and S. D. Levine, *Tetrahedron Lett.* **37**, 4057 (1968).

104. J. Kollonitsch, S. Marburg and L. M. Perkins, *J. Am. Chem. Soc.* **92**, 4489 (1970).

105. J. A. Thomas, D. R. Morris and L. P. Hager, *J. Biol. Chem.* **245**, 3129 (1970). J. Pommier, D. Deme and J. Nunez, *Eur. J. Biochem.* **37**, 406 (1973).

106. S. L. Neidleman, A. I. Cohen and L. Dean, *Biotechnol. Bioeng.* **11**, 1227 (1969).

107. M. C. R. Franssen and H. C. van der Plas, *Rec. J. Roy. Neth. Chem. Soc.* **103**, 99 (1984).

108. For reviews see: S. L. Neidleman and J. Geigert, *Trends Biotechnol.* **1**, 21 (1983). S. L. Neidleman and J. Geigert, *Endeavour* **11**, 5 (1987). S. L. Neidleman and J. Geigert, *Biohalogenation Principles, Basic Roles and Applications*. Ellis Horwood, Chichester (1986); S. L. Neidleman and J. Geigert, *Biochem. Soc. Symp.* **48**, 39 (1984).

109. S. L. Neidleman and J. Geigert, *J. Ann. Proc. Phytochem. Soc. Eur.* **26**, 267 (1985).

110. S. W. May and S. R. Padgette, *Biotechnology* **1**, 677 (1983).

111. B. Boothroyd, E. J. Napier and G. A. Somerfield, *Biochem. J.* **80**, 34 (1961).

112. S.-K. Lin, M. Tin-Wa and E. H. Taylor, *J. Pharm. Sci.* **64**, 2021 (1975).

113. P. J. Davis, D. Wiese and J. P. Rosazza, *J. Chem. Soc., Perkin Trans. 1*, 1 (1977).
114. P. J. Davis, J. C. Glade, A. M. Clark and R. V. Smith, *Appl. Environ. Microbiol.* **38**, 891 (1979).
115. F. S. Sariaslani and J. P. Rosazza, *Appl. Environ. Microbiol.* **46**, 468 (1983).
116. G.-S. Wu, T. Nabih, L. Youel, W. Peczynska-Czoch and J. P. Rosazza, *Antimicrobiol. Agents Chemother.* **14**, 601 (1978).
117. F. Eckenrode and J. P. Rosazza, *J. Nat. Prod.* **45**, 226 (1982).
118. R. V. Smith and P. J. Davis, *Appl. Environ. Microbiol.* **35**, 738 (1978).
119. J. P. Rosazza, A. W. Stocklinski, M. E. Gustafson, J. Adrian and R. V. Smith, *J. Med. Chem.* **18**, 791 (1975).
120. J. P. Rosazza, M. Kammer, L. Youel, R. V. Smith, P. W. Erhardt, D. H. Truong and S. W. Leslie, *Xenobiotica* **7**, 133 (1977).
121. G. Meunier and B. Meunier, *J. Am. Chem. Soc.* **107**, 2558 (1985).
122. F. S. Sariaslani and J. P. N. Rosazza, *Enzyme Microbiol. Technol.* **6**, 242 (1984).

Index of Species, Compounds and Methods

Page numbers in **bold** refer to major entries